Lecture Notes in Computer Science

Lecture Notes in Artificial Intelligence 15956

The series Lecture Notes in Artificial Intelligence (LNAI) was established in 1988 as a topical subseries of LNCS devoted to artificial intelligence.

The series publishes state-of-the-art research results at a high level. As with the LNCS mother series, the mission of the series is to serve the international R & D community by providing an invaluable service, mainly focused on the publication of conference and workshop proceedings and postproceedings.

Tanya Braun · Benjamin Paaßen ·
Frieder Stolzenburg
Editors

KI 2025: Advances in Artificial Intelligence

48th German Conference on AI, Potsdam, Germany, September 16–19, 2025 Proceedings

Springer

Editors
Tanya Braun
University of Münster
Münster, Germany

Benjamin Paaßen
Bielefeld University
Bielefeld, Germany

Frieder Stolzenburg
Harz University of Applied Sciences
Wernigerode, Germany

ISSN 0302-9743 ISSN 1611-3349 (electronic)
Lecture Notes in Artificial Intelligence
ISBN 978-3-032-02812-9 ISBN 978-3-032-02813-6 (eBook)
https://doi.org/10.1007/978-3-032-02813-6

LNCS Sublibrary: SL7 – Artificial Intelligence

This Springer imprint is published by the registered company Springer Nature Switzerland AG
The registered company address is: Gewerbestrasse 11, 6330 Cham, Switzerland

Preface

This proceedings volume contains the papers presented at the 48th German Conference on Artificial Intelligence (KI 2025), held in Potsdam, Germany, during September 16–19, 2025, and co-located with INFORMATIK 2025, EnviroInfo 2025, and #SKILL 2025.

KI 2025 was the 48th German Conference on Artificial Intelligence (AI) organized in cooperation with the German Section on Artificial Intelligence of the German Informatics Society (GI-FBKI, Fachbereich Künstliche Intelligenz der Gesellschaft für Informatik (GI) e.V.). The German AI Conference basically started 50 years ago with the first meeting of the national special interest group on AI within the GI on October 7, 1975. KI is one of the major European AI conferences and traditionally brings together academic and industrial researchers from all areas of AI, providing an ideal place for exchanging news and research results on theory and applications. While KI is primarily attended by researchers from Germany and neighboring countries, it warmly welcomes international participation.

The technical program of KI 2025 featured paper presentations as well as three keynotes, one tutorial, a doctoral symposium, and five workshops. Overall, KI 2025 received 74 submissions from authors in 20 countries, which were single-blind reviewed, in all but very few cases by three Program Committee members each. The Program Committee, comprising 57 experts from 6 countries, accepted 15 full papers (29% of the submissions in that category), 8 technical communications, and 4 extended abstracts. Overall, the acceptance rate across paper categories was 36%.

As a highlight of this year's edition of the KI conference, the main program featured a panel discussion regarding the current state of AI regulation with four renowned experts on the field. We were honored that three prominent researchers kindly agreed to give very interesting keynote talks (alphabetical order):

- Jesse Dinneen (who kindly stepped in for Louisa Specht-Riemenschneider)
- Barbara Hammer
- Rineke Verbrugge

The program also contains a special contribution on the origins of AI research in Germany by Rudolf Seising.

As Program Committee chairs, we would like to thank our speakers for their interesting and inspirational talks, the Workshop and Tutorial Chairs Magda Gregorova, Philipp Väth, and Tamara Bergerhoff, and the Doctoral Symposium Chair Gesina Schwalbe. Our special gratitude goes to the Program Committee, whose sophisticated and conscientious judgement ensured the high quality of the KI conference. Without their substantial voluntary work, this conference would not have been possible.

The following tutorials and workshops took place at KI 2025:

- Humanities-Centered AI (CHAI) organized by Sylvia Melzer, Stefan Thiemann, Hagen Peukert, and Magnus Bender

- Smarter Workplaces: The Role of AI in Promoting Occupational Safety and Health (AI4POSH) organized by Thea Radüntz
- 11th Workshop on Formal and Cognitive Reasoning (FCR) organized by Özgür Lütfü Özçep, Nele Rußwinkel, Kai Sauerwald, and Diedrich Wolter
- Workshop SIG Knowledge Management (FG WM) organized by Johannes Wichmann, Lisa Grumbach, and Pascal Reuß
- Second Workshop on AI in Production organized by Judith Knoblach, Erik Voigt, Christoph Wehner, Martin Krockert, and Lukas Bahr
- Tutorial: Towards FAIR4ML organized by Rohitha Ravinder, Nelson Quiñones, Dietrich Rebholz-Schuhmann, and Leyla Jael Castro

Last not least, we would like to thank our sponsors for their support:

- German Section on Artificial Intelligence of the German Informatics Society (https://fb-ki.gi.de)
- Springer (https://www.springer.com)

September 2025

Tanya Braun
Benjamin Paaßen
Frieder Stolzenburg

Organization

Program Committee

Martin Attmüller	Osnabrück University, Germany
Kerstin Bach	Norwegian University of Science and Technology, Norway
Joachim Baumeister	denkbares GmbH, Germany
Magnus Bender	Aarhus University, Denmark
Christoph Benzmüller	Otto-Friedrich-Universität Bamberg, Germany
Tamara Bergerhoff	Technische Hochschule Würzburg-Schweinfurt, Germany
Ralph Bergmann	University of Trier, Germany
Tanya Braun	University of Münster, Germany
Karin Breckner	University of Applied Sciences Upper Austria, Austria
Ulf Brefeld	Leuphana Universität Lüneburg, Germany
Philipp Cimiano	Bielefeld University, Germany
Stefan Edelkamp	CTU Prague, Czech Republic
Johannes Fähndrich	Hochschule für Polizei Baden-Württemberg, Germany
Johannes Fürnkranz	Johannes Kepler University Linz, Austria
Marcel Gehrke	University of Hamburg, Germany
Magda Gregorova	TH Würzburg-Schweinfurt, Germany
Jürgen Grossmann	Fraunhofer Institute for Open Communication Systems FOKUS, Germany
Jonas Philipp Haldimann	TU Wien, Austria
Heinz Handels	University of Lübeck and DFKI Lübeck, Germany
Tom Hanika	University of Hildesheim, Germany
Andreas Hotho	University of Würzburg, Germany
Xiaoyi Jiang	University of Münster, Germany
Gabriele Kern-Isberner	Technische Universität Dortmund, Germany
Matthias Klusch	German Research Center for Artificial Intelligence (DFKI), Germany
Franziska Klügl	Örebro University, Sweden
Stefan Kopp	Bielefeld University, Germany
Samuel Kounev	University of Würzburg, Germany
Anna Krause	University of Würzburg, Germany
Markus Krötzsch	TU Dresden, Germany

Michael Leuschel	University of Düsseldorf, Germany
Lukas Malburg	University of Trier and German Research Center for Artificial Intelligence (DFKI), Germany
Mirjam Minor	Goethe University Frankfurt, Germany
Till Mossakowski	University of Magdeburg, Germany
Ralf Möller	University of Hamburg, Germany
Bernhard Nebel	Freiburg University, Germany
Özgür Lütfü Özcep	University of Hamburg, Germany
Benjamin Paaßen	Bielefeld University, Germany
Florian Rabe	FAU Erlangen-Nürnberg, Germany
Simon Razniewski	TU Dresden, Germany
Sebastian Rudolph	TU Dresden, Germany
Jürgen Sauer	University of Oldenburg, Germany
Kai Sauerwald	University of Hagen, Germany
Ute Schmid	University of Bamberg, Germany
Claudia Schon	Hochschule Trier, Germany
Lutz Schröder	Friedrich-Alexander-Universität Erlangen-Nürnberg, Germany
René Schumann	HES-SO University of Applied Sciences Western Switzerland, Switzerland
Gesina Schwalbe	University of Lübeck, Germany
Sibylle Schwarz	Hochschule für Technik, Wirtschaft und Kultur Leipzig, Germany
Friedhelm Schwenker	Ulm University, Germany
Dietmar Seipel	University of Würzburg, Germany
Daniel Sonntag	German Research Center for Artificial Intelligence (DFKI), Germany
Alexander Steen	University of Greifswald, Germany
Frieder Stolzenburg	Hochschule Harz, Germany
Heiner Stuckenschmidt	University of Mannheim, Germany
Michael Thielscher	University of New South Wales, Australia
Matthias Thimm	FernUniversität in Hagen, Germany
Philipp Väth	TH Würzburg-Schweinfurt, Germany
Uwe Waldmann	Max Planck Institute for Informatics, Germany
Diedrich Wolter	University of Lübeck, Germany
Stefan Woltran	TU Wien, Austria
Stefan Wrobel	Fraunhofer Institute for Intelligent Analysis and Information Systems and University of Bonn, Germany

Additional Reviewers

Abramov, Olga
Antakli, Andre
Barz, Michael
Battefeld, Dominik
Bengel, Lars
Borst, Vanessa
Brand, Florian
Bühl, Johannes
Fischer, Elisabeth
Grewenig, Lisa
Hadry, Marius
Heinrich, Fabian
Kadir, Md Abdul
Kenneweg, Svenja
Krause, Stefanie
Lackner, Marie-Louise
Lasy, Ilya
Lenz, Mirko
Leppich, Robert
Matthiesen, Jennifer
Mauri, Marcel
Michalczyk, Jakub
Nardi, Oliviero
Roßbach, Jan
Rudolph, Yannick
Schlör, Daniel
Schröder, Florian
Schubert, Julian
Shrestha, Abhishek
Sitapara, Pratik
Speller, Jan
Stenger, Michael
Stodden, Regina
Töberg, Jan-Philipp
Viehmann, Johannes
Vu, Fabian
Weidner, Daniel
Zeyen, Christian

Contents

Technical Communications

Full Technical Papers

Augmenting Systematic Literature Reviews: A Human-AI Collaborative Framework

Constantin Brîncoveanu(✉), K. Valerie Carl, Aaron Witzki, and Oliver Hinz

Goethe University Frankfurt, Chair of Information Systems and Information Management, Theodor-W.-Adorno-Platz 4, 60323 Frankfurt/Main, Germany
{brincoveanu,kcarl,witzki,ohinz}@wiwi.uni-frankfurt.de

Abstract. While Systematic Literature Reviews (SLRs) are integral to research by synthesizing existing knowledge and guiding future inquiry, the exponential increase in academic publications presents significant challenges to traditional, manual review methods, notably regarding scalability, efficiency, and researcher workload. Recent advancements in Artificial Intelligence (AI), particularly Large Language Models (LLMs), offer promising avenues for augmenting the SLR process. Nonetheless, integrating AI into literature reviews introduces methodological complexities, including maintaining accuracy, minimizing biases, and preserving scholarly rigor. To address these challenges, this paper introduces a structured AI-augmented SLR framework, systematically integrating AI capabilities into Wolfswinkel et al.'s [37] established Grounded Theory Literature Review Method. Our framework incorporates AI-driven relevance assessments, automated selection processes, and thematic content analysis, underpinned by rigorous human oversight to ensure reliability and interpretative validity. We empirically illustrate and evaluate our framework through a comparative study, replicating and extending a previously published human SLR. The evaluation assesses AI performance using key metrics such as type I and type II error rates across varying confidence thresholds. Results demonstrate substantial efficiency gains and effective accuracy in AI-assisted selection, highlighting the importance of carefully calibrated thresholds and continued human oversight. Our study contributes practical guidelines for effectively balancing AI automation with human scholarly judgment, offering a replicable methodological approach for researchers seeking to leverage AI capabilities without compromising methodological quality or academic integrity.

Keywords: Systematic Literature Review · Artificial Intelligence · Large Language Models · Research Augmentation · Human-AI Collaboration

1 Introduction

Systematic Literature Reviews (SLRs) are fundamental to research, providing structured methods for synthesizing existing knowledge, uncovering research

T. Braun et al. (Eds.): KI 2025, LNAI 15956, pp. 3–17, 2026.
https://doi.org/10.1007/978-3-032-02813-6_1

gaps, and setting directions for future investigations [36,37]. However, traditional SLR methods face increasing challenges due to the exponential growth in academic literature [18]. Conventional manual approaches encompass comprehensive database searches, extensive article screening, detailed thematic analysis, and synthesis of insights, which are resource-intensive, laborious, and time-consuming [6]. As publication volumes surge, researchers increasingly struggle to maintain rigor and comprehensiveness within reasonable timeframes, highlighting the pressing need for more efficient methodologies [24,35]. Moreover, rapidly evolving theories and technologies further underline the need for efficient SLR methodologies, as timely synthesis of literature is essential to maintain relevance and support ongoing research advancements [24].

Artificial Intelligence (AI), particularly Large Language Models (LLMs), present promising avenues for augmenting and enhancing traditional SLR processes, as they can automate critical components, such as article retrieval, relevance assessment, duplicate removal, and preliminary thematic analysis [12,28, 33,38]. These capabilities offer significant potential to enhance the efficiency, consistency, and scalability of SLR processes, as recent hybrid methodologies combining AI-driven techniques with human oversight have demonstrated substantial reductions in manual effort and improved workflow effectiveness (e.g., in screening of articles [38], or data extraction [28]). Nevertheless, integrating AI into scholarly review processes introduces methodological challenges regarding accuracy, reliability, potential biases, and the preservation of scholarly rigor [25,30,35]. Despite notable progress, most AI-assisted literature reviews still lack structured methodological frameworks explicitly designed to integrate AI systematically with transparent human oversight, often remaining qualitative or exploratory in nature [24,30].

Addressing these methodological challenges, this paper proposes an SLR framework which is augmented by AI. Drawing upon the established Grounded Theory Literature Review Method by Wolfswinkel et al. [37], our framework systematically incorporates AI assistance at critical stages of the SLR process. These include automated article retrieval, relevance classification using LLMs, and AI-supported thematic analysis through open and axial coding. In this way, we conduct research on the capacities of AI, covering the automation of predefined, repetitive tasks [32], the analysis of large data sources [1,10], and the clusters of retrieved information [17]. The framework emphasizes rigorous human oversight and iterative validation at every step, ensuring methodological robustness and scholarly rigor while effectively leveraging AI's capabilities.

Therefore, this study advances the field by systematically integrating AI into a structured SLR framework. We illustrate the practical application and evaluate the methodological rigor of the proposed AI-augmented SLR framework through an empirical comparative case study, replicating and extending a previously conducted human-led SLR [5]. By directly comparing AI-assisted approaches to traditional manual methods within a clearly defined research context, we systematically assess the accuracy, thematic alignment, efficiency gains, and practical trade-offs associated with AI integration.

2 Theoretical Background

With knowledge production accelerating at a tremendous speed while remaining fragmented and interdisciplinary, SLRs are more relevant than ever, for researchers and practitioners alike [31]. Established concept-centric and input-process-output frameworks have significantly shaped how researchers systematically select and synthesize academic literature for SLRs [21,36]. Additional methodological refinements include the five-phase Grounded Theory Literature Review Method by Wolfswinkel et al. [37], which emphasizes rigorous iterative thematic analysis. Despite these methodological advances, researchers increasingly struggle with practical challenges arising from exponential growth in academic publications, difficulties in unbiased study selection, and maintaining reproducibility [35,37].

Many of these challenges can be addressed by integrating AI into the SLR process, since AI (and particularly LLMs) has already demonstrated transformative capabilities across various analytical tasks and contexts, including text summarization [2], sentiment analysis [7,20], thematic classification [9,16,22], and general creative contexts [14]. Recent research has begun to explore these applications within literature reviews by demonstrating the potential of AI to support tasks such as identifying and filtering relevant evidence [24], thematic coding [29], real-time literature searching [13], and drafting summarizations [15], showing that these techniques can significantly reduce manual workload while maintaining acceptable accuracy [27,28]. Additionally, recent classifications have highlighted numerous ready-to-use AI tools supporting the entire SLR process, from search to presentation [3].

However, these existing solutions typically lack systematic empirical evaluation regarding methodological rigor, epistemic implications, and effectiveness compared to traditional human-led approaches, especially for complex tasks like conceptual synthesis and thematic analysis. More recently, attention has shifted toward generative AI and LLMs, such as ChatGPT, to support deeper conceptual analysis and synthesis in literature reviews [4,30,38]. Exploratory studies have illustrated generative AI's potential, while emphasizing limitations like hallucinations, interpretive inaccuracy, and the necessity of sustained human oversight [11,30,38]. Although hybrid workflows combining human judgment with LLM-assisted screening have shown promise [38], fully autonomous reviews remain impractical due to ongoing challenges in interpretive fidelity and methodological rigor. Alongside technical advancements, critical perspectives have examined the epistemic and methodological implications of AI adoption in scholarly review processes [25,34]. These discussions highlight that while AI offers potential in streamlining review tasks, it also risks shifting foundational scholarly values, including interpretive transparency, argumentative coherence, and epistemic accountability [25,34,35]. Structured frameworks clearly defining human-AI collaboration have thus been advocated to maintain scholarly rigor and reproducibility [35].

Extending beyond that, we propose an AI-augmented SLR framework, integrating AI capabilities across the full SLR process, from search and selection through thematic analysis and synthesis. Our approach explicitly delineates human and AI responsibilities, emphasizing rigorous validation through quantitative and qualitative empirical comparisons with a traditional human-led SLR.

3 Methodology

We propose an AI-augmented SLR framework based on Wolfswinkel et al.'s approach [37], comprising the five phases Define, Search, Select, Analyze, and Present. The framework systematically incorporates AI, particularly LLMs, to enhance efficiency while maintaining rigorous human oversight. Figure 1 summarizes the proposed approach, highlighting AI's role in each sub-stage. While AI provides substantial assistance throughout various review stages, human researchers maintain ultimate responsibility for critical interpretation, validation, and methodological rigor. Therefore, scholars employing this framework should proactively and critically evaluate AI-generated outputs, ensuring accuracy and appropriateness, and remain mindful not to inadvertently shift workload toward tasks requiring more substantial human judgment.

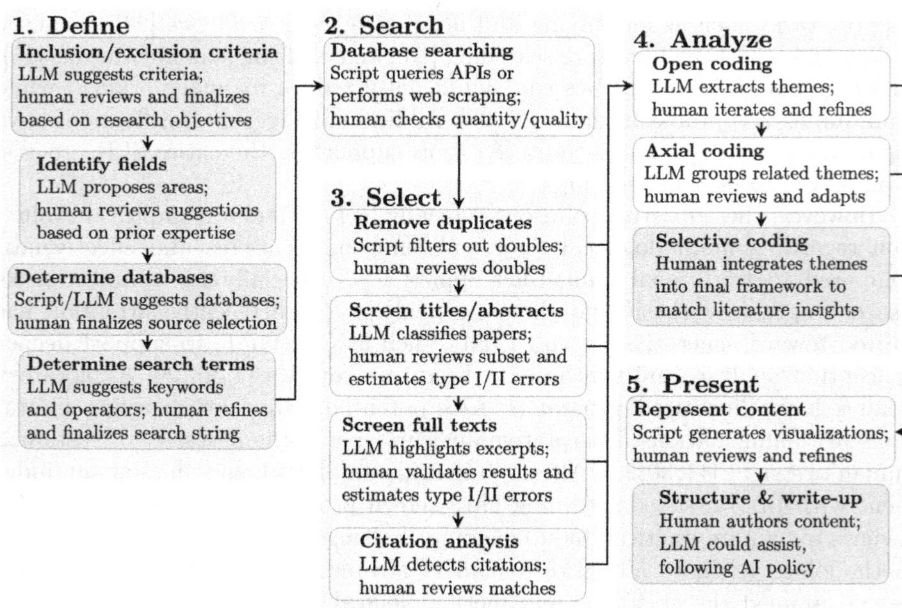

Fig. 1. AI-augmented SLR framework (adapted from [37]). Mainly human-driven (blue), mainly machine-driven (green), and collaborative (yellow) activities.(Color figure online)

The *Define* phase establishes the foundation for the SLR by setting precise boundaries and objectives [37]. This phase is crucially human-driven to ensure that the review aligns closely with the study's research questions and scholarly standards. Initially, researchers define clear and explicit inclusion and exclusion criteria that guide the review process. While LLMs or AI tools could suggest potentially relevant criteria based on patterns found in prior SLRs, these AI-generated suggestions serve only as preliminary input. Human researchers critically review, modify, and finalize these criteria based on expert knowledge and the specific research objectives. Similarly, identifying relevant research fields and selecting appropriate academic outlets and databases remains primarily human-driven. Although AI may perform bibliometric or literature analyses to highlight potentially relevant research fields or frequently used academic databases, the final decision on the scope and sources remains under direct human oversight. Human researchers validate AI-generated insights against domain knowledge, relevance, and practical considerations. Finally, in determining specific search terms, researchers maintain ultimate control. AI-supported tools and LLMs can propose synonyms, variations, or alternative keyword combinations to optimize search queries, but human researchers carefully assess these recommendations through iterative refinement and quality testing, ensuring comprehensive yet focused retrieval of relevant literature.

In the *Search* phase [37], AI-driven automation performs database queries and literature retrieval through the use of APIs or web scraping scripts, using selected academic databases. Automated searches help enhance comprehensiveness, ensuring consistent retrieval across multiple sources and substantially reducing manual workload. However, human oversight remains critical, e.g., controlling completeness.

In the *Select* phase [37], AI-assisted filtering supports the systematic refinement of the retrieved corpus to efficiently identify relevant literature. Initially, automated scripts identify and remove duplicate records across multiple databases, after which human reviewers conduct a brief validation to correct potential inaccuracies in automated duplicate detection. Subsequently, LLMs perform preliminary relevance assessments based on titles and abstracts, providing detailed justifications accompanied by confidence ratings (percentages) for each paper. To quantitatively assess the accuracy of AI-driven classifications, human reviewers then independently validate a representative subset (approximately 20%), akin to assessing intercoder reliability (ICR) commonly used in qualitative research [26]. Such validation ensures transparency, systematicity, and robustness in the relevance assessment, facilitating calibration of AI selection thresholds to balance inclusiveness and specificity. Additionally, LLMs further augment full-text relevance screening by automatically extracting relevant text excerpts and assigning individual confidence scores. AI tools can also assist with citation analysis by identifying potentially relevant forward and backward citations based on title and abstract content. Despite this substantial automation, human researchers retain ultimate responsibility for final verification and correctness, ensuring relevance, accuracy, and thematic alignment with the study's objectives.

In the *Analyze* phase [37], AI can facilitate thematic and conceptual synthesis. AI-driven open coding rapidly extracts relevant excerpts and identifies preliminary thematic patterns (open codes) from the literature corpus. Subsequently, axial coding (i.e., grouping related open codes into broader conceptual categories) is supported by semantic clustering, co-occurrence analysis, and AI-driven recommendations for thematic consolidation. Human researchers iteratively review and refine these AI-generated thematic structures to ensure they are both theoretically sound and contextually appropriate. The final step, selective coding, involves the human-led development of a cohesive, overarching conceptual framework, with AI offering supplemental insights by highlighting connections among themes, summarizing key relationships, and suggesting conceptual syntheses to enrich the analytical process.

During the final *Present* phase [37], AI supports clear communication and effective visualization of the review's outcomes. Automated scripts and visualization tools generate illustrative graphics such as thematic maps, e.g. [8], word clouds, e.g. [23], and histograms, e.g. [19], facilitating the identification and communication of literature trends, concept prominence, and thematic relationships. Given that the underlying data (e.g., selected papers, thematic codes, publication years) is available in a structured format, LLMs or AI tools can automatically generate a variety of meaningful visualizations, such as publication trends over time, concept distributions across studies, or thematic co-occurrence maps. Since these visualizations are grounded in validated and structured data, the risk of hallucinations is inherently minimized. In this context, AI does not create new content but rather assists in the automated generation of accurate and customizable visual representations, allowing human researchers to focus on interpretative analysis and presentation quality. Additionally, LLMs assist in structuring the manuscript by suggesting logical narrative sequences, organizing key findings, and drafting initial summaries. Despite these AI contributions, human researchers retain final responsibility for the manuscript's narrative structure, clarity, coherence, and scholarly quality, carefully reviewing and refining AI-generated content to ensure adherence to rigorous academic standards (e.g., in terms of AI authorship and allowed machine contributions). While LLMs can produce plausible-sounding but fabricated content, commonly referred to as hallucinations, our method actively mitigates this risk through structured prompting and grounded generation. By always providing contextual input based on retrieved and validated source texts (akin to retrieval-augmented generation), the LLM's outputs are anchored in real data. Nonetheless, human researchers must still verify the factual accuracy and interpretive validity of any AI-generated contributions to ensure the final synthesis includes only verifiable sources.

4 Evaluation and Comparative Study

To illustrate and assess the applicability of our proposed AI-augmented SLR framework, we conducted a comparative case study with a previously published human-led SLR [5] serving as a detailed baseline. This specific SLR was chosen for its topical relevance, publicly documented process, and the authors' full access to its dataset (including search strategies, selection criteria, and thematic coding). This comprehensive access enabled a fine-grained, step-by-step comparison essential for this initial validation of our detailed framework, an approach often not feasible with more generic benchmarks that may lack such process transparency for a qualitative methodology like Wolfswinkel et al.'s [37].

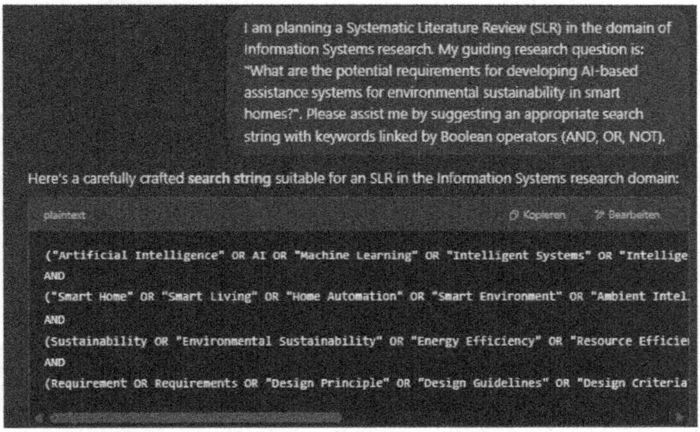

Fig. 2. Example of suggestions by `ChatGPT 4o` for constructing an SLR search string using keywords linked by Boolean operators based on the research question.

The *Define* phase sets a critical foundation for the SLR. While we explicitly replicated the human-led SLR's original research definition, including inclusion/exclusion criteria, research scope, selected databases (ACM Digital Library, AISeL, IEEE Xplore), and exact search terms, to ensure methodological comparability, we also explored potential AI support for augmenting this phase through LLMs: We prompted an LLM to generate a search string structured with Boolean operators (Fig. 2). The resulting output demonstrated the model's capacity to serve as an effective "sparring partner," particularly for individual or early-stage researchers who may lack immediate feedback when developing search strategies. The LLM-generated string offered broad coverage of AI-related concepts, sustainability criteria, smart home contexts, and explicitly stated requirements.

Table 2 highlights key differences between the AI-generated and human-designed search strings. While both addressed core themes such as AI, sustainability, and smart environments, the human-designed string emphasized application contexts and analytical frameworks (e.g., `app*`, `platform*`, `guideline*`), whereas the LLM proposed a wider vocabulary of requirement-related terms and a more expansive understanding of smart environments, including phrases like ambient intelligence and connected home.

These differences underscore the complementary value of AI-generated suggestions: LLMs can enhance terminological breadth and conceptual variety, helping researchers surface relevant alternatives they may not have otherwise considered. However, to maintain methodological rigor, any such suggestions must be critically reviewed, refined, and validated by human researchers to ensure alignment with the specific research goals and domain-specific conventions.

For the *Search* phase, we recommend employing APIs and automated scripts for efficient and consistent literature retrieval across multiple academic databases. Accordingly, we implemented scripts for automatic literature retrieval and tested their functionality, with only minor discrepancies between the sets, attributable to database-specific indexing differences or retrieval settings.

Starting from the *Select* phase, we exclusively used `GPT-4o-mini` via the OpenAI API, prompting the model to return responses in a structured JSON format. Table 3 gives an overview of the prompts employed in our AI-augmented SLR. This ensured all outputs, including relevance justifications and confidence scores, were machine-readable and automatically processable in subsequent steps, eliminating the risk of hallucinations or invented sources. Due to its speed and very low cost, we selected `GPT-4o-mini`, enabling the complete AI-augmented SLR process to be conducted for under \$1. For each paper initially retrieved, the LLM generated a textual justification of its relevance, accompanied by a numeric confidence score (0–100%). Figure 3 illustrates how varying this configurable confidence threshold affects AI classification. At lower thresholds, the AI model becomes overly inclusive, resulting in high false-positive (type I) rates. Conversely, at higher thresholds, the model significantly reduces false positives but at the expense of excluding relevant papers (higher type II errors). A practical compromise was achieved at a threshold of 50%, balancing recall and precision, thereby reducing the manual workload, especially during the title and abstract screening stages. The differences between human-led and AI-assisted selection outcomes (Table 1) result directly from this prioritization of recall. In this case study, the AI-assisted method retained more papers at intermediate steps (826 vs. 190 after title/abstract screening, and 166 vs. 60 after full-text screening). While this broader selection led to lower precision compared to the strictly human approach, it reduced the risk of prematurely excluding relevant studies. It should be noted that assuming human selections as being the ground truth is practical for evaluation purposes but may overlook certain biases or inconsistencies in manual screening. Importantly, our approach retains human validation in the final stages, ensuring overall methodological rigor and interpretative validity.

During the *Analyze* phase, AI substantially supported thematic synthesis by performing initial open coding based on full-text excerpts. Leveraging LLM-generated coding alongside associated confidence ratings, we systematically extracted key thematic concepts from the selected literature. Human researchers subsequently validated, refined, and adjusted these AI-generated codes to ensure conceptual clarity, accuracy, and alignment with domain-specific interpretations. After filtering excerpts by a high-confidence threshold ($\geq$94%), we further pro-

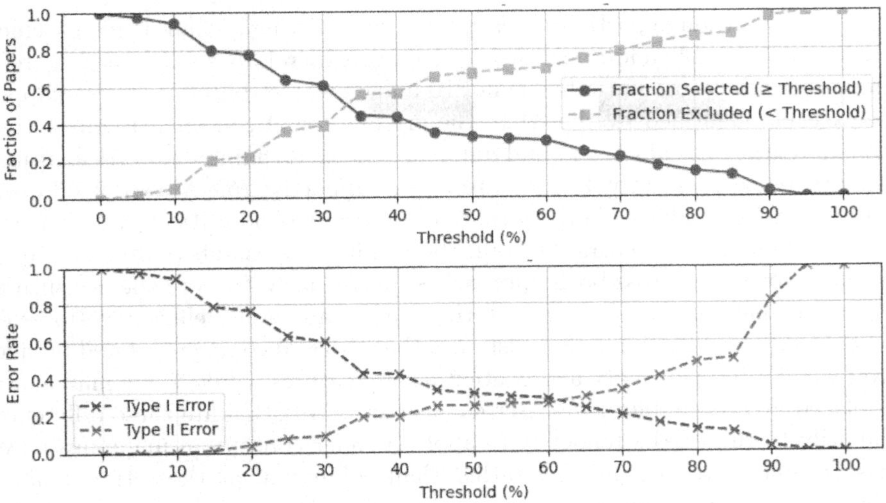

Fig. 3. AI-assisted paper selection and error analysis. Upper panel: Fraction of papers selected (green circles) and excluded (orange squares) at varying AI confidence thresholds. Lower panel: Corresponding type I (false positives) and type II (false negatives) error rates at each threshold, computed against human selections as ground truth. (Color figure online)

cessed the AI-generated open codes using an LLM-driven axial coding procedure, identifying recurrent thematic patterns across the literature corpus. This iterative human-AI collaborative process resulted in a set of AI-suggested conceptual categories. We then critically compared these categories to the established conceptual framework from our previous human-led SLR to assess thematic alignment, novelty, and comprehensiveness, which indicated a strong correspondence between AI-suggested and human-defined thematic patterns, as illustrated in Table 4. Eight of the original nine human-defined themes were effectively covered by the AI-generated categories, demonstrating, in this instance, the AI's ability to capture complex thematic structures accurately. However, the AI's classification highlighted notable points requiring human refinement. First, the AI combined the previously distinct human-defined concepts *Personalized user engagement* and *User interface and experience* into a single broader category labeled *User-Centric Design and Comfort.* This AI-driven merger underscores the importance of human judgment in explicitly determining the appropriate granularity of thematic categories, ensuring theoretical clarity and interpretive coherence. Second, the AI introduced two additional thematic categories not explicitly identified by human reviewers: *Feedback and Monitoring Systems* and *Sustainability and Efficiency in Design.* Upon human evaluation, the latter was assessed as overly generic, effectively encompassing the overall theme of the review itself, thus requiring refinement or potential removal by human experts to maintain conceptual specificity. Conversely, the category *Feedback and Monitor-*

ing Systems, although partially overlapping with existing human-defined themes, contributed valuable additional insights and was therefore retained after careful human consideration and refinement. To allow comparability, we then mapped the excerpts identified by the AI to the original set of concepts. Figure 4 visually summarizes the alignment and differences between human- and AI-identified thematic patterns, clearly demonstrating the supportive role AI can play in conceptual synthesis, alongside the essential interpretive and critical oversight provided by human researchers. Overall, the results show a substantial overlap in thematic coverage across both approaches, particularly for core themes such as predictive analytics, real-time monitoring, and strategies for efficiency. However, some deviations are evident. For instance, the AI-based coding exhibited broader coverage of interoperability and context-aware features, while the human analysis placed greater emphasis on user engagement. We attribute these differences primarily to the LLM's tendency to apply broader conceptual mappings based on surface-level term similarity rather than contextual or theoretical nuance. This may lead to slight over-representation of technical or functional aspects, while more context-sensitive themes (e.g., user engagement or privacy implications) require careful human interpretation and validation. Importantly, these deviations do not constitute methodological flaws but rather highlight the complementary strengths and limitations of AI-driven coding. Our method addresses this challenge by ensuring that AI suggestions are grounded in validated source excerpts and by explicitly integrating human oversight into all stages of coding and theme development. In practice, this means that AI accelerates the identification and initial organization of thematic patterns, while human researchers refine, contextualize, and ensure the interpretative coherence of the final thematic structure.

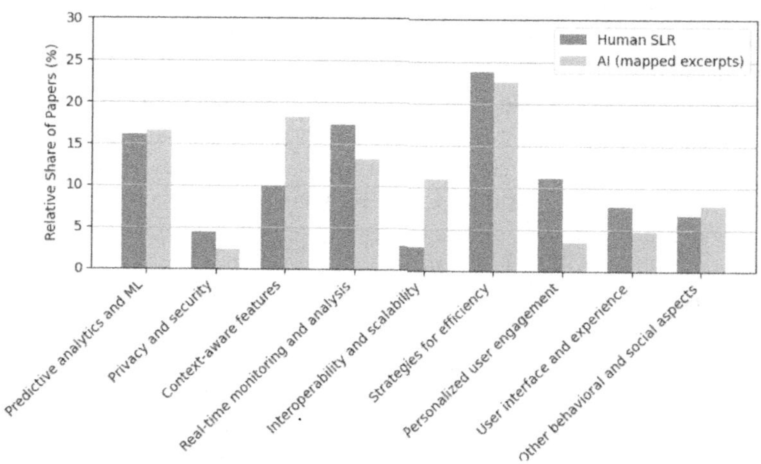

Fig. 4. Comparison of thematic patterns in the human-led SLR and AI-assisted SLR.

5 Discussion and Conclusion

This study presented an AI-augmented SLR framework and illustrated its application through a comparative case study against a traditional human-led SLR. We systematically applied LLMs across key stages of the SLR process, including paper selection, thematic analysis, and conceptual synthesis. The results of this case study suggest that AI can effectively augment multiple stages of the SLR process, particularly in initial screening and thematic analysis. The integration of AI notably reduces manual workload, cognitive burden, and overall review duration, echoing and extending prior exploratory findings [25,35]. At moderate confidence thresholds, AI-assisted selection achieves an optimal balance between type I (false positive) and type II (false negative) errors, closely aligning with human judgment. Furthermore, AI-supported thematic coding yields complementary insights, highlighting concepts and thematic connections that might otherwise remain underexplored in purely manual reviews. Nonetheless, our results underscore the critical importance of continued human oversight, especially given AI's tendency toward conservative relevance assessments, to ensure rigorous methodological quality and interpretive validity. Our study thus aligns with and empirically validates prior qualitative cautions regarding AI's limitations, such as potential biases, reduced transparency, and interpretive risks [25,30].

Our findings have several theoretical implications. First, this study demonstrates that AI tools can significantly augment traditional SLR methodologies, particularly in terms of scalability and efficiency, without compromising methodological rigor. These results substantiate previous qualitative and exploratory claims about AI's potential [24,28,35], providing concrete empirical evidence regarding the practical effectiveness of AI-assisted literature screening and thematic analysis. By systematically integrating AI across all phases, our framework extends beyond previous studies that primarily focused on individual phases [27,38] or provided qualitative demonstrations without systematic validation [30]. Second, extending previous literature [25], the identified complementary thematic insights provided by AI-driven coding suggest that AI can streamline existing methodological procedures and also enhance theoretical discovery. Third, while AI can reshape the epistemic practices of literature reviews, we find that careful human oversight and interpretation remain indispensable to avoid epistemic flattening, aligning with previous findings [25]. Fourth, by explicitly delineating human and AI responsibilities (as depicted in Fig. 1), our framework contributes to clearer methodological standards and transparency in AI-augmented SLRs, addressing calls from prior literature [30,35].

The practical implications of our study are substantial for researchers conducting SLRs. By demonstrating robust accuracy and balanced type I and type II error rates, our case study offers practical guidance for effectively calibrating AI assistance. Applying AI for initial screening and thematic identification can significantly reduce manual workload during these labor-intensive phases. This aligns with previous research [28,38], reinforcing the viability of hybrid workflows that blend AI-driven augmentation with human oversight. Our empirical insights also suggest practices for human-AI collaboration, highlighting the need for iterative validation, threshold calibration, and careful interpretive oversight. Given AI's observed conservative bias, researchers are advised to initially adopt moderate thresholds and subsequently refine their approach iteratively based on intermediate validations. Additionally, the complementary thematic insights identified by AI indicate that researchers can employ AI not merely for efficiency but also as a proactive means of uncovering theoretical insights and enhancing the depth and scope of literature reviews. Lastly, this research offers a notable managerial implication: While practitioners might not necessarily conduct or publish scientific SLRs themselves, the proposed AI-augmented framework could empower them to more effectively and efficiently engage with academic literature. In this way, practitioners can gain enhanced and timely insights into academic research, thus making scholarly knowledge more accessible in practical contexts.

Despite these promising findings, several limitations of our study warrant careful consideration. First, our illustration and evaluation focused on a single SLR domain, potentially limiting generalizability to other research contexts. Future studies should explore the applicability and robustness of our AI-augmented framework across diverse research topics and interdisciplinary domains. Second, LLMs inherently depend on the quality and comprehensiveness of their underlying training data. Thus, biases, inaccuracies, or gaps present in training datasets may propagate into AI-assisted SLR processes, underscoring the critical role of human validation and interpretative oversight. We deliberately used `GPT-4o-mini`, a cost-effective and fast model, making our AI-augmented method highly accessible to a broad range of researchers, even with minimal resources. However, future studies may explore the use of larger (and somewhat costlier) models, potentially improving accuracy or synthesis quality. Third, while our study evaluated AI's impact on efficiency, accuracy, and thematic alignment, additional qualitative investigations into researcher experiences, perceptions of cognitive burden, and trust in AI-generated outcomes would further enrich the understanding of effective human-AI collaboration. Future research should aim to extend our investigation by applying the framework to diverse SLRs across various disciplines to assess its broader applicability and robustness. This would also involve exploring comparisons against other AI-assisted SLR tools or more standardized benchmarks where feasible and appropriate for specific components of our framework, to further contextualize its performance.

In conclusion, this study introduced an AI-augmented SLR framework and provided initial empirical evidence from a comparative case study, suggesting its practical viability and theoretical robustness of integrating AI into SLRs. The proposed framework balances automation with human oversight, significantly enhancing efficiency and thematic exploration while maintaining scholarly rigor. Through iterative validation and human-AI role delineation, researchers can effectively use AI technologies to manage the growing volume of academic literature and deepen thematic insights. We thus contribute meaningfully to ongoing theoretical discussions and practical advancements in SLR methodologies.

Supplementary Material. The appendices for this paper, containing all the referenced tables, are available online at: https://doi.org/10.5281/zenodo.15781057.

Acknowledgments. The COMET project is funded by the Federal Ministry for Economic Affairs and Climate Action (BMWK) as part of the technology program "SmartLivingNEXT - Artificial intelligence for sustainable living and residential environments".

Disclosure of Interests. The authors have no competing interests to declare that are relevant to the content of this article.

References

1. Alahakoon, D., Nawaratne, R., Xu, Y., De Silva, D., Sivarajah, U., Gupta, B.: Self-building artificial intelligence and machine learning to empower big data analytics in smart cities. Inf. Syst. Front. **25**(1), 221–240 (2023). https://doi.org/10.1007/s10796-020-10056-x
2. Ampel, B., Yang, C.H., Hu, J., Chen, H.: Large language models for conducting advanced text analytics information systems research. ACM Trans. Manag. Inf. Syst. **16**(1), 2:1–2:27 (2025). https://doi.org/10.1145/3682069
3. Auth, G., Blei, R., Vogel, R., Jokisch, O.: One for all or the more the better? A survey of ready-to-use AI tools for literature review. In: Refereed Paper Proceedings - KM Conference 2025. Siena, Italy (2025)
4. Benbya, H., Strich, F., Tamm, T.: Navigating generative artificial intelligence promises and perils for knowledge and creative work. J. Assoc. Inf. Syst. **25**(1), 23–36 (2024). https://doi.org/10.17705/1jais.00861
5. Brîncoveanu, C., Carl, K.V., Binz, S., Weiher, M.A., Thomas, O., Hinz, O.: Artificial Intelligence-Based Assistance Systems for Environmental Sustainability in Smart Homes: A Systematic Literature Review on Requirements and Future Directions. In: Lecture Notes in Informatics (LNI) - Proceedings, Volume P-352, pp. 1165–1182. Gesellschaft für Informatik e.V. (2024)
6. Brocke, J., et al.: Reconstructing the giant: on the importance of rigour in documenting the literature search process. In: ECIS 2009 Proceedings (2009)
7. Carvalho, A., Levitt, A., Levitt, S., Khaddam, E., Benamati, J.: Off-the-shelf artificial intelligence technologies for sentiment and emotion analysis: a tutorial on using IBM natural language processing. Commun. Assoc. Inf. Syst. **44**(1) (2019). https://doi.org/10.17705/1CAIS.04443
8. Delavari, V., Shaban, E., Janssen, M., Hassanzadeh, A.: Thematic mapping of cloud computing based on a systematic review: a tertiary study. J. Enterp. Inf. Manag. **33**(1), 161–190 (2019). https://doi.org/10.1108/JEIM-02-2019-0034
9. Deng, C., Ravichandran, T.: Managerial response to online positive reviews: helpful or harmful? Inf. Syst. Res. **35**(4), 1802–1823 (2024). https://doi.org/10.1287/isre.2019.0175
10. Duan, Y., Edwards, J.S., Dwivedi, Y.K.: Artificial intelligence for decision making in the era of big data - evolution, challenges and research agenda. Int. J. Inf. Manag. **48**, 63–71 (2019). https://doi.org/10.1016/j.ijinfomgt.2019.01.021
11. Else, H.: Abstracts written by ChatGPT fool scientists. Nature **613**(7944), 423–423 (2023). https://doi.org/10.1038/d41586-023-00056-7

12. Feuerriegel, S., Hartmann, J., Janiesch, C., Zschech, P.: Generative AI. Bus. Inf. Syst. Eng. **66**(1), 111–126 (2024). https://doi.org/10.1007/s12599-023-00834-7
13. Gwon, Y.N.: The use of generative AI for scientific literature searches for systematic reviews: ChatGPT and Microsoft Bing AI performance evaluation. JMIR Med. Inform. **12**(1), e51187 (2024). https://doi.org/10.2196/51187
14. Heigl, R.: Generative artificial intelligence in creative contexts: a systematic review and future research agenda. Manag. Rev. Q. (2025). https://doi.org/10.1007/s11301-025-00494-9
15. Jenko, N., Ariyaratne, S., Jeys, L., Evans, S., Iyengar, K.P., Botchu, R.: An evaluation of AI generated literature reviews in musculoskeletal radiology. The Surgeon **22**(3), 194–197 (2024). https://doi.org/10.1016/j.surge.2023.12.005
16. Kadhim, A.I.: Survey on supervised machine learning techniques for automatic text classification. Artif. Intell. Rev. **52**(1), 273–292 (2019). https://doi.org/10.1007/s10462-018-09677-1
17. Kumar, S., Lim, W.M., Sivarajah, U., Kaur, J.: Artificial intelligence and blockchain integration in business: trends from a bibliometric-content analysis. Inf. Syst. Front. **25**(2), 871–896 (2023). https://doi.org/10.1007/s10796-022-10279-0
18. Larsen, P., von Ins, M.: The rate of growth in scientific publication and the decline in coverage provided by Science Citation Index. Scientometrics (2010). https://doi.org/10.1007/s11192-010-0202-z
19. Laursen, M., Svejvig, P.: Taking stock of project value creation: a structured literature review with future directions for research and practice. Int. J. Project Manag. **34**(4), 736–747 (2016). https://doi.org/10.1016/j.ijproman.2015.06.007
20. Lee, S.W., Choi, C.W., Kim, D.S., Yeo, W.Y., Kim, J.W.: Multi-category sentiment analysis for social opinion related to artificial intelligence on social media. J. Intell. Inf. Syst. **24**(4), 51–66 (2018). https://doi.org/10.13088/jiis.2018.24.4.051
21. Levy, Y., Ellis, T.J.: A systems approach to conduct an effective literature review in support of information systems research. Inf. Sci. Int. J. Emerg. Transdiscipl. **9**, 181–212 (2006)
22. Minaee, S., Kalchbrenner, N., Cambria, E., Nikzad, N., Chenaghlu, M., Gao, J.: Deep learning–based text classification: a comprehensive review. ACM Comput. Surv. **54**(3), 62:1–62:40 (2021). https://doi.org/10.1145/3439726
23. Moldovan, M.G., Dabija, D.C., Stanca, L., Pocol, C.B.: A qualitative study on the consumer behaviour related to food waste: Romanian perspectives through word cloud and sentiment analysis. Sustainability **16**(10), 4193 (2024). https://doi.org/10.3390/su16104193
24. Napoleão, B., Sarkar, R., Hallé, S., Petrillo, F., Kalinowski, M.: Emerging results on automated support for searching and selecting evidence for systematic literature review updates. In: International Workshop on Methodological Issues with Empirical Studies in Software Engineering (WSESE@ICSE 2024) (2024). https://doi.org/10.1145/3643664.3648202
25. Ngwenyama, O., Rowe, F.: Should we collaborate with AI to conduct literature reviews? Changing epistemic values in a flattening world. J. Assoc. Inf. Syst. **25**(1), 122–136 (2024). https://doi.org/10.17705/1jais.00869
26. O'Connor, C., Joffe, H.: Intercoder reliability in qualitative research: debates and practical guidelines. Int. J. Qual. Methods **19**, 1609406919899220 (2020). https://doi.org/10.1177/1609406919899220
27. Octaviano, F., Felizardo, K.R., Fabbri, S.C.P.F., Napoleão, B.M., Petrillo, F., Hallé, S.: SCAS-AI: a strategy to semi-automate the initial selection task in systematic literature reviews. In: 2022 48th Euromicro Conference on Software Engi-

neering and Advanced Applications (SEAA), pp. 483–490 (2022). https://doi.org/10.1109/SEAA56994.2022.00080

28. Ofori-Boateng, R., Aceves-Martins, M., Wiratunga, N., Moreno-Garcia, C.F.: Towards the automation of systematic reviews using natural language processing, machine learning, and deep learning: a comprehensive review. Artif. Intell. Rev. **57**(8), 200 (2024). https://doi.org/10.1007/s10462-024-10844-w
29. Pan, S.L., Nishant, R., Tuunanen, T., Nah, F.F.H.: Literature review in the generative AI era - how to make a compelling contribution. J. Strateg. Inf. Syst. **32**(3), 101788 (2023). https://doi.org/10.1016/j.jsis.2023.101788
30. Schryen, G., Marrone, M., Yang, J.: Exploring the scope of generative AI in literature review development. Electron. Mark. **35**(1), 13 (2025). https://doi.org/10.1007/s12525-025-00754-2
31. Snyder, H.: Literature review as a research methodology: an overview and guidelines. J. Bus. Res. **104**, 333–339 (2019). https://doi.org/10.1016/j.jbusres.2019.07.039
32. Spring, M., Faulconbridge, J., Sarwar, A.: How information technology automates and augments processes: insights from Artificial-Intelligence-based systems in professional service operations. J. Oper. Manag. **68**(6–7), 592–618 (2022). https://doi.org/10.1002/joom.1215
33. Teubner, T., Flath, C.M., Weinhardt, C., van der Aalst, W., Hinz, O.: Welcome to the era of ChatGPT et al. Bus. Inf. Syst. Eng. **65**(2), 95–101 (2023). https://doi.org/10.1007/s12599-023-00795-x
34. Tingelhoff, F., Brugger, M., Leimeister, J.M.: A guide for structured literature reviews in business research: the state-of-the-art and how to integrate generative artificial intelligence. J. Inf. Technol. **40**(1), 77–99 (2025). https://doi.org/10.1177/02683962241304105
35. Wagner, G., Lukyanenko, R., Paré, G.: Artificial intelligence and the conduct of literature reviews. J. Inf. Technol. **37**(2), 209–226 (2022). https://doi.org/10.1177/02683962211048201
36. Webster, J., Watson, R.: Analyzing the past to prepare for the future: writing a literature review. Manag. Inf. Syst. Q. **26**(2) (2002)
37. Wolfswinkel, J., Furtmueller, E., Wilderom, C.: Using grounded theory as a method for rigorously reviewing literature. Eur. J. Inf. Syst. **22** (2013). https://doi.org/10.1057/ejis.2011.51
38. Ye, A., Maiti, A., Schmidt, M., Pedersen, S.J.: A hybrid semi-automated workflow for systematic and literature review processes with large language model analysis. Future Internet **16**(5), 167 (2024). https://doi.org/10.3390/fi16050167

Balanced Reciprocity for Data Sharing—Axiomatization and Mechanism Design

Björn Filter(✉), Jan Schubsda, Ralf Möller, and Özgür Lütfü Özçep

Institute for Humanities-Centered AI (CHAI), University of Hamburg, Hamburg, Germany
{bjoern.filterbjoern.filter,ralf.moellerbjoern.filter, oezguer.oezcepbjoern.filter}@uni-hamburg.de

Abstract. Collaborative environments, particularly in data-sharing and cooperative settings, require robust mechanisms to fairly allocate rewards among participants. In many scenarios, participants contribute valuable data that improve the collective outcome, yet ensuring that each party is fairly compensated for their contribution remains a complex challenge. Classical cooperative game theory develops mechanisms under the assumption of non-replicable resources, which does not fit the scenario of sharing models or data. Hence, it does not address the potential imbalance in reciprocal benefits among pairs of participants, which can lead to strategic exploitation and, consequently, unfair reward allocations. We explicate fairness w.r.t. balanced reciprocity with an axiom and prove the existence of a reward allocation mechanism that fulfills this axiom, as well as other well-known axioms intended to capture incentivization and fairness. By emphasizing mutual fairness, our approach mitigates the risk of exploitation, fosters stable cooperation, and aligns with the principle of fair exchange. We propose a solution that guarantees proportional benefit allocation while maximizing group welfare under these constraints.

Keywords: Cooperative Game Theory · Reward Allocation · Mechanism Design · Model and Data Sharing · Fairness

1 Introduction

Together with the current breathtaking developments in AI research and applications around the world, the need for a global safe AI ecosystem has been emphasized [1]. The current grievances regarding the copyright issue for the training of large-language models or the exploitation of people for the annotation of training data are some of the examples that underpin this need. The intended ecosystem is meant to be regulated by social mechanisms that (provably) ensure safe and human-values aligned interactions between artificial and human agents. This

T. Braun et al. (Eds.): KI 2025, LNAI 15956, pp. 18–31, 2026.
https://doi.org/10.1007/978-3-032-02813-6_2

paper contributes to the formal foundation of such an ecosystem, focusing on the aspect of fairness in the allocation of rewards between agents.

The allocation of rewards in cooperative settings is a fundamental problem in economics, artificial intelligence, machine learning, and acquisition of research data. Here, multiple participants or players contribute data, computational resources, or expertise to a collective task. A key challenge in such settings is to determine a fair and incentive-compatible allocation of rewards. In data-driven collaborations, for example, participants may contribute data of varying number, quality, and relevance, necessitating a reward allocation mechanism that appropriately reflects their contributions. Applications of such cooperative reward allocation range from federated learning, in which data owners collaborate to train better machine learning models [2,3], over causal inference in multi-agent systems [4] to data acquisition [5].

A crucial challenge in reward allocation for cooperative settings is ensuring that contributions are valued in a way that discourages strategic imbalances and guarantees proportional benefit among participants, in particular, considering the infinite replicability of resources in data-sharing scenarios. Existing solutions, such as the Shapley value [6] and the Banzhaf value [7], provide well-established ways to distribute rewards based on marginal contributions. Various approaches exist that adapt these values to settings with replicable rewards [3,8,9]. However, these methods do not account for imbalances in mutual benefit. Specifically, if one participant contributes significantly to the rewards of another but does not receive a comparable benefit in return, it creates an exploitable asymmetry. This could lead to strategic manipulation, in which, in extreme cases, participants might even free-ride on others' contributions without fair compensation.

To address these limitations, we propose a novel reward allocation framework that incorporates a Balanced Reciprocity Axiom. This axiom ensures that no participant profits from another more than that participant profits in return. This principle is aligned with the economic concept of fair exchange, in which interactions should result in proportional benefits rather than asymmetrical dependencies. Just as in trade, where two parties exchange goods or services of roughly equal value, cooperative interactions should distribute rewards in a way that prevents one party from disproportionately benefiting at the expense of another. Furthermore, the Balanced Reciprocity Axiom prevents the formation of coercive dependencies within a coalition. If a participant has disproportionate influence over another, they may use their leverage to extract concessions by threatening to leave the coalition. By ensuring that the benefit of each player is proportional to their contribution to others, our approach fosters a more stable and mutually beneficial collaborative environment. We show that the Shapley and Banzhaf values are incompatible with our new axiom, and therefore, we present a novel solution function for reward allocation. We show that this function maximizes social welfare under our incentive and fairness constraints.

2 Related Work

Our research aligns with a growing body of work in mechanism design for collaborative data collection, often used for machine learning. Here, it is of interest how multiple parties, each with their own datasets, can collaborate to create joint datasets for training models or making predictions for a shared learning task [2–4,10]. A central challenge in this domain is to develop fair and effective reward mechanisms that evaluate the data contributions of each participant, ensuring that they are adequately recognized while incentivizing participation and maintaining fairness. To address this, numerous reward mechanisms inspired by cooperative game theory have been explored, with the Shapley value and the Banzhaf value being particularly well-studied linear reward functions [8,9,11].

Zou and colleagues propose a surplus allocation method for super-additive games, the proportional surplus division value [12]. In this framework, players possess the power to act as separators, being capable of sabotaging cooperation and reducing the game to its additive core, wherein each participant receives only their stand-alone value. Their proposed solution distributes the surplus so that for any pair of players, the potential loss each would incur from the other defecting as a separator is equal in proportion to their standalone values. This ensures a balanced disincentive structure, stabilizing cooperation through symmetrical threat mitigation.

However, these solution functions assume that rewards are scarce and divisible resources, which may not align with the nature of machine learning models that can be replicated at no cost. Sim and colleagues [3] address this discrepancy by adapting the Shapley value framework to settings where rewards are replicable goods, specifically ML models. They propose a novel reward scheme in which each participant receives an individualized model, with the quality of the model reflecting their contribution to the collaborative training process. This approach ensures fair compensation in the form of model performance proportional to data contributions. The authors further define key properties of the reward mechanism, drawing from cooperative game theory but adapting them to freely replicable model rewards. They first establish axioms that an effective reward scheme should satisfy to incentivize participation. In addition, they outline four fairness criteria to ensure an equitable allocation of rewards and propose a modified version of the Shapley value to meet these criteria. The resulting parametrized Shapley value is not linear w.r.t. the canonical addition and scalar multiplication of games. Those axioms (in a partly adapted form) are the basis for our approach, which adds the axiom of Balanced Reciprocity. As in the case of the parameterized Shapley value, our proposed reward mechanism, which applies the minimality function, is not linear.

Other studies also explore reward mechanisms in federated learning under the assumption of replicable goods. Karimireddy and colleagues [2] propose a data maximizing mechanism in which the value of each data point is assumed to be identical between players. As a result, a player's contribution to a coalition is measured solely by the volume of data provided. However, their primary focus is to maximize data contributions rather than ensuring fairness. In contrast, in a previous work [4] we argued that fairness and data maximization are often at odds with each other. Fairness typically requires limiting access to the best available model or dataset for some participants, restricting them from better models that could otherwise be leveraged as an incentive to encourage further data contributions.

3 Problem Formulation

In many cooperative settings, multiple participants each collect individual data points that are combined into a shared dataset. They are then granted (possibly restricted) access to this dataset or to a machine learning model trained on it. This collaborative approach enables all participants to benefit from the shared resources. The fundamental characteristic that we will focus on in this paper is that every collected data point is of equal worth, meaning that the value function of a coalition depends solely on the number of collected data points. This contrasts with cases in which data points exhibit diminishing returns, redundancy, or dependencies on prior knowledge. Instead, we consider a class of cooperative games where each additional data point contributes an equal and independent increase to the total value of the coalition, potentially without limit.

This setting is relevant in various real-world applications, ranging from scientific research and market intelligence to decentralized information gathering and artificial intelligence training. In such contexts, players are incentivized to collaborate, as pooling their data maximizes the collective benefit, which can then be distributed to all members. This cooperative game is characterized by an additive and monotonically increasing value function, where the value of a coalition is directly proportional to the total number of aggregated data points.

An example of a setting with these characteristics would be the digitization and transcription of historical manuscripts. In order to build a comprehensive digital repository of historical texts, which can be used for research, linguistic analysis, or cultural preservation, data from diverse sources have to be collected. Each participating researcher contributes by providing digitized manuscripts, transcriptions, or metadata, forming a shared dataset from which all members of the coalition can benefit. In this cooperative game, each data point, whether a manuscript scan, transcription, or annotation, provides equal and independent value, since every manuscript (or fragment) contains unique historical information, and its transcription adds to the overall knowledge base. There is no diminishing return because each new data point increases the chances of discovering patterns, linking references, or verifying historical events.

There are numerous similar settings in which each data point is of equal value, such as data collection for traffic, financial market analysis, genetic research, or a database for legal cases.

We will now describe how to model our setting formally as a cooperative game. Let $N \subset \mathbb{N}$ be a finite set of players, and we define $|N| = n$. Let a game v be a real-valued function $v : 2^N \to \mathbb{R}$ that satisfies $v(\emptyset) = 0$ and for all $C \subseteq N$, $v(C) \geq 0$. We call a set $C \subseteq N$ a coalition and $v(C)$ the value of that coalition.

To simplify the notation, we use v_C for $v(C)$, and, in the case of singular member coalitions, we write v_i for $v(\{i\})$. As mentioned above, we presume an additive game. This means that for any $C \subseteq N$, $v_C = \sum_{i \in C} v_i$. Let Γ^N denote the set of all games v with the set of players N.

In our setting, a solution function prescribes an element of $\mathbb{R}^{n \times 2^n}$ for each game $v \in \Gamma^N$. This function maps coalitions to the rewards agents receive. Each column corresponds to a possible coalition $C \subseteq N$ while each row corresponds to a player $i \in N$. For such a solution function

$$\mathcal{M}(v) = \begin{bmatrix} \mathcal{M}(v)_1^{\emptyset} & \mathcal{M}(v)_1^{\{1\}} & \mathcal{M}(v)_1^{\{2\}} & \dots & \mathcal{M}(v)_1^{C} & \dots & \mathcal{M}(v)_1^{N} \\ \vdots & \vdots & \vdots & \ddots & \vdots & \ddots & \vdots \\ \mathcal{M}(v)_n^{\emptyset} & \mathcal{M}(v)_n^{\{1\}} & \mathcal{M}(v)_n^{\{2\}} & \dots & \mathcal{M}(v)_n^{C} & \dots & \mathcal{M}(v)_n^{N} \end{bmatrix} \tag{1}$$

and any $C \subseteq N$, $i \in N$, let $\mathcal{M}(v)_i^C$ denote the reward obtained by player i if coalition C forms. We will write $\mathcal{M}_i^C$ for $\mathcal{M}(v)_i^C$ when the function v is clear from the context. We will call $\mathcal{M}_i^C$ the reward of the player i if coalition C forms.

In this approach, the solution function assigns a reward to every player, including those who are not part of a coalition. We assume that these players operate independently and obtain the value of their individual datasets. Alternatively, one might treat the solution function as a mechanism that interacts solely with the coalition, disregarding players outside the coalition and assigning them a value of zero. However, we decided to adopt the former approach because it accounts for the outcome of every player, not just those within a coalition. Furthermore, we note that the first $n+1$ columns in $\mathcal{M}(v)$ are redundant. These correspond to cases where either no coalition forms or a coalition consists of a single player, both of which yield the same outcome, where each player is working independently. We decided to stick to our definition because it makes the whole approach more homogeneous and eases the notation of the axioms.

4 Axioms for Incentivization and Fairness

An effective reward allocation mechanism in the described setting has to both incentivize players to participate and reward them fairly. To ensure this, we establish a set of axioms that govern valid reward distributions. We begin by defining incentive constraints that ensure feasibility, efficiency, and individual rationality. These constraints guarantee that the solution function remains viable and encourages participation. Building on this foundation, we introduce fairness constraints to prevent unjust disparities in reward distribution. The following sections formalize these principles and their implications.

4.1 Incentive Axioms

To formulate a valid solution function, we state the following incentive constraints, ensuring that a solution function is feasible, efficient, and individually rational. These are common solution concepts from cooperative game theory. For any $v \in \Gamma^N$ fulfilling the conditions stated in Sect. 3, a valid solution function $\mathcal{M}(v)$ must fulfill the following axioms:

R1 **Non-negativity:** Each player must get a non-negative reward: $\forall C \subseteq N$, $\forall i \in C : \mathcal{M}_i^C \geq 0$.

R2 **Feasibility:** The reward for each player in any coalition $C \subseteq N$ cannot be larger than the value achieved by that coalition: $\forall C \subseteq N$, $\forall i \in C : \mathcal{M}_i^C \leq v_C$.

R3 **Weak Efficiency:** In each coalition $C \subseteq N$, the reward received by at least one player $i \in C$ must be as large as the total value that coalition C can achieve: $\forall C \subseteq N \; \exists i \in C : \mathcal{M}_i^C = v_C$.

R4 **Individual Rationality:** Each player must receive a reward that is at least as large as the value that player can achieve by themselves: $\forall C \subseteq N$, $\forall i \in N : \mathcal{M}_i^C \geq v_i$.

R5 **Non-participation:** For any $C \subseteq N$, the reward received by players outside the coalition $i \notin C$ equals the value they achieve by themselves: $\forall C \subseteq N$, $\forall i \notin C : \mathcal{M}_i^C = v_i$.

R1 and R4 are the same solution concepts as in cooperative game theory with non-replicable rewards. R2 and R3 have been adapted from Chalkiadakis and colleagues [13], since in our setting we can give every member of a coalition $C \subseteq N$ a reward of up to v_C. We only require weak efficiency, since strong efficiency would imply having to pay out v_C to every member of the coalition, which would maximize welfare, but impede any fairness considerations. Axiom R5 is necessary in our setting, since a solution function describes rewards for every player regardless of whether they are in the coalition. Thus, it has to be ensured that they are treated as if they work alone.

4.2 Fairness Axioms

Additionally, to ensure a fair allocation of rewards, the solution function must meet the following four fairness axioms:

F1 **Uselessness:** If player i's data does not provide an increase to the value of any coalition (in our setting if $v_i = 0$), then player i should receive a valueless reward: For all $i \in N$,

$$\left(\forall C \subseteq N \setminus \{i\} : v_C = v_{C \cup \{i\}}\right) \Rightarrow \left(\forall C \subseteq N : \mathcal{M}_i^C = 0\right). \tag{2}$$

F2 **Symmetry:** If players i and j contribute identically to any coalition (in our setting if $v_i = v_j$), they should receive equal rewards: For all $i, j \in N$ s.t. $i \neq j$,

$$\left(\forall C \subseteq N \setminus \{i,j\} : v_{C \cup \{i\}} = v_{C \cup \{j\}}\right) \Rightarrow \left(\forall C \subseteq N : \mathcal{M}_i^C = \mathcal{M}_j^C\right). \tag{3}$$

F3 **Strict Desirability:** If the value of at least one coalition improves more by including player i instead of player j, but the reverse is not true, player i should receive a more valuable reward than j. For all $i, j \in N$ s.t. $i \neq j$,

$$\begin{aligned}&\left(\exists B \subseteq N \setminus \{i,j\} : v_{B \cup \{i\}} > v_{B \cup \{j\}}\right) \wedge \\ &\left(\forall C \subseteq N \setminus \{i,j\} : v_{C \cup \{i\}} \geq v_{C \cup \{j\}}\right) \Rightarrow \left(\forall C \subseteq N : \mathcal{M}_i^C > \mathcal{M}_j^C\right).\end{aligned} \tag{4}$$

F4 **Strict Monotonicity:** Suppose the value of at least one coalition containing player i improves (for example, by including more data of i), ceteris paribus, then player i should receive a more valuable reward than before: Let v and v' denote any two value functions over all coalitions $C \subseteq N$ and $\mathcal{M}(v)_i^C$ and $\mathcal{M}(v')_i^C$ be the corresponding values of rewards received by player i in coalition C. For all $C \in N$ and all $i \in C$,

$$\begin{aligned}(\exists i \in C : v'_i > v_i) \wedge \left(\forall j \in C \setminus \{i\} : v'_j = v_j\right) \wedge \left(v'_C > \mathcal{M}(v)_i^C\right) \Rightarrow \\ \mathcal{M}(v')_i^C > \mathcal{M}(v)_i^C.\end{aligned} \tag{5}$$

F1 and F2 are axioms of the Shapley value [6]. Axiom F3 was first introduced by Maschler and Peleg [14] and, in our setting, reduces to the fact that players who contribute larger values should receive larger rewards, compared to players with less valuable contributions. F4 is adapted from Young [15]. These axioms together as conditions for fairness were introduced by Sim and colleagues [3].

We expand these fairness conditions by our so-called Balanced Reciprocity Axiom. It states that no player should profit from another player more than that player profits from the first player. If one player contributes significantly to another player's rewards but does not receive a comparable benefit in return, this creates an imbalance that can lead to strategic exploitation. The Balanced Reciprocity Axiom ensures that rewards are assigned proportionally to mutual benefit, preventing situations where a player can make a free ride on another's contributions without fair compensation.

Suppose agents i and j are members of a coalition C. The difference in the reward that agent i receives in $C \setminus \{j\}$ versus C represents the value i gains from cooperating with j. A natural fairness requirement would be that this gain equals the benefit that j derives from cooperating with i, that is, the difference in j's reward in $C \setminus \{i\}$ versus C. Our Balanced Reciprocity Axiom formalizes precisely this idea. We believe this interpretation of fairness is both important and novel in the context of cooperative game theory.

When agents are rewarded according to this axiom, a second desirable property emerges: Balance of strategic leverage. If agent j threatens to leave the coalition, agent i would lose value (from M_i^C to $M_i^{C\setminus j}$). This gives j potential leverage over i. However, our axiom ensures that i holds the same leverage over j, since j would suffer an equal loss if i were to leave. Thus, no agent is in a position to exploit another through threats or strategic withdrawals.

We formulate this axiom as follows:

F5 **Balanced Reciprocity:** Any agent i should profit from the participation of any other agent j in the same way that j profits from the participation of i. For all $C \subseteq N$ and all $i, j \in C$:

$$\mathcal{M}_i^C - \mathcal{M}_i^{C\setminus\{j\}} = \mathcal{M}_j^C - \mathcal{M}_j^{C\setminus\{i\}}, \tag{6}$$

We consider this axiom as a fairness condition, and hence name it as the fifth fairness axiom F5. For any game $v \in \Gamma^N$ that fulfills the conditions stated in Sect. 3, for a function $\mathcal{M}(v)$ to be a valid solution, the following must hold:

R6 **Fairness:** For all $C \subseteq N$ and $i \in N$, the rewards $\mathcal{M}_i^C$ must satisfy F1 to F5.

5 Solution Function

Our solution function satisfies the axioms of Sect. 4, in particular F5, by explicitly balancing reciprocity among participants. The intuition behind our approach is simple: Each player receives a baseline reward equal to their individual contribution, ensuring individual rationality. Additionally, to reflect the cooperative nature of the setting, each player also benefits from the contributions of others to the extent that they can provide a comparable value in return. For any $C \subseteq N$ and any $i \in N$,

$$\mathcal{M}(v)_i^C = \begin{cases} v_i + \sum_{j \in C \setminus \{i\}} \min(v_i, v_j) & \text{if } i \in C, \\ v_i & \text{else,} \end{cases} \tag{7}$$

Indeed, the solution function is the reward mechanism that we were looking for: It fulfills all standard axioms of incentivization and fairness, as well as the Balanced Reciprocity Axiom (Theorem 1), and it is the only function that maximizes group welfare (Theorem 4 further below).

Theorem 1. *The solution function $\mathcal{M}(v)_i^C$ fulfills R1 to R6.*

Proof. We will show, that $\mathcal{M}(v)$, as described above, satisfies R1 to R5 and F1 to F5. We will assume some arbitrary but fixed function v, which fulfills the criteria listed in Sect. 3.

R1 follows directly from the fact that for all i: $v_i \geq 0$, thus for all $C \subseteq N$ and all $i \in N$: $\mathcal{M}_i^C \geq 0$.
R2 holds because for any $C \subseteq N$ and all $i \in C$: $\mathcal{M}_i^C = v_i + \sum_{j \in C \setminus \{i\}} \min(v_i, v_j) \leq \sum_{j \in C} v_j = v_C$, due to the additive property of v.
R3 holds because: For any $C \subseteq N$, let $i = \arg\max_{j \in C} v_j$. Then, again due to the additive property of v: $\mathcal{M}_i^C = v_i + \sum_{j \in C \setminus \{i\}} \min(v_i, v_j) = \sum_{j \in C} v_j = v_C$.
R4 holds, since $\mathcal{M}_i^C \geq v_i$ for all $C \subseteq N$ and all $i \in N$.
R5 holds, since for all $C \subseteq N$ and all $i \notin C$, $\mathcal{M}_i^C = v_i$.

F1 holds because if $v_i = 0$, then $\mathcal{M}_i^C = 0 + \sum_{j \in C \setminus \{i\}} \min(0, v_j) = 0$.
F2 holds because if $v_i = v_j$, then for all $C \subseteq N$: $\mathcal{M}_i^C = v_i + \sum_{k \in C \setminus \{i\}} \min(v_i, v_k) = v_j + \sum_{k \in C \setminus \{j\}} \min(v_j, v_k) = \mathcal{M}_j^C$.

F3 holds because if $v_i > v_j$, then

$$\mathcal{M}_i^C = v_i + \sum_{k \in C \setminus \{i\}} \min(v_i, v_k) > v_j + \sum_{k \in C \setminus \{j\}} \min(v_j, v_k) = \mathcal{M}_j^C \tag{8}$$

F4 holds because if $(\exists i \in N : v_i' > v_i) \wedge (\forall j \in N \setminus \{i\} : v_j' = v_j)$, then, due to the additivity of v: $v_C' > v_C$. Furthermore: $\mathcal{M}(v')_i^C = v_i' + \sum_{j \in C \setminus \{i\}} \min(v_i', v_j') > v_i + \sum_{j \in C \setminus \{j\}} \min(v_i, v_j) = \mathcal{M}(v)_i^C$.

F5 holds because of the following chain of equations:

$$\begin{aligned}
&\mathcal{M}_i^C - \mathcal{M}_i^{C \setminus \{j\}} \\
&= \left(v_i + \sum_{k \in C \setminus \{i\}} \min(v_i, v_k) \right) - \left(v_i + \sum_{k \in C \setminus \{i,j\}} \min(v_i, v_k) \right) \\
&= \min(v_i, v_j) = \min(v_j, v_i) \\
&= \left(v_j + \sum_{k \in C \setminus \{j\}} \min(v_j, v_k) \right) - \left(v_j + \sum_{k \in C \setminus \{i,j\}} \min(v_j, v_k) \right) \\
&= \mathcal{M}_j^C - \mathcal{M}_j^{C \setminus \{i\}}
\end{aligned} \tag{9}$$

Thus, our mechanism from Eq. (7) fulfills R1 to R6. □

5.1 Comparison to Shapley Value and Banzhaf Value

We will now compare our solution function with the Shapley and Banzhaf values. Sim and colleagues [3] presented a mechanism for a setting with replicable goods, similar to ours, and a solution function based on the Shapley value. As we will demonstrate, such a solution is not possible here, since it is not compatible with our Balanced Reciprocity Axiom.

The Shapley value [6] is defined as follows:

$$Shapley(i) = \frac{1}{n!} \sum_{\pi \in \Pi_N} \left(v\left(S_{\pi,i} \cup \{i\}\right) - v\left(S_{\pi,i}\right) \right), \tag{10}$$

where Π_N is the set of all possible permutations of N and $S_{\Pi,i}$ is the coalition of parties preceding i in permutation π. Similar to this, the Banzhaf value [16] is defined as:

$$Banzhaf(i) = \frac{1}{2^{n-1}} \sum_{C \subseteq N \setminus \{i\}} \left(v_{C \cup \{i\}} - v_C \right), \tag{11}$$

In our setting, the equations reduce to $Shapley(i) = \frac{1}{n!} \sum_{\pi \in \Pi_N} (v_i) = v_i$ and $Banzhaf(i) = \frac{1}{2^{n-1}} \sum_{\pi \in \Pi_N} (v_i) = v_i$, respectively. So, here, the Shapley and Banzhaf values are equal and simply correspond to the values v_i for all $i \in N$.

However, the distribution of rewards according to the Shapley or Banzhaf values would not fulfill R3 since no agent would get the maximum reward. Hence, Shapley or Banzhaf rewards should be scaled in such a way that the agent with the largest contribution gets the maximum reward. How this can be done was shown by Sim and colleagues [3], however, due to our simplified setting, the following scaling is sufficient. Furthermore, to stay in line with our definition of a solution function from Eq. 1, we define our scaled Shapley or Banzhaf for any coalition $C \subseteq N$, distributing a value of v_i to all players $i \notin C$ outside a coalition $C \subseteq N$:

$$\phi(v)_i^C = \begin{cases} v_C \frac{v_i}{v_C^{max}} & \text{if } i \in C, \\ v_i & \text{else,} \end{cases} \tag{12}$$

where $v_C^{max} = \max_{i \in C} v_i$. We will assume that $\forall i \in C : v_i > 0$, thus always $v_C^{max} > 0$. If this were not the case, some agent in C not would contribute any valuable data, and thus should be left out. Thus, similarly to the setting examined by Sim and colleagues, we can state the following result.

Theorem 2. *$\phi(v)_i^C$ fulfills R1 to R5, as well as F1 to F4.*

Proof. Let v be arbitrary but fixed.

- R1 holds, since for all $C \subseteq N$, $i \in C$: $v_C \geq v_C^{max} \geq v_i \geq 0$. Thus, for all $i \in C$, $\phi(v)_i^C = v_C \frac{v_i}{v_C^{max}} \geq 0$.
- R2 holds because for all i in any $C \subseteq N$: $\frac{v_i}{v_C^{max}} \leq 1$, thus $v_c \geq v_c \frac{v_i}{v_C^{max}} = \phi(v)_i^C$.
- R3 holds due to the following observations: For any $C \subseteq N$, let $i = \arg\max_{j \in C} v_j$, i.e. $v_i = v_C^{max}$. Then $\phi(v)_i^C = v_C \frac{v_i}{v_i} = v_C$.
- R4 holds, since $\frac{v_c}{v_C^{max}} \geq 1$ and $v_C \geq v_i$, thus $v_c \frac{v_i}{v_C^{max}} = v_i \frac{v_c}{c_C^{max}} \geq v_i$ for all $i \in C$.
- R5 holds, since for all $C \subseteq N$ and all $i \notin C$, $\phi(v)_i^C = v_i$.
- F1 holds since if $v_i = 0$, then by definition $\phi(v)_i^C = 0$.
- F2 holds because if $v_i = v_j$, then $\phi(v)_i^C = v_C \frac{v_i}{v_C^{max}} = v_C \frac{v_j}{v_C^{max}} = \phi(v)_j^C$.
- F3 holds because if $v_i > v_j$, then $\phi(v)_i^C = v_C \frac{v_i}{v_C^{max}} > v_C \frac{v_j}{v_C^{max}} = \phi(v)_j^C$.
- F4 because if $(\exists i \in N : v_i' > v_i) \wedge (\forall j \in N \setminus \{i\} : v_j' = v_j)$, then, due to the additivity of v: $v_C' > v_C$. Furthermore, $\frac{v_i'}{v'^{max}_C} \geq \frac{v_i}{v_C^{max}}$, since if $v'^{max}_C > v_C^{max}$, this would imply $v'^{max}_C = v_i'$ and therefore $\frac{v_i'}{v'^{max}_C} = 1$, while $\frac{v_i}{v'^{max}_C} \leq 1$. Thus, $\phi(v')_i^C = v_C' \frac{v_i'}{v'^{max}_C} > v_C \frac{v_i}{v_C^{max}} = \phi(v)_i^C$. □

However, $\phi(v)_i^C$ is not compatible with the new Balanced Reciprocity Axiom. In particular, this result shows that our new axiom F5 is not entailed by the other axioms.

Theorem 3. *The scaled Shapley value/Banzhaf value $\phi(v)_i^C$ does not fulfill F5.*

Proof. Let v be arbitrary but fixed. Let $C \subseteq N$ with $|C| \geq 2$. Let $v_i = v_C^{max}$ and v_j be the second highest contribution with $v_i > v_j$, thus $v_{C\setminus\{i\}}^{max} = v_j$. If $\phi(v)_i^C$ fulfilled F5, the following would have to hold:

$$\begin{aligned} \mathcal{M}_i^C - \mathcal{M}_i^{C\setminus\{j\}} &= v_C \frac{v_i}{v_C^{max}} - (v_C - v_j) \frac{v_i}{v_C^{max}} \\ &= v_C \frac{v_j}{v_C^{max}} - (v_C - v_i) \frac{v_j}{v_{C\setminus\{i\}}^{max}} = \mathcal{M}_j^C - \mathcal{M}_j^{C\setminus\{i\}}. \end{aligned} \tag{13}$$

However, this would lead to the following chain of equations:

$$\begin{aligned} v_C \frac{v_i}{v_C^{max}} - (v_C - v_j) \frac{v_i}{v_C^{max}} &= v_C \frac{v_j}{v_C^{max}} - (v_C - v_i) \frac{v_j}{v_{C\setminus\{i\}}^{max}} \\ \Rightarrow \frac{v_j v_i}{v_C^{max}} &= v_C \frac{v_j}{v_C^{max}} - (v_C - v_i) \frac{v_j}{v_{C\setminus\{i\}}^{max}} \\ \Rightarrow \frac{v_j v_i}{v_C^{max}} &= v_C \frac{v_j}{v_C^{max}} - (v_C - v_i) \\ \Rightarrow v_j v_i &= v_C v_j - v_C^{max}(v_C - v_i) \\ \Rightarrow v_j v_i - v_C v_j &= -v_C^{max}(v_C - v_i) \\ \Rightarrow v_j(v_i - v_C) &= v_C^{max}(v_i - v_C) \\ \Rightarrow v_j &= v_C^{max}, \end{aligned} \tag{14}$$

which is a contradiction to our assumption that $v_C^{max} = v_i > v_j$. □

However, in the special case where for some $C \subseteq N$ and for all $i, j \in C$, $v_i = v_j$, both under our solution function and under the scaled Shapley or Banzhaf values, all agents achieve the same (maximal) reward, thus, in this case, the Shapley value, the Banzhaf value, and our mechanism lead to equal outcomes. Furthermore, any mechanism that fulfills R3 and F3 would lead to the same result.

5.2 Achieving Group Welfare

For any coalition $C \subseteq N$, the group welfare is defined as the total sum of rewards received by all players: $\sum_{i \in N} \mathcal{M}_i^C$ for any $C \subseteq N$. Since we allow any coalition $C \subseteq N$, we calculate the group welfare for all players $i \in N$, including those outside the coalition. However, due to R5, for any $C \subseteq N$, all players $i \notin C$ outside C will only receive their values v_i. For any $C \subseteq N$, maximal group welfare is achieved by giving the maximum reward v_C to each player in C. This will usually be incompatible with the fairness incentives. For example, neither the Shapley nor the Banzhaf value achieves maximal group welfare. The same is true for our solution function. This motivates considering group welfare maximization not as an absolute axiom but as an objective function to be maximized w.r.t. axioms considered as constraints (Comparable to linear programming). Now, we can show that our reward function maximizes group welfare under the constraints induced by axioms R1 to R6.

Theorem 4. *The solution function $\mathcal{M}$ defined in Eq. (7) maximizes group welfare while fulfilling R1 to R6.*

Proof. Let v be arbitrary but fixed. We have already seen that $\mathcal{M}$ fulfills R1 to R6. Suppose, there was another mechanism $\mathcal{M}'$, which also fulfilled R1 to R6 but achieved larger group welfare, so for some $C \subseteq N$, $\sum_{i \in C} \mathcal{M}'^C_i > \sum_{i \in C} \mathcal{M}^C_i$. We can disregard all $j \notin C$ here, since due to R5, for all $j \notin C$, $\mathcal{M}'^C_j = \mathcal{M}^C_j = v_j$. Then there must be at least one player $k \in C$, who receives a larger reward under $\mathcal{M}'$ compared to under $\mathcal{M}$, i.e., $\mathcal{M}'^C_k > \mathcal{M}^C_k$. Let $j = \arg\max_{i \in C} v_i$. Since $\mathcal{M}^C_j = v_C$, it must hold $k \neq j$, otherwise $\mathcal{M}'$ would violate R2. Consider now $C' = C \setminus \{j\}$. The following must hold:

$$\mathcal{M}'^C_k - \mathcal{M}'^{C'}_k = \mathcal{M}'^C_j - \mathcal{M}'^{C \setminus \{k\}}_j = \mathcal{M}^C_j - \mathcal{M}^{C \setminus \{k\}}_j = v_k. \tag{15}$$

Here, the first equation follows from the fact that $\mathcal{M}'$ fulfills F5. The second equation follows from the fact that both $\mathcal{M}$ and $\mathcal{M}'$ fulfill F3 and R3. Thus, under both j must get the maximum reward. The third equation follows directly from the definition of $\mathcal{M}$. Therefore: $\mathcal{M}'^C_k - \mathcal{M}'^{C'}_k = v_k = \mathcal{M}^C_k - \mathcal{M}^{C'}_k$, which leads to $\mathcal{M}'^{C'}_k > \mathcal{M}^{C'}_k$. We can now apply the same argument as before: Let $j' = \arg\max_{i \in C'} v_i$. Since $\mathcal{M}^{C'}_{j'} = v_{C'}$, it must hold $k \neq j'$, otherwise $\mathcal{M}'$ would violate R2. Set now $C'' = C' \setminus \{j'\}$ and proceed as before. By applying the same procedure $|C| - 1$ times, we will finally reach a coalition $\hat{C} = \{k\}$, such that: $\mathcal{M}'^{\hat{C}}_k > \mathcal{M}^{\hat{C}}_k = v_k$. But this would contradict R2. Thus, such a mechanism $\mathcal{M}'$ cannot exist and $\mathcal{M}$ is welfare maximizing while achieving R1 to R6. □

6 Conclusion and Future Work

We propose a novel framework for reward allocation in additive cooperative games with equally valuable data contributions. Grounded in cooperative game theory, our approach enforces incentive and fairness constraints. A key contribution is the Balanced Reciprocity Axiom, which prevents strategic imbalances and ensures proportional distribution of mutual benefits.

We formally establish that our proposed solution function satisfies all specified constraints, providing a principled mechanism for cooperation which is both fair and incentive-compatible. In contrast, we demonstrate that neither the Shapley value nor the Banzhaf value can accommodate the Balanced Reciprocity Axiom, highlighting a fundamental limitation of these classical solution concepts in our setting. Furthermore, we show that our solution function maximizes group welfare while preserving fairness and incentivization.

The results of this paper rely on the assumption that games are additive. In the case of classical cooperative game theory, where resources are assumed to be scarce and divisible, the theory of reward functions becomes trivial: this is reflected in the trivialization of the Shapley value and the Banzhaf value. However, in our case, where we assume infinitely replicable data, still relevant aspects of fair reward allocation in the form of the Axiom of Balanced Reciprocity emerge. Future research should drop the additivity constraint and extend the framework laid down in this paper to settings with heterogeneous value functions, including cases with diminishing returns, redundant data, or context-dependent valuations. An open question is whether the Balanced Reciprocity Axiom remains compatible with the other fairness axioms in these more complex environments. Additionally, exploring dynamic coalition formation, where players enter and exit over time, would provide a more realistic model of real-world cooperation. Developing a reward allocation mechanism that accounts for these dynamics remains an important direction for future work.

References

1. Conitzer, V., et al.: Social Choice Should Guide AI Alignment in Dealing with Diverse Human Feedback. arXiv e-prints arXiv:2404.10271 (2024)
2. Karimireddy, S.P., Guo, W., Jordan, M.I.: Mechanisms that Incentivize Data Sharing in Federated Learning. Papers 2207.04557, arXiv.org (2022)
3. Sim, R.H.L., Zhang, Y., Chan, M.C., Low, B.K.H.: Collaborative machine learning with incentive-aware model rewards. In: International Conference on Machine Learning, pp. 8927–8936. PMLR (2020)
4. Filter, B., Möller, R., Özçep, Ö.L.: Mechanisms for data sharing in collaborative causal inference. In: German Conference on Artificial Intelligence (Künstliche Intelligenz), pp. 86–98. Springer (2024)
5. Duan, L., Kubo, T., Sugiyama, K., Huang, J., Hasegawa, T., Walrand, J.: Incentive mechanisms for smartphone collaboration in data acquisition and distributed computing. In: 2012 Proceedings IEEE INFOCOM, pp. 1701–1709. IEEE (2012)
6. Shapley, L.S.: A value for n-person games. In: Kuhn, H.W., Tucker, A.W. (eds.) Contributions to the Theory of Games II, pp. 307–317. Princeton University Press, Princeton (1953)
7. Lehrer, E.: An axiomatization of the banzhaf value. Int. J. Game Theory **17**, 89–99 (1988)
8. Wang, T., Rausch, J., Zhang, C., Jia, R., Song, D.: A principled approach to data valuation for federated learning. Federated Learning: Privacy and Incentive, pp. 153–167 (2020)
9. Wang, J.T., Jia, R.: Data banzhaf: a robust data valuation framework for machine learning. In: International Conference on Artificial Intelligence and Statistics, pp. 6388–6421. PMLR (2023)
10. Xu, X., Lyu, L., Ma, X., Miao, C., Foo, C.S., Low, B.K.H.: Gradient driven rewards to guarantee fairness in collaborative machine learning. Adv. Neural Inf. Process. Syst. **34**, 16104–16117 (2021)
11. Jia, R., et al.: Towards efficient data valuation based on the shapley value. In: The 22nd International Conference on Artificial Intelligence and Statistics, pp. 1167–1176. PMLR (2019)
12. Zou, Z., van den Brink, R., Funaki, Y.: Sharing the surplus and proportional values. Theory Decis. 1–33 (2020)
13. Chalkiadakis, G., Elkind, E., Wooldridge, M.: Computational Aspects of Cooperative Game Theory. Morgan & Claypool Publishers, San Rafael (2011)
14. Maschler, M., Peleg, B.: A characterization, existence proof and dimension bounds for the kernel of a game. Pac. J. Math. **18**(2), 289–328 (1966)

15. Young, H.P.: Monotonic solutions of cooperative games. Int. J. Game Theory **14**(2), 65–72 (1985)
16. Banzhaf, J.: Weighted voting doesn't work: a mathematical analysis. Rutgers Law Rev. **19**(2), 317–343 (1965)

Toward Simple and Robust Contrastive Explanations for Image Classification by Leveraging Instance Similarity and Concept Relevance

Yuliia Kaidashova, Bettina Finzel(✉), and Ute Schmid

Cognitive Systems, University of Bamberg, Weberei 5, 96047 Bamberg, Germany
yuliia.kaidashova@stud.uni-bamberg.de,
{bettina.finzel,ute.schmid}@uni-bamberg.de
https://www.uni-bamberg.de/kogsys/

Abstract. Understanding why a classification model prefers one class over another for an input instance is the challenge of contrastive explanation. This work implements concept-based contrastive explanations for image classification by leveraging the similarity of instance embeddings and relevance of human-understandable concepts used by a fine-tuned deep learning model. Our approach extracts concepts with their relevance score, computes contrasts for similar instances, and evaluates the resulting contrastive explanations based on explanation complexity. Robustness is tested for different image augmentations. Two research questions are addressed: (1) whether explanation complexity varies across different relevance ranges, and (2) whether explanation complexity remains consistent under image augmentations such as rotation and noise. The results confirm that for our experiments higher concept relevance leads to shorter, less complex explanations, while lower relevance results in longer, more diffuse explanations. Additionally, explanations show varying degrees of robustness. The discussion of these findings offers insights into the potential of building more interpretable and robust AI systems.

Keywords: Explainable Artificial Intelligence · Contrastive Explanation · Complexity · Robustness · Concept-based Explanation

1 Introduction

Artificial Intelligence systems have produced impressive results because of the rapid development of deep learning, which has resulted in significant technological advancement and broad adoption. But a big problem with deep learning models is that they frequently act as "black boxes". This may result in an inability to evaluate such systems' reliability, endangering adoption by human users due to a lack of trust in AI [2,3,5,12,25].

The goal of Explainable Artificial Intelligence (xAI) is to shed light on the inner workings of these "black-box" models [12,16]. Among the diverse range

T. Braun et al. (Eds.): KI 2025, LNAI 15956, pp. 32–46, 2026.
https://doi.org/10.1007/978-3-032-02813-6_3

of xAI approaches [2,3,27,29], *concept-based explanations* (see, e.g., Poeta et al. [22]) currently stand out for their focus on interpretability. We introduce *concept-based contrastive explanations* that explain a model's decision using human-understandable concepts that distinguish instances from different classes.

As the human is considered in this work as the recipient of explanations, we refer to the term *concept* from a cognitive psychology perspective. A concept is a mental representation that helps people categorize and name objects, events, or ideas based on shared features [18,21]. Concepts further play a crucial role in cognition by allowing humans to recognize patterns, make decisions, and infer the properties of new instances based on prior knowledge [18,21].

One important aspect related to a human's processing of information is the evaluation of an explanation's complexity (see, e.g., Schwalbe & Finzel for an overview of methods and metrics [27]). As demonstrated by Kulesza et al. [15] an explanation should be as concise and complete as possible, given that it contains the relevant information to be well understood and processed by human users.

In this work, explanation complexity is measured by the length of an explanation which is determined by the number of concepts included. The shorter (simpler) explanations are preferred, as they likely improve human understanding in cases that may be easily confused otherwise [10,31].

Another important criterion is explanation robustness, which refers to the stability of an explanation when the input is slightly modified [3,4,27]. According to Arcaini et al., a common way to test robustness is image augmentations, which are small, controlled modifications applied to input images [4]. Gaussian noise is one of the typical augmentation techniques, where adaptations are normally distributed across the image to simulate sensor noise. Image rotation is very common, too (e.g., by 1 to 180° to test the model's consistency under geometric transformations) among other techniques such as brightness variation, blur, and compression [4]. Robust explanations should remain consistent across similar inputs, maintaining the same set of concepts and preserving the explanation length [4]. This work focuses on alterations by introducing Gaussian noise and by applying either 10° or 180° rotation to input images as all these modifications are expected to impact the recognition of material and concept-specific spatial orientation [4].

Further, the domain of image classification is a suitable use case for investigating and improving the interpretability of learned models with concept-based contrastive explanations. Even though models like Convolutional Neural Networks (CNNs) are found to be very accurate, they often make decisions based on small details and diverse pixel values, which makes it difficult to understand why they chose a certain label for an image. Concept-based contrastive explanations can help by showing which visual parts of the image were important for the model's decision, and how human-understandable representations of such visual parts made the model choose one class over another. Linking these concepts and contrasts with the relevance of each visual part, allows for an explanation that takes into account the different contribution strengths of concepts that result from the model's weight learning.

To explore the potential of concept-based contrastive explanations in image classification, this work focuses on explanation complexity and robustness and considers different relevance ranges for explanation generation. More specifically, relevance ranges are determined based on quantiles over the relevance distribution of relevance-ranked concepts. Accordingly, the following research questions are formulated for the evaluation of our novel approach:

R1 Does the length of concept-based explanations vary for similar images within different relevance ranges?
R2 Does the length of concept-based explanations remain stable under image augmentations such as noise, 10 ° rotation, and 180° rotation?

Subsequently, we introduce the related work that inspired our approach (Sect. 2), introduce our methodology (Sect. 3), evaluate and discuss it (Sect. 4 and Sect. 5) and conclude with a summary of our findings in Sect. 6.

2 Related Work

Linking cognitive research with xAI, the term *concept* typically refers to higher-level, semantically meaningful attributes learned by a model, which can be interpreted and understood by humans [22]. In the explainability literature, *concepts* are defined in various ways, including, but not limited to symbolic concepts, unsupervised concept bases, prototypes, and textual concepts [9,22]. This work focuses on symbolic concepts, which are human-defined symbols representing interpretable abstractions of a feature, such as *red*, *striped* or *handle* [9,22]. Such concepts are well-suited for model interpretability since they are naturally consistent with human understanding.

To fully leverage the usefulness of concepts in explaining model predictions, it is essential to understand what concept-based explanations are. One of the categories of concept-based explanations is the *node-concept association*, which refers to explicitly linking internal units (e.g., neurons or filters) in a neural network to specific concepts [9,22]. This can be achieved either post hoc, by analyzing which concepts maximally activate certain units, or during training, by requiring a unit to predict a concept [22]. This approach improves model transparency by revealing whether and which internal representations correspond to human-understandable terms, thereby allowing explainees to better understand what a network has learned [1,9,22].

Contrastive explanations aim to explain why a model chose one class over another by contrasting outcomes for instances a model has been applied to (see, e.g., Stepin et al. [29]). Unlike traditional explanations that focus on a single class, contrastive explanations identify which features would need to be altered in order to predict a different class or which features are missing to support the alternative class. In the context of xAI, understanding how humans naturally interpret explanations is crucial [5,8,25]. Research in philosophy and cognitive science highlights that people do not typically ask for explanations in isolation "Why P?", but rather in relation to an alternative that did not occur (see, e.g.,

Miller [20]). This form of contrastive explanation is typically structured as "Why P rather than Q?", where P is the observed fact and Q is the imagined foil, an alternative that was expected or considered but did not occur. The foil may not be explicitly stated in the question, but humans are generally capable of inferring it from context, such as tone, expectations, or background knowledge [20]. One of the very first contrastive approaches was introduced, e.g., in Winston's seminal work on *near misses* that are differently classified explanatory examples lacking a minimal set of properties. [24,31].

Lipton emphasizes that a contrastive explanation is often cognitively easier to produce and more relevant to human reasoning than exhaustive explanations of a fact [17]. For example, when classifying an image as a teapot rather than a vase, mentioning a "spout" might be sufficient to explain the contrast, even if it does not completely explain why some vessel is a teapot (see Finzel et al. [9]).

Building on the theoretical foundations of contrastive explanation, several approaches in machine learning have proposed methods to generate such explanations for neural networks. One known area of research is using input space perturbations to create contrastive explanations [22,29]. For example, the work of Dhurandhar et al. [7] introduced the concept of *pertinent negatives*, which are features that are deemed irrelevant by a model and need to be included to get another prediction. Their method generates contrastive explanations by identifying minimal changes to an input that would alter the classification, effectively answering the question: "Why is input x classified in class y?". This is achieved by finding points close to the original input that lie just outside the decision boundary of the current class but within the boundary of a contrasting class. These points help highlight what is missing in the input to prevent classification into the foil class, thus providing a contrastive explanation through absence.

Prabhushankar et al. [23] propose a method for contrastive explanations in neural networks that focuses on identifying salient features that separate two classes. Their approach constructs explanations by comparing the internal activations of the neural network for two target classes: the predicted class and an alternative class. By analyzing differences in feature importance between these classes, the method highlights the contrastive attribution, that is the set of features that pushes the model towards one class and away from another.

While both methods ([7,23]) effectively identify differences between classes, they primarily focus on plain features rather than human-understandable concepts. This can limit the interpretability for users, as the explanations lack semantic clarity. Including recognizable concept-based elements could significantly improve human perception and trust in the explanations.

Recent works by Finzel et al. [9,10] propose a framework for contrastive explanations based on the extraction of human-understandable concepts from node-concept associations to explain model decisions in image classification. The concept-based and relational explainer introduced in [9] utilizes a sampling method that selects concepts for explanation generation based on quantiles over the relevance distribution of concepts, where concepts are ranked by the strength of their relevance. The conceptual framework presented in [10] suggests to compute the similarity of *embeddings* (a model's internal vectorized representation of input data) for the generation of concept-based contrastive explanations as embeddings can yield more meaningful similarity scores compared to plain input.

This work presents an actual implementation of these ideas utilizing embeddings for similarity computation and adds an evaluation of the robustness and interpretability of *contrastive* explanations dependent of different *relevance ranges* (different quantiles over the relevance distribution) for node-concept associations. It builds upon the state of the art in concept-based and contrastive explanation and particularly contributes an evaluation based on measurable metrics for robustness and an empirical investigation of the relevance range's effect on explanation simplicity.

3 Concept-Based Contrastive Explanations

Our methodology for generating human-understandable contrastive explanations is applied to image classification with a CNN. The model is based on the *VGG16* architecture as introduced by Simonyan & Zisserman [28], in particular a CNN pre-trained on ImageNet. CNNs consist of convolutional layers that automatically extract visual features from images, followed by classification layers that use these features to predict class labels.

For this research, a publicly available subset of *ImageNet* was used (see Deng et al. [6]) called the *ImageNet Large Scale Visual Recognition Challenge* (ILSVRC) 2012–2017 image classification and localization dataset (published by Russakovsky et al. [26]). ILSVRC uses *WordNet* from Princeton University [19] to construct a semantic hierarchy that helps in selecting similar classes that still have meaningful distinctions. For this work, we selected indirect child classes of a common parent node to enable reasonable contrastive explanations, ensuring that the classes share some visual similarities while still being distinct. In particular, we set up the data and model to perform classification between *teapot* and *vase* images. Contrasting these classes is a cognitively interesting case, due to their similarity, yet, heterogeneity in design [9]. Following the *WordNet* hierarchy, the path is as follows: *artifact* → *instrumentality* → *container* → *vessel* → *teapot/vase*. The number of images per class varies slightly. For fine-tuning the CNN model (with 2 output neurons), we used 622 images of teapots and 636 images of vases. The test split includes 678 teapot images and 663 vase images.

To generate contrastive explanations that highlight the distinguishing features between visually and semantically related classes, Algorithm 1 was applied (see also its illustration in Fig. 1).

First, the model M is fine-tuned on instances from the training data D belonging either to a target class t_1 or the contrastive class t_0 (e.g., teapot vs. vase) to prepare the explanation generation for the fine-tuned model M' (see Algorithm 1, line 8). Second, embeddings (Em_0 or Em_1) are extracted for instances from the test set D' that are correctly classified by M', utilizing only feature extraction layers of M' (line 9). Third, an example's embedding

$Em_1(x_i)$ is selected, either by choosing the image instance x_i arbitrarily or by human choice (line 10).

Algorithm 1: Concept-based Contrastive Explanation

```
1  Input:
2  (M, D, D', C_i), where M is the model, D is the training set, D' is the test set, and C_i is the
   set of concepts for each class (t_1 for the target, t_0 otherwise)
3  Output:
4  E: concept-based contrastive explanation
5  Begin:
6    maxSim ← −1
7    bestMatch ← ∅
8    M' ← FineTune(M, D)
9    Em_0, Em_1 ← embedding(M', D'), embeddings of correctly classified instances
10   Em_1(x_i) ← selectExampleEmbedding(Em_1)
11   For each Em_0(x_j) ∈ Em_0 do
12     s ← CosSim(Em_0(x_j), Em_1(x_i)), where x_i of class t_1, x_j instance of class t_0
13     If s > maxSim then
14       maxSim ← s
15       bestMatch ← x_j
16   C_1 ← ranking(extract(x_i, M', D'))
17   C_0 ← ranking(extract(bestMatch, M', D'))
18   E ← contrastive(naming(difference(C_1, C_0)))
19   Return E
20 End
```

Fourth, the cosine similarity is computed between the embedding of the input image $Em_1(x_i)$ and the embeddings of all images from the opposite class $Em_0(x_j)$. The best matching contrastive image x_j is the one with the highest cosine similarity between embeddings, that is the most semantically and visually similar image to x_i from the opposite class (lines 11–15). Next (see lines 16–17), concept extraction is performed by capturing internal relevance attributions from predefined convolutional layers (e.g., layer 40) with the help of Concept Relevance Propagation (CRP). CRP as introduced by Achtibat et al. [1] links neuron-level signals derived during the forward pass to concepts by performing a masked backward pass. To isolate a specific concept, CRP selectively masks irrelevant pathways during propagation, allowing only the neurons associated with that concept to retain relevance. This yields concept-specific relevance scores that quantify how much each concept (neuron) contributed to the prediction. In our approach, all concepts are sorted in descending order based on their contribution to the prediction (also lines 16–17). In particular, two concept sets (C_1 and C_0) are constructed, one for instances of t_1 and one for t_0. Next, unique concepts are identified, defined as those that occur in only one sample from either class. Concepts that appear in both classes are excluded to ensure that the contrastive explanation is focused on class-distinctive features. This step is coupled with optional naming (automated or manual) of features to turn them into human-understandable concepts (line 18). The final explanation is formatted like so:

> The model classified the image as a **Target class** instead of a **Contrastive class** because it contains the concepts $c_{1,1}$, $c_{1,2}$, ..., $c_{i,n}$, and does not contain the concepts $c_{0,1}$, $c_{0,2}$, ..., $c_{j,m}$.

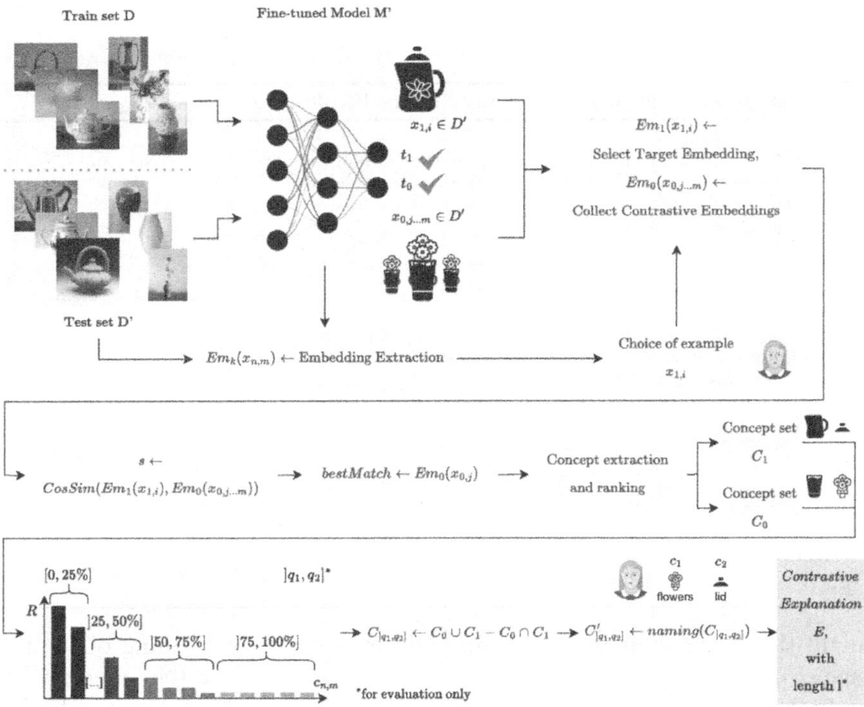

Fig. 1. Illustration of the overall concept-based contrastive explanation approach following the steps of Algorithm 1 and extended by the mention of relevance ranges and explanation length as evaluation variables (best viewed in color).

Here, $c_{\mathrm{i,n}}$ refers to the unique concepts found in the target class, while $c_{\mathrm{j,m}}$ corresponds to the unique concepts found in the contrastive class. For a concrete example, see Fig. 2. The contrast helps to understand why one class is preferred over another and to see the main concept differences between classes[1].

For the evaluation, the sorted concepts are grouped into categories representing different levels of relevance, from most to least influential. These relevance levels form the basis for evaluating explanation length and robustness. Explanation lengths are compared across the different relevance levels to examine whether highly relevant concepts tend to produce shorter and more focused explanations than less relevant ones (see next Section for a detailed description).

[1] Note that optional concept naming is supported by techniques such as reference sampling based on activation maximization or similar methods [1]. Unnamed concepts may shed light on ambiguities or complexities in finding a suitable name. Otherwise, the model needs to be provided with pre-defined concepts [22].

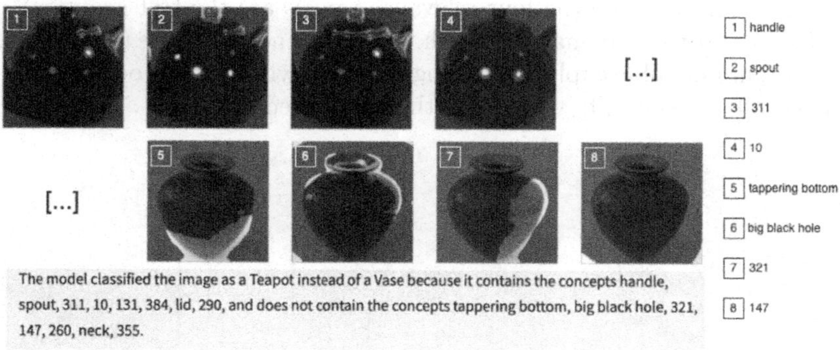

Fig. 2. Contrastive explanation for teapot and vase (best viewed in color).

4 Evaluation and Results

The goal of R1 is to examine whether the length of concept-based explanations differs depending on the relevance strength of the concepts. To address this, explanations are analyzed in distinct relevance ranges. To compute the ranges, all concepts are first sorted in descending order of relevance for a given prediction, where *very strong* relevance is the top 25% of total relevance, *strong* are concepts that lie between 25% and 50%, *low* are concepts that account for the 50% to 75% and *very low* is the remaining 25% of the cumulative relevance. The evaluation of robustness for R2 is based on comparing the original images and their augmented versions within relevance ranges, assessing whether the length of concept-based explanations remains stable. We formulate the following hypotheses, accordingly:

H1 Regarding explanation length, it is expected that similar images with concepts of *very strong* relevance lead to shorter explanations compared to concepts with *strong*, *low*, and *very low* relevance. Furthermore, it is expected that concepts with *strong* relevance result in longer explanations than those with *very strong* relevance, but shorter explanations than concepts with *low* and *very low* relevance. Similarly, concepts with *low* relevance are expected to produce longer explanations than those with *very strong* and *strong* relevance, but shorter explanations than those with *very low* relevance. Finally, concepts with very *low* relevance are expected to lead to the longest explanations compared to all other relevance levels.

H2 Regarding robustness, it is expected that within the relevance ranges the explanation length remains consistent between the original explanation (produced using the unaltered image) and the explanation generated after applying an augmentation technique to the input image (e.g., noise, rotation).

The evaluation of explanation length for R1 is performed with an ANOVA test as introduced by Fisher [11] that assesses whether the means of multiple groups differ significantly, where the four relevance ranges are the test groups. We can see in Fig. 3 that the explanation length varies depending on the relevance range. The maximum possible explanation length is 512, which would occur if one class contained all 512 concepts while the other contained none[2].

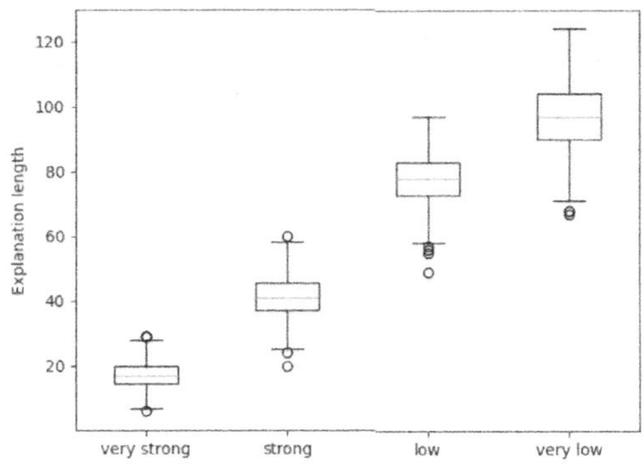

Fig. 3. Explanation length across relevance ranges (n = 615 teapots, 328 vases).

For *very strong* relevance, the median explanation length is around 17. The length in the 25th to 75th percentile ranges from 14 to 21. For some explanations the length ranges from 8 to 30. Only a few explanations have a length below 10 and above 30. Considering the *strong* relevance, the median explanation length is around 40. Most values lie between 35 and 45, while overall, the explanation length from this group mainly varies from 25 to 60, with some outlier samples having a length below 25 and above 60. For *low* relevance, the median explanation length increases to 78, with most explanations ranging from 70 to 85. The length mainly varies from 55 to 98, with only few outliers having a length below 55. Considering the *very low* relevance, we can see that the median value of explanation length is around 100. For most samples the explanation length is between 90 and 110, with a general variation between 70 to 123 and some outliers below 70. Also, the results of the ANOVA test show a high F-value of 354.41 and a statistically significant p-value below 0.001.

The robustness evaluation for R2 is performed using a paired t-test as introduced by Student [30], comparing the means of sampled explanation lengths for original images and their augmented versions (Rotation 180°, Rotation 10°, Gaussian Noise) for each relevance range (*very strong*, *strong*, *low*, and *very low*).

For the augmentation of 180° rotation, we can see that the distribution of explanation lengths is visually quite similar between original and rotated images across all relevance levels (please, view Fig. 4 in the Appendix of the extended paper version [14]). The paired t-tests (see Table 1) support this observation: For *very strong*, *strong*, and *low* relevance, the differences are not statistically

[2] The number of extractable concepts depends on the architecture of the model in use.

significant ($p > 0.05$). Only the *very low* relevance category shows a significant difference ($p = 0.0002$), where rotated images tend to have slightly longer explanations.

For the augmentation of 10° rotation, Fig. 5 (see extended paper version [14]) reveals more noticeable differences. For *very strong* relevance, explanation length appears slightly reduced for rotated images, though not significantly. However, for *strong* and *low* relevance, explanation lengths are significantly shorter after rotation ($p = 0.0006$ and $p = 0.0175$ respectively). For *very low* relevance, no significant difference is observed ($p = 0.7401$), with nearly identical distributions.

For the augmentation with Gaussian noise, the effect of adding it is especially evident in Fig. 6 (see extended paper version [14]). Across all relevance levels, explanation lengths are consistently and significantly lower for noisy images compared to the original ones. This is confirmed by the very low p-values (all <0.001). The boxplots clearly show downward shifts in the distributions, most noticeably in the *very strong* and *strong* relevance categories.

Table 1. Results of t-tests for different augmentations across relevance ranges

Relevance Range	180° Rotation		Gaussian Noise		10° Rotation	
	t	p	t	p	t	p
Very strong	1.7630	0.0784	25.7762	**0.0000**	5.6708	**0.0000**
Strong	−1.9153	0.0559	18.1813	**0.0000**	−3.4474	**0.0006**
Low	1.7312	0.0839	13.2428	**0.0000**	−2.3825	**0.0175**
Very low	3.7505	**0.0002**	13.9907	**0.0000**	−0.3318	0.7401

5 Discussion

The results derived from t-testing the effect of different augmentations, in combination with the outcomes of the ANOVA test, demonstrate a clear relationship between relevance strength and explanation length. Explanations that were based on highly relevant concept groups (e.g., *very strong*) tended to be shorter, while explanations involving less relevant concept groups (e.g., *very low*) were significantly longer. This trend suggests that when a model identifies highly relevant concepts, explaining its decision can be done more briefly and efficiently. In contrast, the longer explanations can be seen as an indicator for contrastive explanations being less effective, when the included concepts received less relevance by the learned model. This also supports the initial claim that highly relevant concepts are more discriminative, fewer of them are needed to form a confident prediction and explanation. On the other hand, longer explanations in the *very low* relevance group may reflect model uncertainty, concept redundancy, or the lack of a dominant set of concepts. These findings support H1, as we clearly see that lower relevance was associated with longer explanations.

H2 stated that the explanation length would remain the same for original images and their augmented versions within a certain relevance range. However, the results showed that this assumption only holds partially.

For rotation by 180°, which is a strong geometric transformation, the effect on explanation length was relatively small. Only the *very low* relevance range showed a significant change, while the more important relevance levels remained stable. This suggests that the most relevant concepts used by the model were largely preserved even when the image was flipped upside down. It seems that the model benefited from a degree of orientation invariance, possibly due to highly relevant concepts not changing in appearance if flipped (such as a vertical handle). The slightly varying results in the lower relevance ranges might reflect small changes in concept activations. Overall, these findings mostly support H2 for this type of augmentation.

In contrast, rotation by 10°, although much more subtle, led to more noticeable changes in explanation length. Significant differences in the *very strong*, *strong*, and *low* relevance categories indicate that even slight misalignments can alter the model's internal concept processing. Some explanations may become shorter, others longer, possibly depending on which concepts gain or lose relevance after the rotation. Unexpectedly, the *very low* relevance group remained stable, which may imply that such minor transformations primarily affected the core discriminative concepts. These results suggest that even small augmentations can influence the structure of the explanation, which is challenging H2 in this case. Comparing both rotation cases, the results nevertheless confirm that robustness under augmentation highly depends on the geometrical properties of learned concepts.

For Gaussian noise, hypothesis H2 clearly does not hold. Explanation lengths increased significantly across all relevance ranges when noise was added. Although the semantic content of the image remained the same, the model reacted strongly to this low-level perturbation. It appeared to lose focus on a small set of discriminative concepts and instead activated a broader range of concepts, leading to longer and more diffuse explanations. This indicates a lack of robustness in how the used model handles noisy inputs and suggests that its concept-based reasoning becomes less certain under such conditions.

In summary, hypothesis H2 was partially supported. While explanation lengths remained mostly stable under strong global changes like 180° rotation, even slight rotations or noise could lead to substantial shifts in explanation structure. This shows that the produced contrastive explanations are somewhat sensitive to image augmentations, especially those affecting low-level visual details.

While the results offer valuable insights into the behavior of concept-based contrastive explanations, several limitations should be acknowledged. First, the evaluation of explanation length was based on the assumption that shorter explanations are preferable, as they are generally easier for humans to interpret and understand. While this assumption is supported by prior work on explanation comprehensibility [20], explanation length alone may not fully capture the quality or faithfulness of an explanation [27]. A short explanation might omit relevant information, while a longer one might still be meaningful if it includes necessary or supporting information [15]. Second, the robustness analysis was limited to three types of augmentations: rotation by 10°, rotation by 180°, and Gaussian noise. While they reflect global and local image changes, other relevant transformations, like occlusion, color shifts, or brightness [4,10], could be considered in the future. Thus, conclusions about the model's robustness may not generalize to all forms of input variation. Additionally, the analysis focused on statistical changes in explanation length, but did not examine which concepts were added, removed, or altered under augmentation. This limits the interpretability of changes and assessment of whether the model shifted to semantically meaningful alternative concepts or just produced noisier explanations. However, this open question may be an interesting future evaluation step, especially if examined on richer data sets (e.g., showing the same object from different angles) and extended by a human study to measure understanding in explainees.

6 Conclusion

This work introduced a concept-based contrastive explanation method for image classification, aiming to clarify why a model prefers one class over another using human-understandable concepts. Two key aspects were examined: the relationship between explanation length and concept relevance, and the robustness of explanations under image augmentations.

Experiments showed that highly relevant concepts led to shorter, more concise contrastive explanations, while less relevant ones produced longer, more diffuse outputs. This finding supports our hypothesis that relevance strength significantly influences explanation complexity, and that contrastive approaches can benefit from exploiting this relationship.

Robustness analysis revealed that explanation length remained stable under strong global transformations (e.g., 180° rotation), but changed notably with small local perturbations (e.g., 10° rotation, Gaussian noise), particularly within concept relevance ranges. This suggests that while explanations exhibit invariance to coarse transformations, the model is sensitive to fine-grained input variations, underscoring the role of geometrical properties of concepts.

Future extensions could leverage the finding that contrastive explanations, grounded in instance similarity and concept relevance, offer interpretable class distinctions. Several promising directions arise from this insight.

Models could be fine-tuned to classify instances with concept inputs disabled when those concepts were deemed irrelevant in a contrastive explanation task (i.e., mistakenly identified as important) [9,10], potentially reducing *Clever-Hans behavior* (see, e.g., Hernández-Orallo [13]) and improving generalization.

Rather than focusing solely on explanation length, future work could examine how individual concepts change under perturbations, identifying additions, removals, or alterations and their semantic relevance. Robustness testing could be expanded to include more realistic perturbations to better reflect real-world scenarios. Further, alternative explanation quality metrics like faithfulness, completeness, and user-perceived interpretability could be introduced and assessed via user studies [9,27] to better evaluate our method's practical utility.

Overall, this work contributes to explainable AI by providing a concept-based contrastive explanation method for image classification, laying the groundwork for future research on explanation complexity and robustness.

References

1. Achtibat, R., et al.: From attribution maps to human-understandable explanations through concept relevance propagation. Nat. Mach. Intell. **5**(9), 1006–1019 (2023). https://doi.org/10.1038/s42256-023-00711-8
2. Adadi, A., Berrada, M.: Peeking inside the black-box: a survey on explainable artificial intelligence (XAI). IEEE Access **6**, 52138–52160 (2018). https://doi.org/10.1109/ACCESS.2018.2870052
3. Ali, S., et al.: Explainable artificial intelligence (XAI): what we know and what is left to attain trustworthy artificial intelligence. Inf. Fus. **99**, 101805 (2023). https://doi.org/10.1016/J.INFFUS.2023.101805
4. Arcaini, P., Bombarda, A., Bonfanti, S., Gargantini, A.: Dealing with robustness of convolutional neural networks for image classification. In: 2020 IEEE International Conference on Artificial Intelligence Testing (AITest), pp. 7–14 (2020). https://doi.org/10.1109/AITEST49225.2020.00009
5. Bruckert, S., Finzel, B., Schmid, U.: The next generation of medical decision support: a roadmap toward transparent expert companions. Front. Artif. Intell. **3**, 507973 (2020). https://doi.org/10.3389/FRAI.2020.507973
6. Deng, J., Dong, W., Socher, R., Li, L., Li, K., Fei-Fei, L.: ImageNet: a large-scale hierarchical image database. In: 2009 IEEE Conf. on Comp. Vis. and Patt. Recogn. (CVPR 2009), 20-25 June 2009, Miami, Florida, USA, pp. 248–255. IEEE Computer Society (2009). https://doi.org/10.1109/CVPR.2009.5206848
7. Dhurandhar, A., et al.: Explanations based on the missing: towards contrastive explanations with pertinent negatives. In: Bengio, S., Wallach, H.M., Larochelle, H., Grauman, K., Cesa-Bianchi, N., Garnett, R. (eds.) Adv. in Neur. Inf. Proc. Sys. 31, NeurIPS 2018, December 3-8, 2018, Montréal, Canada, pp. 590–601 (2018)
8. Doshi-Velez, F., Kim, B.: Towards a rigorous science of interpretable machine learning (2017). https://arxiv.org/abs/1702.08608
9. Finzel, B., Hilme, P., Rabold, J., Schmid, U.: Telling more with concepts and relations: exploring and evaluating classifier decisions with CoReX. arXiv preprint arXiv:2405.01661 (2024). https://arxiv.org/abs/2405.01661
10. Finzel, B., Knoblach, J., Thaler, A.M., Schmid, U.: Near hit and near miss example explanations for model revision in binary image classification. In: Julián, V., Camacho, D., Yin, H., Alberola, J.M., Nogueira, V.B., Novais, P., Tallón-Ballesteros, A.J. (eds.) Intelligent Data Engineering and Automated Learning - IDEAL 2024 - 25th Int. Conf., Valencia, Spain, November 20-22, 2024, Proceedings, Part II. LNCS, vol. 15347, pp. 260–271. Springer (2024). https://doi.org/10.1007/978-3-031-77738-7_22
11. Fisher, R.A.: Statistical Methods for Research Workers, pp. 66–70. Springer, New York (1992). https://doi.org/10.1007/978-1-4612-4380-9_6

12. Gunning, D., Aha, D.W.: Darpa's explainable artificial intelligence (XAI) program. AI Mag. **40**(2), 44–58 (2019). https://doi.org/10.1609/AIMAG.V40I2.2850
13. Hernández-Orallo, J.: Gazing into clever HANs machines. Nat. Mach. Intell. **1**(4), 172–173 (2019). https://doi.org/10.1038/S42256-019-0032-5
14. Kaidashova, Y., Finzel, B., Schmid, U.: Toward simple and robust contrastive explanations for image classification by leveraging instance similarity and concept relevance (2025). https://arxiv.org/abs/2506.23975
15. Kulesza, T., Stumpf, S., Burnett, M.M., Yang, S., Kwan, I., Wong, W.: Too much, too little, or just right? Ways explanations impact end users' mental models. In: Kelleher, C., Burnett, M.M., Sauer, S. (eds.) 2013 IEEE Symposium on Vision Languages and Human Centric Computing. San Jose, CA, USA, September 15-19, 2013, pp. 3–10. IEEE Computer Society (2013). https://doi.org/10.1109/VLHCC.2013.6645235
16. van Lent, M., Fisher, W., Mancuso, M.: An explainable artificial intelligence system for small-unit tactical behavior. In: Proceedings of the 2004 National Conference Artificial Intelligence, pp. 900–907. AAAI Press; MIT Press (2004)
17. Lipton, P.: Contrastive explanation. R. Inst. Philos. Suppl. **27**, 247–266 (1990). https://doi.org/10.1017/S1358246100005130
18. Margolis, E., Laurence, S.: The ontology of concepts—Abstract objects or mental representations? Noûs **41**(4), 561–593 (2007). https://doi.org/10.1111/j.1468-0068.2007.00663.x
19. Miller, G.A.: WordNet: a lexical database for English. Commun. ACM **38**(11), 39–41 (1995). https://doi.org/10.1145/219717.219748
20. Miller, T.: Explanation in artificial intelligence: insights from the social sciences. Artif. Intell. **267**, 1–38 (2019). https://doi.org/10.1016/J.ARTINT.2018.07.007
21. Palmer, D.C.: Psychological essentialism: a review of E. Margolis and S. Laurence (eds.), Concepts: Core readings (2002)
22. Poeta, E., Ciravegna, G., Pastor, E., Cerquitelli, T., Baralis, E.: Concept-based explainable artificial intelligence: a survey (2023). https://arxiv.org/abs/2312.12936
23. Prabhushankar, M., Kwon, G., Temel, D., AlRegib, G.: Contrastive explanations in neural networks. In: IEEE International Conference on Image Processing. ICIP 2020, pp. 3289–3293. IEEE (2020)
24. Rabold, J., Siebers, M., Schmid, U.: Generating contrastive explanations for inductive logic programming based on a near miss approach. Mach. Learn. **111**(5), 1799–1820 (2022). https://doi.org/10.1007/s10994-021-06048-w
25. Rudin, C.: Stop explaining black box machine learning models for high stakes decisions and use interpretable models instead. Nat. Mach. Intell. **1**(5), 206–215 (2019). https://doi.org/10.1038/S42256-019-0048-X
26. Russakovsky, O., et al.: ImageNet large scale visual recognition challenge. Int. J. Comput. Vision **115**(3), 211–252 (2015). https://doi.org/10.1007/s11263-015-0816-y
27. Schwalbe, G., Finzel, B.: A comprehensive taxonomy for explainable artificial intelligence: a systematic survey of surveys on methods and concepts. Data Min. Knowl. Discov. **38**(5), 3043–3101 (2024). https://doi.org/10.1007/S10618-022-00867-8
28. Simonyan, K., Zisserman, A.: Very deep convolutional networks for large-scale image recognition. In: Bengio, Y., LeCun, Y. (eds.) 3rd International Conference on Learning Representations, ICLR 2015, San Diego, CA, USA, May 7-9, 2015 (2015). http://arxiv.org/abs/1409.1556

29. Stepin, I., Alonso, J.M., Catalá, A., Pereira-Fariña, M.: A survey of contrastive and counterfactual explanation generation methods for explainable artificial intelligence. IEEE Access **9**, 11974–12001 (2021). https://doi.org/10.1109/ACCESS.2021.3051315
30. Student: The probable error of a mean. Biometrika **6**(1), 1–25 (1908). https://doi.org/10.2307/2331554
31. Winston, P.H.: Learning structural descriptions from examples. In: The Psychology of Computer Vision, pp. 157–210. McGraw-Hill (1975)

Intermediate-Task Transfer Learning for Bioacoustic Data

Hannes Kath[1,2(✉)], Thiago S. Gouvêa[1], and Daniel Sonntag[1,2]

[1] German Research Center for Artificial Intelligence (DFKI), Oldenburg, Germany
{hannes.kath,thiago.gouvea,daniel.sonntag}@dfki.de
[2] University of Oldenburg, Applied Artificial Intelligence (AAI), Oldenburg, Germany

Abstract. The accelerating loss of biodiversity underscores the urgent need for up-to-date and reliable quantitative data to support evidence-based biosphere management. Passive acoustic monitoring (PAM) has emerged as a key technology for biosphere management, scalable wildlife monitoring in particular. While PAM facilitates large-scale data collection, efficiently analyzing the vast amounts of recorded data remains a significant challenge. This work explores the use of transfer learning and systematically investigates state-of-the-art models, comparing fine-tuning these models with using frozen layer weights, and evaluating the application of intermediate-task transfer learning. In intermediate-task transfer learning, a model pre-trained on data less related to the target data is first re-trained on a larger dataset more closely related to the target, before being re-trained on the target data itself. Our results show that fine-tuning improves performance compared to using frozen model weights, and that intermediate-task transfer learning is only beneficial for models trained on data significantly different from PAM data. These findings pave the way for developing a real-world, efficient PAM data analysis tool.

Keywords: Passive Acoustic Monitoring · Transfer Learning · Fine-Tuning

1 Introduction

Biodiversity is defined by the Convention on Biological Diversity (CBD) as 'the variability among living organisms from all sources, including terrestrial, marine, and other aquatic ecosystems, and the ecological complexes of which they are part; this includes diversity within species, between species, and of ecosystems' [43]. Conservation of biodiversity is one of the greatest challenges facing humanity today, as it underpins essential global processes such as climate regulation, food security and ecosystem stability. This is reflected in the UN sustainable development goals 14 (life below water) and 15 (life on land) [40].

Evidence-based management of ecosystems requires quantitative, reliable biodiversity data. While biodiversity encompasses the variety of all living organisms, wildlife specifically refers to the diversity of animal species within ecosystems.

T. Braun et al. (Eds.): KI 2025, LNAI 15956, pp. 47–61, 2026.
https://doi.org/10.1007/978-3-032-02813-6_4

While traditional wildlife quantification methods rely on experts conducting field observations and counting species instances (e.g., [5,27]), comprehensive biodiversity assessments in bioreserves require scalable and accessible solutions for efficient data acquisition and analysis.

Passive acoustic monitoring (PAM) is a powerful, scalable technology for wildlife monitoring data acquisition, utilizing audio recorders in bioreserves to enable researchers and biodiversity managers to collect extensive acoustic data with minimal disturbance to natural habitats [36,37]. PAM enables the continuous recording of environmental sounds (soundscapes), which encode valuable information about animal behaviour, species richness and ecosystem health. Decoding this information enables evidence-based ecosystem management, rapid biodiversity assessment and fundamental research [30,35]. While the development of recording devices for PAM, such as AudioMoth [16], enables cost-effective and scalable acquisition of audio data, analysing the vast amount of data generated by these systems still poses significant challenges, limiting the adoption of PAM methods for biodiversity monitoring.

Acoustic indices are commonly used to quantify and summarize features of large soundscape datasets [4,23,34]. However, these methods are controversial and have been shown to misrepresent biodiversity in some cases [3,31]. To overcome these limitations and extract more precise, ecologically relevant information, it is essential to focus on alternative approaches that detect and identify discrete sound events at the species level.

This paper advances current research on leveraging transfer learning (TL) to improve sound event detection for PAM data. Current approaches either fine-tune models trained on data less related to PAM (e.g., [9,10]) or use embeddings from models with frozen weights more closely related to PAM (e.g., [12,19]). As fine-tuning generally improves model performance [44], a trend also observed for PAM [9], we present experiments fine-tuning state-of-the-art bioacoustic models, comparing accuracy and training effort between fine-tuning and frozen model weights. Furthermore, this paper presents the adaptation of intermediate-task transfer learning from natural language processing [26] to PAM, comparing various state-of-the-art models trained on different intermediate-task datasets and subsequently fine-tuned on target datasets of varying sizes. All experiments utilize real-world multi-label PAM datasets for both intermediate-task re-training and re-training on the target task.

2 Related Work

The related work focuses on sound event detection and transfer learning in the context of bioacoustic data.

2.1 Sound Event Detection

In the field of machine learning, sound event detection refers to the detection and identification of discrete events within acoustic data, typically performed by computing spectrograms and using convolutional neural networks (CNNs) [15,25]. Due to the potential simultaneous occurrence of sounds from multiple species in a soundscape, species detection is a multi-label sound event detection task, meaning that multiple labels can be assigned to a single audio segment.

Training sound event detection models requires annotated training data, such as species identity and times of occurrence. As obtaining or generating these annotations can take over 10 min of analysis labour per minute of recorded data [22], real-world applications are often constrained by the scarcity of annotated multi-label PAM datasets [18]. Instead, single-label datasets containing focal animal recordings, such as Xeno-Canto[1], are commonly used to train species detection models for PAM datasets – for example, BirdNet [18], a model specifically designed for bird species recognition. Focal recordings differ from PAM data as they are directional, professional-grade recordings actively aimed at the sound source, resulting in high-quality recordings with a high signal-to-noise ratio.

2.2 Transfer Learning

Practical sound event detection for PAM datasets requires few-shot learning techniques. TL is a key few-shot learning technique where a model is initially trained on a large source dataset and subsequently re-trained on a smaller target dataset. Different layers of deep neural networks encode different levels of abstraction: For CNNs, early layers capture low-level features such as Gabor-like gray-scale and RGB filters. Subsequent layers capture progressively higher-level features, with the output layer being directly activated by the semantic labels used for training [1]. Greater similarity between the target task and the source task leads to better adoption of the learned features, resulting in improved TL performance. Therefore, TL is usually most effective when using a model trained on source data similar to the target data [1,24]. For re-training on the target data, the last or last few layers are replaced with classification layers that are trainable on the target dataset. Weights for the new layers are randomly initialized and learned from the target data, while weights for the pre-trained layers can be either frozen at their pre-trained values or fine-tuned, starting from their pre-trained values [1,44]. Fine-tuning typically improves performance but demands more computational resources. Freezing layer weights, however, may cause performance drops due to fragile co-adaptation, where features interact in complex ways that upper layers cannot relearn, and representation specificity, where features from the frozen layers are too specific to the source data and do not transfer well to the target data [44]. Maximizing performance is therefore achieved by maintaining model depth, modifying only the classification layer, and fine-tuning the entire model, while minimal computational resources are required by freezing pre-trained weights and training only the replaced layers [1,44].

Deep learning pipelines for bioacoustic data processing typically follow the standard sound event detection approach, converting audio into spectrograms and analyzing these visual representations with CNNs. Due to the lack of large

[1] https://xeno-canto.org/.

annotated PAM datasets, models trained on datasets with varying similarity to the target data have been used for TL in bioacoustics. ImageNet [7], a large dataset on the visual modality, serves as source data for many pre-trained models explored in bioacoustic sound event detection, including variants of ResNet [14] and VGG [32] to detect frogs [12,20], mammals [12], monkeys [9], whales [19], and birds [9,10,12,19,20]. AudioSet [11], a large dataset in the target modality (audio) but from a different domain (YouTube sound clips), serves as source data for VGGish[2], YAMNet[3], and Audio-MAE [17], which have been used as TL models for PAM data [12,19]. Various unlabelled audio datasets from different domains serve as source data for Wav2Vec [2], which demonstrates strong performance as a TL model for bird detection [38]. Bird recording databases, such as Xeno-Canto[4], from the target modality (audio) and a related domain (single-label focal bird recordings), serve as source data for BirdNet [18] and Perch [13], which consistently show superior performance as TL models for PAM data [12,19]. Due to the scarcity of public multi-label datasets, sound event detection using PAM data is mostly treated as either a single-class task (e.g., [9,10]) or a multi-class task (e.g., [12]). While treating sound event detection as a multi-label task previously required creating real multi-label PAM datasets [19,20] or synthetic ones [19,38], recent studies leverage newly published multi-label datasets [19]. Re-training TL models varies depending on the specific objectives. To directly compare the utility of pre-trained model embeddings and reduce training effort, studies replace and retrain only the classification layer while keeping the pre-trained model weights frozen [12,20]. Comparing embeddings from different layers of the same model confirms the general finding that earlier embeddings perform better when the source data is less similar to the target data (e.g., ImageNet), while later embeddings perform better when the source data is more similar (e.g., bird vocalisations) [19]. Fine-tuning the entire model outperforms freezing the model weights, with performance gains increasing as more training data becomes available [9]. Therefore, implementations seeking the highest accuracy fine-tune the entire model [10,38].

Intermediate-task transfer learning introduces an additional training stage, where a model pre-trained on source data is first fine-tuned on an intermediate dataset before being fine-tuned on the target data. The method was originally developed in the field of natural language processing to fine-tune large language models such as BERT [8] or RoBERTa [21] for downstream applications (e.g., [26,28,41]). Especially when limited training data is available for the target task, fine-tuning can be brittle and depends on the pre-trained encoder parameters being sufficiently close to an optimal configuration for the target task. An intermediate fine-tuning stage on a closely related, ample dataset improves model robustness and performance [26].

[2] https://tfhub.dev/google/vggish/1.
[3] https://tfhub.dev/google/yamnet/1.
[4] https://xeno-canto.org/.

3 Methodology

This section outlines the datasets used, the models evaluated, and the training procedures applied.

3.1 Datasets

In this study, three real-world PAM datasets with manually created multi-label annotations and high class imbalance are used. Pre-processing standardizes the datasets by generating 2-second audio snippets and resampling to 48 kHz. The datasets used are:

1. AnuraSet [6], which consists of 27 h of frog and toad vocalizations and features 42 species. It includes a predefined 70/30 training/evaluation split.
2. The american avian dawn dataset (AADD) [42], which consists of 11 h of avian, mammalian, and amphibian vocalizations. We exclude all species with fewer than 200 occurrences in the 2-second audio snippets, resulting in 20 species. Since no training/evaluation split is provided by the authors, we randomly select 30 % of the raw 5-minute files for evaluation before pre-processing, ensuring that all snippets from a given file are either in the training or evaluation set.
3. BirdSet PER [29], which consists of 21 h of bird vocalizations. We discard all species with fewer than 1000 occurrences in the 2-second audio snippets, resulting in 21 species. Since no predefined training/evaluation split is provided by the authors, we split the raw 1-h files into 5-min segments and randomly select 30 % of these segments for evaluation before pre-processing, ensuring that all snippets from a segment are either in the training or evaluation set.

3.2 Models

In this study, we evaluate the performance of three widely used models pre-trained on datasets with different levels of similarity to the target PAM data. Standardized datasets are transformed to match the model's input requirements, such as computing spectrograms for image-based models. The models used are:

1. ResNet152v2 [14], pre-trained on ImageNet. It serves as the AnuraSet baseline [6], but shows highly variable performance in other experiments. It excels when fine-tuned [9], but underperforms when used with frozen weights [19].
2. VGG16 [32], also pre-trained on ImageNet. It serves as a baseline in many experiments (e.g., [9,19,38]). Performance improves when embeddings further from the classification layer are used during re-training with frozen layers [19], suggesting that fine-tuning helps adjust overly specific final-layer features.
3. BirdNet [18], based on EfficientNet-B1 [39] and pre-trained on bird vocalizations. It consistently outperforms other models as a TL method for PAM [12,19], although it has not yet been fine-tuned for TL tasks.

4. Google Perch [13], also based on EfficientNet-B1 and trained on a similar bird vocalization dataset as BirdNet. It shows comparable performance as a TL model [12]. However, due to its similarity to BirdNet and limited attention in the literature, it was excluded from this study.

3.3 Training and Evaluation

The last layer of the transfer learning model is replaced with a fully connected layer, where each output node indicates the presence or absence of a species from the training dataset. Training updates either only this layer (frozen) or the entire model (fine-tuned). Since we use multi-label data, we apply binary cross-entropy loss with logistic activation, optimized using the Adam optimizer. All models are trained for 50 epochs on either the intermediate or target task using a learning rate of 1e-4. For intermediate-task transfer learning, a batch size of 32 was used, while for re-training on the target task, a batch size of 2 was employed.

All experiments use five random seeds to ensure result robustness and reliability. All results presented are from the evaluation set, which was not used during training. The macro F1 score is used as the performance metric to assign equal weight to each class. We report the mean and standard error of the mean (SEM) over five independent runs for robust evaluation.

The experimental analysis is conducted using Python 3.10, employing key libraries such as Librosa for audio processing and TensorFlow for model training and evaluation. All experiments are performed on a computational setup comprising an NVIDIA Tesla V100 GPU with 16 GB of memory and four supporting CPU cores.

4 Experiments

The following analyses compare the top-performing transfer learning models for PAM data, investigate the effects of fine-tuning versus using frozen model weights, evaluate the impact of sample size on retraining, and explore the potential of intermediate-task transfer learning. The transfer learning models used are ResNet152v2 and VGG16, pre-trained on ImageNet, and BirdNet, pre-trained on bird vocalizations. The sample sizes investigated are 10, 20, 50, 100, 200, 500, 1000, 2000, and 5000. The samples are randomly selected, ensuring that all classes are represented. For investigating intermediate-task transfer learning, none, one, or both of the datasets not used for evaluation were employed. As these experiments aim to lay the groundwork for a real-time PAM analysis tool that combines transfer learning and active learning, the computation time is also assessed in addition to model performance.

4.1 Transfer Learning Model

Figure 1 shows the performance of each dataset for each model without intermediate-task transfer learning. Comparing the performance of the transfer learning

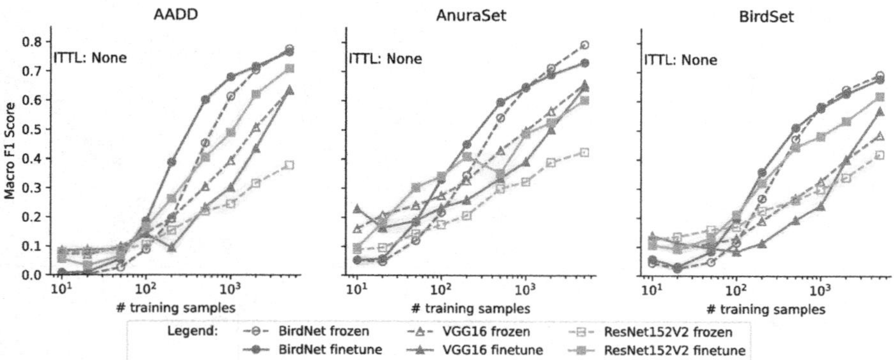

Fig. 1. Mean ± SEM of performance across five independent runs for different transfer learning models, comparing retraining only the last layer (frozen) and the full-model (finetune), as a function of the number of training samples.

models, three distinct phases are observed based on the number of training samples: For small dataset sizes (10 to approximately 50 samples), VGG16 outperforms, followed by ResNet152v2 and BirdNet. As the dataset size increases (approximately 20 to 50 samples), ResNet152v2 surpasses VGG16. For dataset sizes larger than 100 samples, BirdNet's performance significantly increases, surpassing the other models. Figure 2 and 3 show that these phases are largely independent of intermediate-task transfer learning.

4.2 Retraining

Figure 1 shows the performance of transfer learning models retrained with frozen model weights and fine-tuned model weights. The transfer learning models exhibit different behaviours for the retraining methods: Fine-tuning ResNet152v2 generally outperforms using frozen weights. However, for the AADD and BirdSet datasets with small sample sizes up to approximately 50 samples, frozen weights outperform fine-tuning. Figures 2 and 3 show that this behaviour is largely independent of intermediate-task transfer learning. VGG16, using frozen model weights, outperforms fine-tuning for most dataset sizes. However, for large datasets (approximately 2000 to 5000 samples), fine-tuning is equal to or better than using frozen weights. Figures 2 and 3 show that this behaviour changes when intermediate-task transfer learning is applied, with fine-tuning clearly outperforming the use of frozen model weights. BirdNet with fine-tuning performs better for most datasets. However, for large datasets (approximately 1000 to 2000 samples and beyond), using frozen model weights results in comparable or even improved performance. Figures 2 and 3 show that, when intermediate-task transfer learning is applied, fine-tuning outperforms the use of frozen model weights, even for large datasets in most cases.

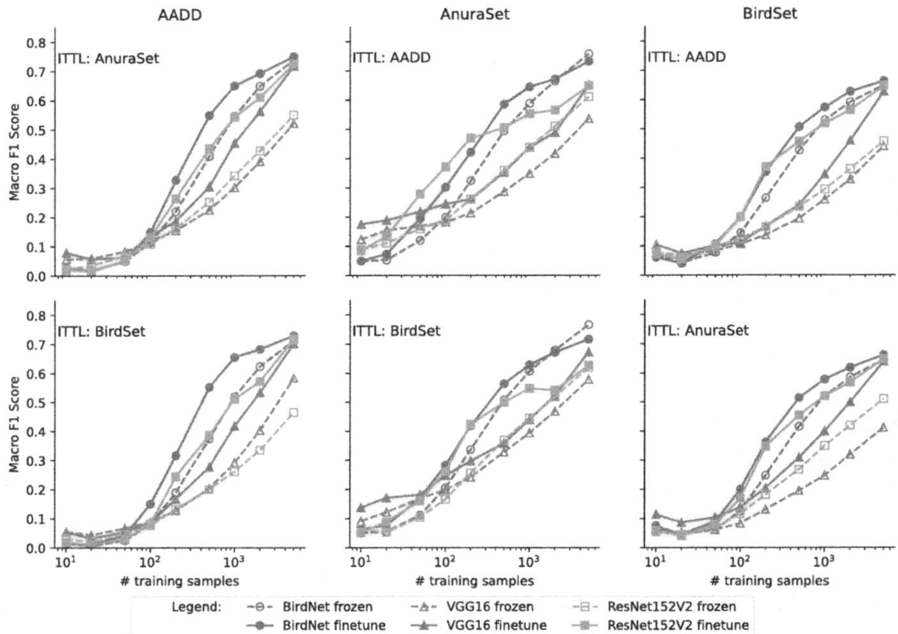

Fig. 2. Mean ± SEM of performance across five independent runs for different transfer learning models using intermediate task transfer learning with one different PAM dataset, comparing retraining only the last layer (frozen) and the full-model (finetune), as a function of number of training samples.

4.3 Sample Size Performance

Figure 1 also shows the performance of transfer learning models for different sample sizes. All models show increasing performance with larger training datasets, regardless of the retraining procedure. Figures 2 and 3 show that this behaviour is independent of intermediate-task transfer learning.

4.4 Intermediate-Task Transfer Learning

Figure 4 illustrates the impact of intermediate-task transfer learning on the transfer learning models. The results for BirdNet are visualized in the first row of Fig. 4, indicating that BirdNet does not benefit from intermediate-task transfer learning, which either leads to equivalent or lower performance compared to the original model. The second row of Fig. 4 shows the results for VGG16. While the original model performs better with frozen layer weights, this behaviour changes when intermediate-task transfer learning is applied. All models re-trained on intermediate tasks show improved performance when further fine-tuned on the target data, surpassing the original model's baseline for the AADD and BirdSet datasets. For AnuraSet, re-training on intermediate tasks followed by fine-tuning

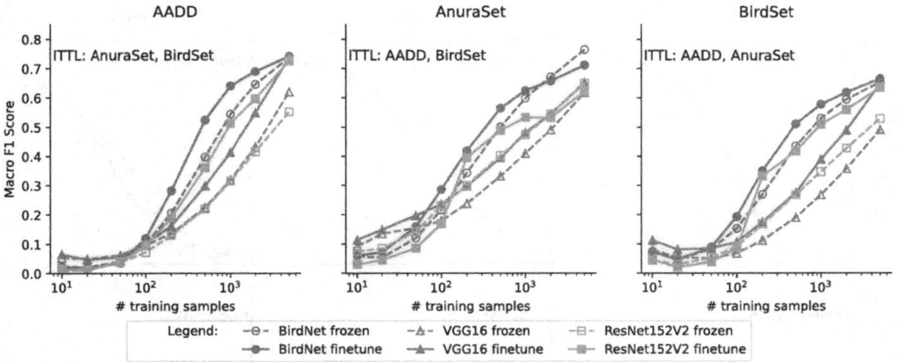

Fig. 3. Mean ± SEM of performance across five independent runs for different transfer learning models using intermediate task transfer learning with two different PAM dataset, comparing retraining only the last layer (frozen) and the full-model (finetune), as a function of number of training samples.

on the target data outperforms the fine-tuned original model but remains inferior to the original model with frozen weights. The last row of Fig. 4 shows the results for ResNet152v2. When using frozen model weights, models trained on intermediate tasks generally outperform the original model. Fine-tuning generally outperforms frozen models. The original model performs similarly to models trained on intermediate tasks, with no clearly superior variant.

Table 1. Training time for the intermediate-task transfer learning process [h:mm]

Model	Intermediate-task transfer learning data					
	AADD	AnuraSet	BirdSet	AADD + AnuraSet	AADD + BirdSet	AnuraSet + BirdSet
ResNet152v2	1:49	3:36	2:33	5:21	4:20	6:07
VGG16	0:54	1:51	1:18	2:42	2:11	3:15
BirdNet	0:29	0:59	0:42	1:27	1:10	1:40

4.5 Computation Time

Table 1 presents the computation time required to re-train each model on the intermediate tasks. Re-training BirdNet requires the least amount of time, while re-training VGG16 takes approximately twice as long as BirdNet, and re-training ResNet152v2 takes roughly three times the time required by BirdNet.

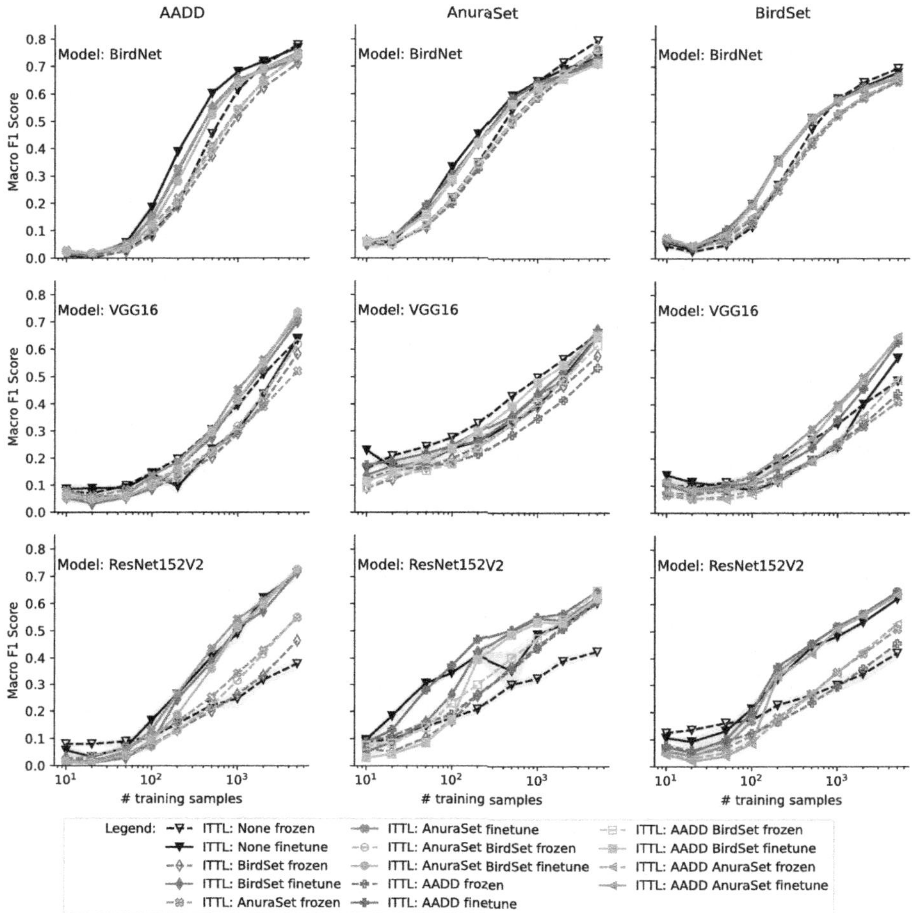

Fig. 4. Mean ± SEM of performance across five independent runs for one transfer learning model using the raw model or one or two PAM datasets for intermediate task transfer learning with different PAM datasets, comparing retraining only the last layer (frozen) and the full-model (finetune), as a function of the number of training samples.

Figure 5 shows the time required to re-train the models as a function of the number of training samples. The required training time clearly increases with the dataset size, regardless of the model architecture or training process. Models trained on intermediate tasks modify only the weights, not the architecture, which is why only the different model architectures are compared. Re-training only the classification layer while keeping the original model weights frozen significantly reduces training time. When comparing the training time for fine-tuning and using frozen model weights, VGG16 requires the least time, followed by BirdNet, and ResNet152v2 requiring the most time.

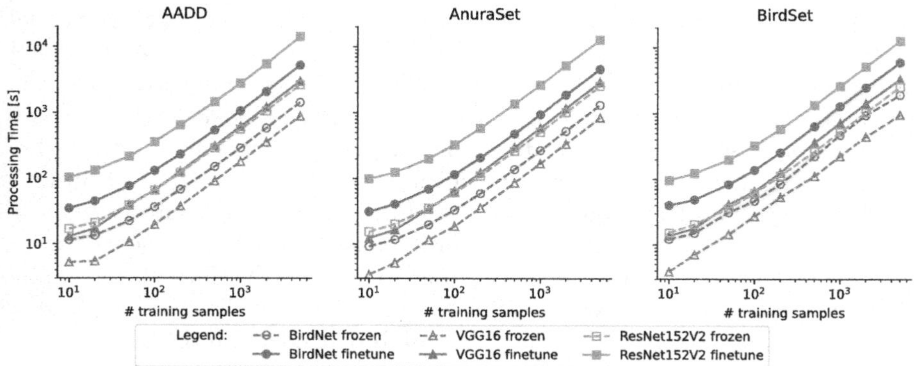

Fig. 5. Mean ± SEM of computation time across five independent runs for different transfer learning models, comparing retraining only the last layer (frozen) and the full-model (finetune), as a function of the number of training samples.

5 Discussion

This work utilizes three real-world multi-label PAM datasets to compare the state-of-the-art models BirdNet, ResNet152v2 and VGG16, exploring the benefits of fine-tuning versus retraining only the classification layer while keeping the other model weights frozen, the impact of adding training samples to the labeled dataset, and the value of intermediate-task transfer learning.

Our results show that while BirdNet performs best when the number of training samples exceeds 100, the other models are superior for smaller training sets. These results suggest that BirdNet requires more training data to leverage its architecture and pretraining, as it depends on data diversity for effective generalization. With smaller datasets, other models like VGG16 and ResNet152v2 perform better due to their more general pretraining, making them more effective with limited data.

Comparing fine-tuning with the use of frozen model weights, results show that fine-tuning generally improves model performance. The exceptions for VGG16 and ResNet152v2 are that, for small datasets, using frozen model weights outperforms the fine-tuned model. While this could be attributed to overfitting the model to the small dataset, a more reasonable explanation is that the features learned on ImageNet are too distinct from the target features. Small datasets may not be able to fully adapt these features, causing the model to get stuck in less useful representations. This is supported by the observation that, for these models trained on intermediate tasks, where the representations are already aligned with useful features for PAM data, fine-tuning consistently outperforms frozen model weights. The exception for BirdNet is that, for large datasets, using frozen weights outperforms fine-tuning, though not significantly. This could be because the relevant features are already present in the BirdNet embedding but are hard to interpret. Correctly interpreting these features through classifier training requires a large dataset.

Intermediate-task transfer learning improves performance for the models VGG16 and ResNet152v2, whereas for BirdNet, the performance using the original model weights is equal to or higher. These results demonstrate the general potential of intermediate-task transfer learning for domains where no pre-trained model is available, providing a means to adapt models pre-trained on distantly related domains. The results indicate that the training data of BirdNet, vocal bird recordings, are similar enough to PAM data that relevant features are already captured, making intermediate-task transfer learning not beneficial.

Results on computation time show that increasing the dataset size significantly increases the required computation time. Additionally, the results show that the training time for a given architecture is dependent on the batch size. As Table 1 shows, BirdNet clearly requires the least amount of time when a batch size of 32 is used, while Fig. 5 shows that VGG16 requires the least amount of time when a batch size of 2 is used.

Based on our results, we hypothesize that fine-tuning generally outperforms frozen weights, despite higher computational cost. Furthermore, the effectiveness of intermediate-task transfer learning increases with domain shift, suggesting broad cross-domain applicability.

6 Future Work

While current pipelines for analyzing PAM datasets using state-of-the-art models rely on the use of frozen model weights [12,19], we show that fine-tuning TL models is beneficial for PAM data analysis. Future work includes the development of a real-life tool that leverages this knowledge. While our experimental setup selected training samples randomly, active learning has demonstrated significant potential for PAM data analysis [19], sparking interest in comparing fine-tuning the model using actively selected data. For this human-in-the-loop approach, the system must be real-time capable, and the present work provides initial results regarding the computation time for model re-training. Furthermore, we demonstrate that our models trained on vocal bird recordings cannot be improved through intermediate-task transfer learning as employed here. This leads us to conclude that the development of a more general PAM foundation model, explicitly trained on a variety of multi-label PAM data, is necessary to advance PAM data analysis. Future work also includes intermediate-task transfer learning for other sustainability data sources and AI-based application systems [33].

Acknowledgements. This research is part of the Computational Sustainability & Technology project area[5], and has been supported by the Ministry for Science and Culture of Lower Saxony (MWK), the Endowed Chair of Applied Artificial Intelligence, Oldenburg University, and DFKI.[5] https://cst.dfki.de/

References

1. Azizpour, H., Razavian, A.S., Sullivan, J., Maki, A., Carlsson, S.: From generic to specific deep representations for visual recognition. In: CVPR Workshop, pp. 36–45. IEEE Computer Society (2015). https://doi.org/10.1109/CVPRW.2015.7301270
2. Baevski, A., Zhou, Y., Mohamed, A., Auli, M.: wav2vec 2.0: a framework for self-supervised learning of speech representations. In: NeurIPS, pp. 12449–12460 (2020)
3. Bicudo, T., Llusia, D., Anciães, M., Gil, D.: Poor performance of acoustic indices as proxies for bird diversity in a fragmented Amazonian landscape. Eco. Inform. **77**, 102241 (2023). https://doi.org/10.1016/j.ecoinf.2023.102241
4. Campos, I.B., Fewster, R., Truskinger, A., Towsey, M., et al.: Assessing the potential of acoustic indices for protected area monitoring in the Serra do Cipó National Park. Brazil. Ecol. Indicators **120**, 106953 (2021). https://doi.org/10.1016/j.ecolind.2020.106953
5. Carthy, R., Oli, M., Wooding, J., Berish, J., Meyer, W.: Analysis of gopher tortoise population estimation techniques. Tech. rep, US Army Engineer Research and Development Center, Construction Engineering (2005)
6. Cañas, J.S., Toro-Gómez, M.P., Sugai, L.S.M., Benítez Restrepo, H.D., et al.: A dataset for benchmarking neotropical anuran calls identification in passive acoustic monitoring. Sci. Data **10**(1), 771 (2023). https://doi.org/10.1038/s41597-023-02666-2
7. Deng, J., Dong, W., Socher, R., Li, L.J., et al.: ImageNet: a large-scale hierarchical image database. In: CVPR, pp. 248–255. IEEE (2009)
8. Devlin, J., Chang, M., Lee, K., Toutanova, K.: BERT: pre-training of deep bidirectional transformers for language understanding. In: NAACL-HLT, pp. 4171–4186. Association for Computational Linguistics (2019). https://doi.org/10.18653/V1/N19-1423
9. Dufourq, E., Batist, C., Foquet, R., Durbach, I.: Passive acoustic monitoring of animal populations with transfer learning. Eco. Inform. **70**, 101688 (2022). https://doi.org/10.1016/j.ecoinf.2022.101688
10. Florentin, J., Dutoit, T., Verlinden, O.: Detection and identification of European woodpeckers with deep convolutional neural networks. Ecol. Informatics **55** (2020). https://doi.org/10.1016/J.ECOINF.2019.101023
11. Gemmeke, J.F., Ellis, D.P.W., Freedman, D., Jansen, A., et al.: Audio Set: an ontology and human-labeled dataset for audio events. In: ICASSP, pp. 776–780 (2017). https://doi.org/10.1109/ICASSP.2017.7952261
12. Ghani, B., Denton, T., Kahl, S., Klinck, H.: Global birdsong embeddings enable superior transfer learning for bioacoustic classification. Sci. Rep. **13**(1), 22876 (2023). https://doi.org/10.1038/s41598-023-49989-z
13. Google research: Google bird vocalization classifier: a global bird embedding and classification model (2023). https://tfhub.dev/google/bird-vocalization-classifier/4
14. He, K., Zhang, X., Ren, S., Sun, J.: Identity mappings in deep residual networks. In: Computer Vision ECCV 2016, pp. 630–645. Springer International Publishing, Cham (2016). https://doi.org/10.1007/978-3-319-46493-0_38
15. Hershey, S., Chaudhuri, S., Ellis, D.P., Gemmeke, J.F., et al.: CNN architectures for large-scale audio classification. In: ICASSP, pp. 131–135 (2017). https://doi.org/10.1109/ICASSP.2017.7952132
16. Hill, A., Prince, P., Piña-Covarrubias, E., Doncaster, C., et al.: AudioMoth: evaluation of a smart open acoustic device for monitoring biodiversity and the environment. Methods Ecol. Evol. **9** (2017). https://doi.org/10.1111/2041-210X.12955

17. Huang, P.Y., Xu, H., Li, J., Baevski, A., et al.: Masked autoencoders that listen. In: NeurIPS, pp. 28708–28720 (2022)
18. Kahl, S., Wood, C.M., Eibl, M., Klinck, H.: BirdNet: a deep learning solution for avian diversity monitoring. Eco. Inform. **61**, 101236 (2021). https://doi.org/10.1016/j.ecoinf.2021.101236
19. Kath, H., Serafini, P.P., Campos, I.B., Gouvêa, T.S., Sonntag, D.: Leveraging transfer learning and active learning for data annotation in passive acoustic monitoring of wildlife. Eco. Inform. **82**, 102710 (2024). https://doi.org/10.1016/j.ecoinf.2024.102710
20. LeBien, J., Zhong, M., Campos-Cerqueira, M., Velev, J.P., et al.: A pipeline for identification of bird and frog species in tropical soundscape recordings using a convolutional neural network. Eco. Inform. **59**, 101113 (2020). https://doi.org/10.1016/j.ecoinf.2020.101113
21. Liu, Y., Ott, M., Goyal, N., Du, J., et al.: RoBERTa: a robustly optimized BERT pretraining approach. CoRR abs/1907.11692 (2019)
22. Lüers, B., Serafini, P.P., Campos, I.B., Gouvêa, T.S., Sonntag, D.: BirdNET-annotator: AI-assisted strong labelling of bird sound datasets. In: 3rd Annual AAAI Workshop on AI to Accelerate Science and Engineering (AI2ASE). Vancouver, Canada (2024)
23. Machado, R.B., Aguiar, L., Jones, G.: Do acoustic indices reflect the characteristics of bird communities in the savannas of Central Brazil? Landsc. Urban Plan. **162**, 36–43 (2017). https://doi.org/10.1016/j.landurbplan.2017.01.014
24. Mensink, T., Uijlings, J.R.R., Kuznetsova, A., Gygli, M., Ferrari, V.: Factors of influence for transfer learning across diverse appearance domains and task types. IEEE Trans. Pattern Anal. Mach. Intell. **44**(12), 9298–9314 (2022). https://doi.org/10.1109/TPAMI.2021.3129870
25. Nolasco, I., Singh, S., Morfi, V., Lostanlen, V., et al.: Learning to detect an animal sound from five examples. Eco. Inform. **77**, 102258 (2023). https://doi.org/10.1016/j.ecoinf.2023.102258
26. Phang, J., Févry, T., Bowman, S.R.: Sentence encoders on stilts: supplementary training on intermediate labeled-data tasks. CoRR abs/1811.01088 (2018)
27. Plumptre, A.J.: Monitoring mammal populations with line transect techniques in African forests. J. Appl. Ecol. **37**(2), 356–368 (2000)
28. Pruksachatkun, Y., Phang, J., Liu, H., Htut, P.M., et al.: Intermediate-task transfer learning with pretrained models for natural language understanding: when and why does it work? CoRR abs/2005.00628 (2020)
29. Rauch, L., Schwinger, R., Wirth, M., Heinrich, R., et al.: BirdSet: a multi-task benchmark for classification in avian bioacoustics. CoRR abs/2403.10380 (2024). https://doi.org/10.48550/ARXIV.2403.10380
30. Ross, S.R.J., O'Connell, D.P., Deichmann, J.L., Desjonquères, C., et al.: Passive acoustic monitoring provides a fresh perspective on fundamental ecological questions. Funct. Ecol. **37**(4), 959–975 (2023)
31. Sethi, S.S., Bick, A., Ewers, R.M., Klinck, H., et al.: Limits to the accurate and generalizable use of soundscapes to monitor biodiversity. Nat. Ecol. Evol., 1–6 (2023). https://doi.org/10.1038/s41559-023-02148-z
32. Simonyan, K., Zisserman, A.: Very deep convolutional networks for large-scale image recognition. In: ICLR (2015)
33. Sonntag, D.: Computational Sustainability and Technology (CST). Tech. rep, German Research Center for AI (2025)

34. Sueur, J., Farina, A., Gasc, A., Pieretti, N., Pavoine, S.: Acoustic indices for biodiversity assessment and landscape investigation. Acta Acust. Acust. **100**(4), 772–781 (2014). https://doi.org/10.3813/AAA.918757
35. Sueur, J., Pavoine, S., Hamerlynck, O., Duvail, S.: Rapid acoustic survey for biodiversity appraisal. PLoS ONE **3**(12), e4065 (2008)
36. Sugai, L., Llusia, D.: Bioacoustic time capsules: using acoustic monitoring to document biodiversity. Ecol. Ind. **99**, 149–152 (2019)
37. Sugai, L.S.M., Silva, T.S.F., Ribeiro, J.W., Llusia, D.: Terrestrial passive acoustic monitoring: review and perspectives. Bioscience **69**(1), 15–25 (2019). https://doi.org/10.1093/biosci/biy147
38. Swaminathan, B., Jagadeesh, M., Vairavasundaram, S.: Multi-label classification for acoustic bird species detection using transfer learning approach. Eco. Inform. **80**, 102471 (2024). https://doi.org/10.1016/j.ecoinf.2024.102471
39. Tan, M., Le, Q.V.: EfficientNet: rethinking model scaling for convolutional neural networks. In: ICML, vol. 97, pp. 6105–6114. PMLR (2019)
40. United Nations: Transforming our world: the 2030 agenda for sustainable development. Resolution A/RES/70/1, United Nations General Assembly (2015). https://sdgs.un.org/2030agenda, adopted on 25 September 2015
41. Wang, A., Hula, J., Xia, P., Pappagari, R., et al.: Can you tell me how to get past sesame street? sentence-level pretraining beyond language modeling. In: ACL, pp. 4465–4476. Association for Computational Linguistics (2019). https://doi.org/10.18653/V1/P19-1439
42. Weldy, M.J., Denton, T., Fleishman, A.B., Tolchin, J., et al.: Audio tagging of avian dawn chorus recordings in California, Oregon and Washington. Biodivers. Data J. **12**, e118315 (2024)
43. Wilson, E., University, H. (eds.): Biodiversity. The National Academies Press, Washington, DC (1988)
44. Yosinski, J., Clune, J., Bengio, Y., Lipson, H.: How transferable are features in deep neural networks? In: NeurIPS, pp. 3320–3328 (2014)

On the Domain Robustness of Contrastive Vision-Language Models

Mario Koddenbrock(✉), Rudolf Hoffmann, David Brodmann, and Erik Rodner(✉)

KI-Werkstatt/Fachbereich 2, University of Applied Sciences Berlin, Wilhelminenhofstr. 75A, 12459 Berlin, Germany
{Mario.Koddenbrock,Rudolf.Hoffmann,David.Brodmann, Erik.Rodner}@htw-berlin.de

Abstract. In real-world vision-language applications, practitioners increasingly rely on large, pretrained foundation models rather than custom-built solutions, despite limited transparency regarding their training data and processes. While these models achieve impressive performance on general benchmarks, their effectiveness can decline notably under specialized domain shifts, such as unique imaging conditions or environmental variations. In this work, we introduce `DeepBench`, a framework designed to assess domain-specific robustness of vision-language models (VLMs). `DeepBench` leverages a large language model (LLM) to generate realistic, context-aware image corruptions tailored to specific deployment domains without requiring labeled data. We evaluate a range of contrastive vision-language architectures and architectural variants across six real-world domains and observe substantial variability in robustness, highlighting the need for targeted, domain-aware evaluation. `DeepBench` is released as open-source software (`DeepBench` is available at https://github.com/ml-lab-htw/deepbench) to support further research into domain-aware robustness assessment.

Keywords: Vision-Language Models · Robustness Evaluation · Domain Shift · Benchmark

1 Introduction

Ensuring robust performance in computer vision applications has been a longstanding challenge, with numerous benchmarks and evaluation strategies proposed for common corruptions and adversarial attacks [9,19,32,41]. While such work has significantly advanced our understanding of robustness under controlled conditions, it often overlooks the *domain-specific* factors that arise in real-world scenarios. Consequently, even high-performing models can fail when facing new contexts—such as specialized manufacturing processes or unique medical imaging conditions—that diverge from the distribution seen during training [36].

This challenge becomes increasingly important with the growing adoption of large VLMs trained on massive, weakly curated datasets scraped from the

T. Braun et al. (Eds.): KI 2025, LNAI 15956, pp. 62–76, 2026.
https://doi.org/10.1007/978-3-032-02813-6_5

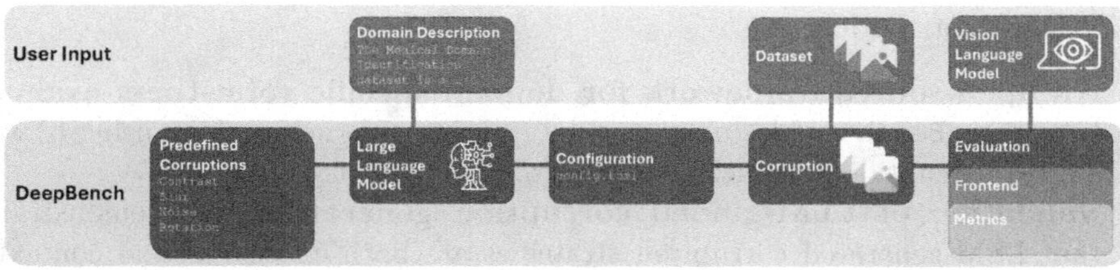

Fig. 1. Overview of the DeepBenchframework. A predefined set of corruption methods and a domain-specific description are used to prompt a Large Language Model (LLM), which outputs a tailored set of image corruptions. These are applied to a dataset, and model performance is evaluated under various perturbations. Results are stored and visualized through an interactive frontend.

internet [5]. Although these models excel at generic tasks (e.g., zero-shot image classification and retrieval), they may exhibit reduced reliability on domains where object appearance, environmental variability, or imaging protocols differ sharply from the data seen during pretraining [43]. Notably, many VLMs expose little information about the exact nature of their training data or data augmentation strategies, making it difficult to anticipate whether they will maintain accuracy and consistency when deployed in unfamiliar environments.

To tackle this gap, we introduce `DeepBench` (Fig. 1), a general-purpose framework that evaluates domain-specific robustness *without* requiring labeled data. At its core `DeepBench` is an LLM-guided corruption generation pipeline: given a high-level domain description (such as imaging conditions or expected variability), `DeepBench` selects tailored perturbations from a predefined set to mimic real-world challenges. These corruptions, combined with an unsupervised consistency metric called *label flip probability*, facilitate a thorough robustness assessment. By quantifying how often a model changes its prediction under perturbation, label flip probability offers a practical way to evaluate robustness even when ground-truth labels are scarce or expensive to obtain.

We focus on three contrastive VLMs—`CLIP` [30], `SigLIP` [42], and `ALIGN` [20], each pretrained on large-scale image-text pairs. Our evaluation covers six real-world domains, ranging from everyday photography to industrial and medical applications. In addition to standard accuracy-based metrics, we assess prediction consistency under corruption using label flip probability. We also analyze architectural variants of `CLIP`, finding that robustness improves with larger model capacity, finer patch resolution, and the use of QuickGELU activations—though even the strongest models exhibit domain-specific weaknesses.

The `DeepBench` framework is open-source and designed to integrate with custom datasets or workflows. Its modular structure comprises a configurable backend for corruption generation and metric computation, alongside a user-facing interface for browsing and visualizing results. By emphasizing domain realism and data-scarce conditions, `DeepBench` highlights robustness failure modes that more generic benchmarks miss [23].

Our main contributions are:

1. **An open-source framework for domain-specific robustness evaluation.** `DeepBench` enables unsupervised robustness benchmarking using LLM-guided corruption generation tailored to deployment scenarios.
2. **Validation of LLM-guided corruption generation.** We demonstrate that LLM-generated corruption strategies are both consistent and context-aware across diverse domains, enabling effective simulation of real-world robustness challenges.
3. **A systematic analysis of contrastive vision-language models under domain shift.** We benchmark leading contrastive VLMs across six real-world domains, showing that `CLIP` achieves the highest overall robustness, while performance varies substantially by task and corruption type.
4. **Insights into architectural design for robustness.** We analyze architectural variants and show that model capacity, patch size, input resolution, and activation function all influence robustness—highlighting `CLIP` ViT-L/14 with QuickGELU as the most resilient configuration.

2 Related Work

General Robustness in Computer Vision: Robustness has long been a central topic in computer vision, driven by the need to ensure stable performance under real-world variability. Efforts to measure and improve robustness against common corruptions and adversarial attacks date back to benchmarks like ImageNet-C and ImageNet-P [19], as well as work on adversarial examples [16]. Later studies revealed that even mild natural distribution shifts can substantially degrade performance [32]. While these efforts provide valuable insights, they largely focus on *general-purpose* corruptions, overlooking how domain-specific factors can reduce model reliability in real-world applications.

Robust Learning Approaches: Improving robustness often involves modifying the training process, for instance via adversarial training [27], distributionally robust optimization (DRO) [11], or advanced data augmentation [10,26,41]. Such methods typically aim to harden models against diverse perturbations. However, they are no guarantee for resilience and should therefore be assessed post-hoc, regardless of the training method.

Robustness in Vision-Language Models: VLMs such as `CLIP`, `ALIGN`, and `SigLIP` have demonstrated strong zero-shot capabilities, leveraging large-scale image-text pretraining [12,40,42]. While open-source variants like `OpenCLIP` [30] provide transparency, many pretrained datasets remain opaque [5], complicating bias and robustness analysis. Despite their scale, VLMs are prone to spurious correlations [36] and often underperform in domain-specific settings. Existing solutions—such as fine-tuning [18], ensembling [22], or test-time adaptation [28]—typically require labeled or generic data. Our work introduces a label-free pipeline to evaluate VLMs under realistic, domain-specific corruptions.

Domain-Specific Evaluations: Despite recent advances in general robustness methods, specialized application domains such as industrial inspection [4] or scientific imaging [35] often present unique distribution shifts that generic techniques fail to capture [13]. In these contexts, labeled or source data may be scarce or inaccessible [25]. Our framework addresses this by enabling realistic, zero-shot robustness assessments without requiring labels, providing practical insights into VLM performance under domain-specific shifts.

3 DeepBench: A Framework for Domain-Specific Robustness Evaluation

3.1 Mathematical Framework and Robustness Metrics

Using pretrained models without any fine-tuning is increasingly common, particularly for multimodal systems controlled solely through prompting. In classical computer vision, one typically considers a pretrained model f mapping the input space $\mathcal{X}$ (e.g., images) to an output space $\mathcal{Y}$ (e.g., a softmax over M classes with $\mathcal{Y} = [0,1]^M$). In the special case of pretrained contrastive VLMs, f consists of an image encoder $\phi_I : \mathcal{X} \to \mathcal{Z}$ and a text encoder $\phi_T : \mathcal{Y} \to \mathcal{Z}$ mapping to an shared embedding space $\mathcal{Z} \subset \mathbb{R}^d$. f then performes prediction via some similarity measure sim : $\mathcal{Z} \times \mathcal{Z} \to \mathbb{R}$ (typically cosine similarity) between the image embedding of a sample $\boldsymbol{x}$ and the embeddings of all possible labels:

$$f(\boldsymbol{x}) = \arg\max_{y \in \mathcal{Y}} \operatorname{sim}\Big(\phi_I(\boldsymbol{x}), \phi_T(y)\Big) \ . \tag{1}$$

An application domain $\mathcal{D} \subseteq \mathcal{X}$ reflects the conditions under which images are acquired, inducing a domain-specific data distribution $p(\boldsymbol{x}, y \mid \mathcal{D})$. In the classical (supervised) evaluation, the overall performance of model f in $\mathcal{D}$ is defined using a example-based 0âĂŞ1 loss function L, such as the misclassification error $L(z, z') = [z \neq z']$:

$$\varepsilon(f, \mathcal{D}) = \mathbb{E}_{\boldsymbol{x},y \sim p(\boldsymbol{x},y \mid \mathcal{D})} \left(L(y, f(\boldsymbol{x})) \right) . \tag{2}$$

Since the joint distribution $p(\boldsymbol{x}, y \mid \mathcal{D})$ is unknown, it is commonly (e.g. [7]) approximated by a test set $\{(\boldsymbol{x}_i, y_i)\}_{i=1}^m$ using Dirac delta functions:

$$p(\boldsymbol{x}, y \mid \mathcal{D}) = \frac{1}{m} \sum_{i=1}^{m} \delta(y - y_i)\, \delta(\boldsymbol{x} - \boldsymbol{x}_i) \ . \tag{3}$$

In contrast, vicinal learning [33,34], replaces the Dirac delta functions with a smoothing function $\mathcal{K}$, which can, for example, consider a smooth region around each test example $\boldsymbol{x}_i$:

$$p(\boldsymbol{x}, y \mid \mathcal{D}) = \frac{1}{m} \sum_{i=1}^{m} \delta(y - y_i) \left(\int \mathcal{K}(\boldsymbol{x}, \boldsymbol{x}_i)\, \mathrm{d}\boldsymbol{x} \right) \ . \tag{4}$$

The work of [7] showed that for learning with this scheme and a Gaussian kernel $\mathcal{K}$, classical Tikhonov regularization evolves. However, a Gaussian assumption often does not align with complex data, such as images, and it completely ignores additional domain knowledge that might be available. One type of domain knowledge that is natural to model involves invariant transformations $c(\cdot, \boldsymbol{\theta})$ of inputs that do not alter the output.

Therefore, instead of placing a Dirac delta at $\boldsymbol{x}_i$, it is more natural to marginalize over all possible transformation parameters $\boldsymbol{\theta}$ that do not lead to output changes and then approximate this marginalization with empirical samples of these parameters:

$$\mathcal{K}(\boldsymbol{x}, \boldsymbol{x}_i) = \mathbb{E}_{\boldsymbol{\theta} \sim p(\boldsymbol{\theta} \mid \mathcal{D})} \left(\delta(\boldsymbol{x} - c(\boldsymbol{x}_i, \boldsymbol{\theta})) \right) \approx \frac{1}{K} \sum_{k=1}^{K} \delta(\boldsymbol{x} - c(\boldsymbol{x}_i, \boldsymbol{\theta}_k)) \ , \tag{5}$$

where $(\boldsymbol{\theta}_k)_{k=1}^{K}$ are sampled from the domain-specific distribution $p(\boldsymbol{\theta} \mid \mathcal{D})$. Plugging this approximation into our performance measure from (2), we obtain the classical evaluation on a set of augmented test samples:

$$\varepsilon(f, \mathcal{D}) \approx \frac{1}{mK} \sum_{i=1}^{m} \sum_{k=1}^{K} L(f(c(\boldsymbol{x}_i, \boldsymbol{\theta}_k)), y_i) = \frac{1}{mK} \sum_{i=1}^{m} \sum_{k=1}^{K} [f(c(\boldsymbol{x}_i, \boldsymbol{\theta}_k)) \neq y_i] \tag{6}$$

Extension to Unsupervised Robustness: In many real-world applications, ground-truth labels are unavailable. In such cases, robustness can be quantified by measuring the consistency of the model's predictions under corruption. Instead of using supervised loss L, we consider whether the prediction changes when a corruption is applied, i.e., we replace the label y in Eq. 2 with the clean prediction of image $\boldsymbol{x}$. The *label flip probability (FP)* of the model f is defined as

$$\mathrm{FP}^f(\boldsymbol{x}) = \mathbb{E}_{\boldsymbol{\theta} \sim p(\boldsymbol{\theta} \mid \mathcal{D})} \left[f\big(c(\boldsymbol{x}, \boldsymbol{\theta})\big) \neq f(\boldsymbol{x}) \right] \ . \tag{7}$$

Transformed into the embedding space, a label flip means that applying a corruption $c(\cdot, \boldsymbol{\theta})$ shifts the image embedding such that there exists a label $y' \neq f(\boldsymbol{x})$ satisfying:

$$\mathrm{sim}(\phi_I(c(\boldsymbol{x}, \boldsymbol{\theta})), \phi_T(y')) > \mathrm{sim}(\phi_I(c(\boldsymbol{x}, \boldsymbol{\theta})), \phi_T(f(\boldsymbol{x}))) \ . \tag{8}$$

To approximate the expectation from Eq. 7, we simply evaluate on the sampled transformation parameters $(\boldsymbol{\theta}_k)_{k=1}^{K}$:

$$\mathrm{FP}^f(\boldsymbol{x}) \approx \frac{1}{K} \sum_{k=1}^{K} [f\big(c(\boldsymbol{x}, \boldsymbol{\theta}_k)\big) \neq f(\boldsymbol{x})] \tag{9}$$

and derive a dataset-level flip rate FP^f, by averaging Eq. 9 over all test samples. Finally, we get the *mean flip rate* relative to an baseline model as

$$\mathrm{mFR} = \frac{\mathrm{FP}^f}{\mathrm{FP}^{\mathrm{baseline}}} . \tag{10}$$

Mean and Relative Corruption Error (mCE, rCE). When ground-truth labels are available, we first establish a baseline using the *clean balanced accuracy*—balanced accuracy computed on the unperturbed test set. Balanced accuracy, defined as the macro-average of per-class recall, compensates for class-imbalance bias. Building on this baseline, we adopt the error-based robustness metrics introduced by Hendrycks et al. [19].

We consider $b : \mathcal{X} \to \mathcal{Y}$ a baseline model and $\tilde{\epsilon}(f, c, \mathcal{D})$ the error rate of f on $\mathcal{D}$, when corruption c is applied (averaged over all corresponding configurations). Let $\mathcal{C}$ now be a set of different types of corruptions (i.e. blur, contrast, lightning etc.), then the *mean Corruption Error (mCE)* is defined as

$$\mathrm{mCE} = \frac{1}{|\mathcal{C}|} \sum_{c \in \mathcal{C}} \frac{\tilde{\epsilon}(f, c, \mathcal{D})}{\tilde{\epsilon}(b, c, \mathcal{D})} . \tag{11}$$

Let further $\tilde{\epsilon}(f, \mathcal{D})$ be the clean error rate of f on $\mathcal{D}$, then the *relative Corruption Error (rCE)* is defined as:

$$\mathrm{rCE} = \frac{1}{|\mathcal{C}|} \sum_{c \in \mathcal{C}} \frac{\tilde{\epsilon}(f, c, \mathcal{D}) - \tilde{\epsilon}(f, \mathcal{D})}{\tilde{\epsilon}(b, c, \mathcal{D}) - \tilde{\epsilon}(b, \mathcal{D})} . \tag{12}$$

3.2 Evaluation Framework

`DeepBench` systematically evaluates model robustness across specialized domains using a modular architecture that combines automated corruption generation, metric-based evaluation, and interactive result visualization. At its core, the backend processes high-level domain descriptions and employs an LLM—default is `GPT-4o` [29]—to generate a tailored set of corruptions from a predefined list. This list, detailed in the supplementary material (see Sect. 6), includes corruption methods like geometric distortions, noise, color shifts, or simulated environmental effects. To guide the LLM's reasoning, we use structured prompts consisting of a role definition, a domain description, a list of available corruptions with brief explanations, and explicit output format instructions (details can be found in the the supplementary material. The generated output is parsed into a configuration file that specifies which corruptions to apply and with which parameters. The overall workflow is shown in Fig. 1.

Each image in the selected dataset is processed under all defined corruption conditions, and model performance is measured using metrics such as balanced accuracy and label flip probability (see Sect. 3.1). Results are stored in a MongoDB database and visualized through a browser-based frontend that supports interactive comparisons across models and domains. The framework is designed for extensibility: researchers can easily add new corruptions and configurations via a simple Python interface, enabling adaptation to emerging robustness challenges with minimal effort.

3.3 Application Scenarios and Datasets

`DeepBench` currently provides six diverse application scenarios, each representing a distinct real-world domain with its own imaging conditions, data distributions, and robustness challenges. Further application scenarios can be easily integrated into our open-source framework by specifying a high-level domain description (see Fig. 1), which informs the corruption generation pipeline. Providing details such as image source, expected variability, and deployment context, an LLM translates it into a tailored set of corruption transformations.

1. **Medical Domain Identification:** We use eight datasets covering diverse medical imaging types, including chest X-rays [37], hand X-rays for bone age assessment [17], knee X-rays for osteoarthritis diagnosis [31], dental X-rays [2], ultrasound images for nerve segmentation [3], breast ultrasound images [1], liver fibrosis ultrasound images [21], and mammography images [24]. Each image is labeled according to its medical domain, with 126âĂŞ999 samples per class.
2. **Autonomous Driving (KITTI)** [14]**:** KITTI (Karlsruhe Institute of Technology and Toyota Technological Institute) is an autonomous driving dataset, consisting of hours of traffic scenarios recorded with a variety of sensor modalities, including high-resolution RGB images. These images were cropped around non-overlapping object bounding boxes and re-labeled into four categories: *Car*, *Person*, *Tram*, *Truck*, with 156âĂŞ4660 samples per class.
3. **Manufacturing Quality Control (MVTec AD)** [4]**:** The MVTec anomaly detection dataset (MVTec AD) is repurposed as a 15-class object classification task over industrial items, with 60âĂŞ391 samples per class.
4. **People and Face Recognition (FER13)** [15]**:** The Facial Expression Recognition 2013 (FER13) dataset contains 35,887 grayscale 48×48 face images labeled with seven emotion categories, with 751âĂŞ1093 samples per class.
5. **Satellite and Aerial Imaging (AID)** [39]**:** The Aerial Image Dataset (AID) offers 15 scene categories (e.g., farmland, airport) with 800 high-res RGB images per class from Google Earth.
6. **Handheld Image Recognition (Food-101)** [6]**:** Food-101 is a dataset including 101,000 images across 101 food categories (1,000 per class), showing a range of dishes under varied real-world conditions.

3.4 Built-In Vision Language Models

`DeepBench` integrates with widely used APIs like Hugging Face [38] and OpenCLIP [8], providing access to a broad range of pretrained vision-language models

(VLMs). This study focuses on three representative models—CLIP, SigLIP, and ALIGN—chosen for their architectural diversity and popularity. Unless noted otherwise, all models are evaluated at an input resolution of 224×224.

New models can be added via two mechanisms: (1) specifying the model name in the configuration file for API-based loading, or (2) implementing a lightweight Python wrapper for custom architectures. This modular design allows easy extension to new inference pipelines and VLM variants.

1. CLIP [30] (OpenAI) is a contrastive model trained on 400M image-text pairs using separate Transformer encoders for vision and text. It aligns both modalities in a shared embedding space for zero-shot classification.
2. SigLIP [42] (Google Research) builds on CLIP but replaces the softmax loss with a sigmoid-based variant, improving performance in multi-label settings.
3. ALIGN [20] (Google Brain) is trained on over 1B noisy image-text pairs. It uses an EfficientNet-L2 image encoder and a BERT-style text encoder, emphasizing scale and weak supervision over architectural novelty.

4 Experiments

This study is structured around three key research questions. Below is a brief overview of the main findings; detailed analyses and results are provided in the subsections that follow.

1. **Can LLMs generate consistent and context-aware corruption strategies?** (Sect. 4.1)
   Yes; our approach produces interpretable and domain-relevant corruptions across diverse application settings.
2. **How robust are foundation VLMs under domain-specific corruptions?** (Sect. 4.2)
   CLIP is most robust overall; ALIGN is weakest; SigLIP shows domain-specific strengths but lower stability.
3. **How does model architecture affect robustness?** (Sect. 4.3)
   Transformer models with larger capacity and QuickGELU activation show the strongest overall robustness. ViT-L/14 (QuickGELU) consistently leads across domains. Smaller patch sizes and higher input resolution can improve robustness in some cases, but effects vary by domain.

4.1 Can LLMs Generate Consistent and Context-Aware Corruption Strategies?

Answer: Yes, LLMs consistently produce stable and context-relevant corruption sets, supporting automated domain-aware robustness evaluation.

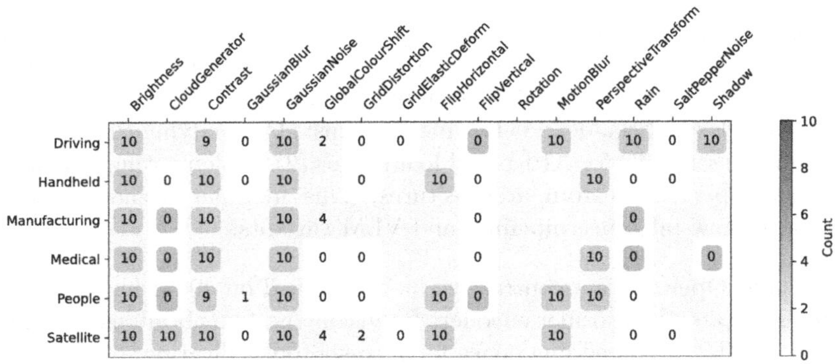

Fig. 2. Heatmap of selection frequencies showing how often each corruption was chosen in 10 independent `GPT-4o` runs. As a sanity check, we defined a simple *whitelist* of obviously relevant corruptions and a *blacklist* of clearly implausible ones for every domain. Green cells belong to the whitelist, whereas red cells belong to the blacklist. All whitelist/blacklist rules are summarised in the supplementary material (see Sect. 6). (Color figure online)

Experimental Setup: To assess whether large language models (LLMs) can generate stable and context-aware corruption strategies, we prompted `GPT-4o` (temperature = 0) with a domain description and a predefined list of corruptions. The complete list, with the descriptions and examples can be found in the supplementary material (see Sect. 6). For each of the six application from 3.3, we repeated the prompting process 10 times and recorded which corruptions were selected. This allowed us to evaluate both the stability and contextual relevance of the LLM outputs. Corruption sets for downstream robustness evaluation were determined via majority vote across these runs.

Results: Figure 2 shows the selection frequencies of each corruption method across all domains. We observe that core transformations such as *Brightness*, *Contrast*, and *ImageRotation* were selected consistently across domains, indicating stable LLM behavior. Additionally, the emergence of domain-specific corruptions—e.g., *CloudGenerator* for Satellite imagery and *GridDistortion* for Manufacturing—demonstrates that the model adapts its choices to the provided context. This confirms that LLMs can serve as a reliable component in generating tailored robustness benchmarks.

A qualitative analysis of the textual justifications generated for each corruption is provided in the supplementary material (see Sect. 6), where we assess the reasoning quality and its alignment with domain-specific visual challenges.

4.2 How Robust are Foundation VLMs Under Domain-Specific Corruptions?

Answer: Across all evaluated domains, `CLIP` consistently achieves the best mean robustness. However, in certain specialized domains `SigLIP` exhibits superior relative robustness, highlighting the importance of domain-specific evaluations

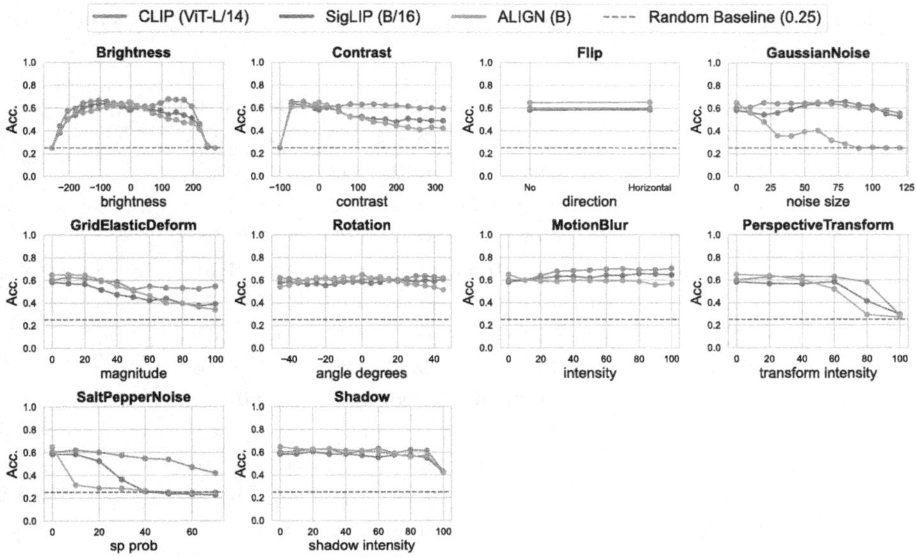

Fig. 3. Balanced accuracy under various corruptions in a **People Recognition** task. `ALIGN` leads in clean accuracy, however, drops even under moderate noise, while `CLIP` remains more stable overall.

Experimental Setup: We evaluate three large contrastive VLMs—`CLIP`, `SigLIP`, and `ALIGN`—under a zero-shot protocol across six real-world domains (see Sect. 3.3). For each domain, a tailored corruption set is applied (Sect. 4.1), encompassing geometric, photometric, and environment-specific disturbances. We test multiple corruption intensities and primarily measure *balanced accuracy*. In addition, we calculate aggregate robustness metrics including mCE, rCE, and mFR as defined in Sect. 3.1. Figure 3 presents representative results for the **People Recognition** domain. Detailed per-domain plots for all six scenarios can be found in the supplementary material (see Sect. 6).

Results: `CLIP` consistently achieves the lowest mCE across all domains, indicating strong overall robustness. However, some domains exhibit nuanced trends: in **Medical Diagnosis** and **Manufacturing Quality**, `SigLIP` surpasses `CLIP` in rCE. Notably, `ALIGN`, despite achieving high clean accuracy, experiences severe performance degradation under mild noise corruptions (rCE up to 14.78). Figure 3 highlights this effect in **People Recognition**, where `ALIGN`'s accuracy sharply declines with minimal noise, approaching random-chance levels, whereas `CLIP` and `SigLIP` remain robust. Additionally, brightness and contrast changes only slightly affect `CLIP`, but significantly degrade `ALIGN` performance beyond moderate distortion levels.

Table 1. Per-domain clean balanced accuracy (Acc.) and mCE for architectural variants of `CLIP` relative to `CLIP (ViT-L/14)`

	Driving		Handheld		Manufacturing	
	Acc. ↑	mCE ↓	Acc. ↑	mCE ↓	Acc. ↑	mCE ↓
ResNet101	0.66	1.27	0.72	1.48	0.77	0.83
ResNet50	0.66	1.25	0.73	1.52	0.65	1.05
ViT-B/16	0.74	1.01	0.81	1.25	0.65	0.98
ViT-B/32	**0.82**	**0.84**	0.71	1.51	0.61	1.10
ViT-L/14 (QGelu)	0.74	0.87	0.86	**0.88**	**0.78**	**0.74**
ViT-L/14 (baseline)	0.69	1.00	0.85	1.00	0.62	1.00
ViT-L/14-336	0.74	0.85	**0.87**	0.90	0.70	0.91
	Medical		**People**		**Satellite**	
	Acc. ↑	mCE ↓	Acc. ↑	mCE ↓	Acc. ↑	mCE ↓
ResNet101	0.63	1.61	0.62	1.35	0.75	1.49
ResNet50	0.66	1.64	0.68	1.28	0.70	1.65
ViT-B/16	0.78	1.38	0.66	1.22	0.80	1.23
ViT-B/32	0.81	1.32	**0.70**	1.11	0.75	1.40
ViT-L/14 (QGelu)	**0.97**	**0.70**	0.60	1.20	0.85	**0.84**
ViT-L/14 (baseline)	0.85	1.00	**0.70**	1.00	0.84	1.00
ViT-L/14-336	0.86	0.91	0.68	**0.97**	**0.87**	0.92

4.3 How Does Model Architecture Affect Robustness?

Answer: Transformer-based architectures generally offer superior robustness compared to ResNet baselines. ViT-L/14 (QuickGELU) consistently ranks best or near-best across domains in terms of robustness. While smaller patch sizes can improve robustness in some cases, this effect is not consistent across all domains. Similarly, higher input resolution leads to lower robustness on every domain.

Experimental Setup: We compare a set of `CLIP` model variants with architectural differences in backbone type (ResNet vs. Vision Transformer), patch size, input resolution, and activation function. All models were trained by OpenAI on the same dataset, ensuring consistent pretraining conditions.

Each model is evaluated in a zero-shot setting across six real-world application domains (Sect. 3.3), using the tailored domain-specific corruption sets from Sect. 4.1. For each corruption type and intensity, we compute balanced accuracy. We also report aggregate robustness metrics: mCE, rCE, and mFR, as defined in Sect. 3.1.

Results: Table 1 shows clean accuracy and mCE for each architecture across all six domains. Several consistent patterns emerge:

- **Transformer vs. ResNet:** Vision Transformer (ViT) models substantially outperform ResNet-based architectures in both clean accuracy and robust-

ness. ResNet-50 and ResNet-101 show the lowest accuracy and highest average mCE across domains.
- **Patch size:** ViT-B/16 does not consistently outperform ViT-B/32. In both clean accuracy and mCE, there is no clear advantage for one of the two architectures.
- **Resolution:** ViT-L/14@336 achieves lower mCE than ViT-L/14 (baseline) in all domains, indicating improved robustness with higher input resolution. Gains in clean accuracy are observed in 5 of the 6 application scenarios.
- **Activation function:** The ViT-L/14 variant with QGelu activation achieves better clean accuracy and mCE then the ViT-L/14 baseline in 5 out of 6 cases.
- **Best overall model:** ViT-L/14 (QGelu) emerges as the most robust architecture in our comparison, achieving the lowest mCE in 4 and highest clean accuracy in 2 application scenarios.

5 Conclusion and Future Work

In this work, we introduced `DeepBench`, a modular, label-free framework for evaluating the domain-specific robustness of VLMs. Unlike traditional robustness benchmarks that focus on synthetic or generic corruptions, `DeepBench` employs large language models to generate realistic, scenario-specific corruptions, enabling scalable, zero-shot robustness assessments without requiring labeled data.

Our experiments across six real-world domains highlight several insights. Among contrastive VLMs, `CLIP` consistently achieves superior robustness compared to `SigLIP` and `ALIGN`. Architectural choices also significantly influence robustness: Transformer models with finer patches, higher resolution, and QuickGELU activation generally outperform other variants. These findings emphasize the importance of targeted evaluations beyond conventional benchmarks.

Limitations. `DeepBench` currently uses a fixed set of corruption primitives and relies on LLM-generated context interpretations. Our analysis focuses primarily on image classification for contrastive models and does not yet fully address autoregressive VLMs, whose robustness heavily depends on prompt design and output parsing.

Future Work. Future work will extend `DeepBench` beyond classification tasks, incorporate adaptive corruption sampling, and include temporal or multi-frame corruptions suitable for video and robotics scenarios. We also plan to systematically evaluate autoregressive models (e.g., `LLaVA`), investigating the impact of prompt strategies and output handling on their robustness.

6 Supplementary Material

Owing to the strict page limit in the published proceedings, a complete set of figures, tables, code listings, and additional analyses is available in the supplementary material that accompanies the preprint at: https://arxiv.org/abs/2506.23663.

Acknowledgments. This research was funded by the *Deutsche Forschungsgemeinschaft (DFG, German Research Foundation)* via the Project ApplFM (528483508) and the *Institut für angewandte Forschung Berlin (IFAF, Berlin Institute for Applied Research)* via Project TrustAdHocAI (TAHAI).

Disclosure of Interests. The authors declare no competing financial or non-financial interests related to this work. This research was conducted independently and is not affiliated with or endorsed by any of the organizations that developed the evaluated models or datasets.

References

1. Al-Dhabyani, W., Gomaa, M., Khaled, H., Fahmy, A.: Dataset of breast ultrasound images. Data Brief **28**, 104863 (2020)
2. Ali, R.B., Ejbali, R., Zaied, M.: Detection and classification of dental caries in x-ray images using deep neural networks. In: International Conference on Software Engineering Advances (ICSEA), p. 236 (2016)
3. Baby, M., Jereesh, A.: Automatic nerve segmentation of ultrasound images. In: 2017 International Conference of Electronics, Communication and Aerospace Technology (ICECA), vol. 1, pp. 107–112. IEEE (2017)
4. Bergmann, P., Fauser, M., Sattlegger, D., Steger, C.: MVTEC AD–A comprehensive real-world dataset for unsupervised anomaly detection. In: Proceedings of the IEEE/CVF Conference on Computer Vision and Pattern Recognition, pp. 9592–9600 (2019)
5. Bommasani, R., et al.: On the opportunities and risks of foundation models. arXiv preprint arXiv:2108.07258 (2021)
6. Bossard, L., Guillaumin, M., Van Gool, L.: Food-101 – Mining discriminative components with random forests. In: Fleet, D., Pajdla, T., Schiele, B., Tuytelaars, T. (eds.) ECCV 2014. LNCS, vol. 8694, pp. 446–461. Springer, Cham (2014). https://doi.org/10.1007/978-3-319-10599-4_29
7. Chapelle, O., Weston, J., Bottou, L., Vapnik, V.: Vicinal risk minimization. In: Advances in Neural Information Processing Systems, vol. 13 (2000)
8. Cherti, M., et al.: Reproducible scaling laws for contrastive language-image learning. In: Proceedings of the IEEE/CVF Conference on Computer Vision and Pattern Recognition, pp. 2818–2829 (2023)
9. Croce, F., et al.: RobustBench: a standardized adversarial robustness benchmark. arXiv preprint arXiv:2010.09670 (2020)
10. Cubuk, E.D., Zoph, B., Mane, D., Vasudevan, V., Le, Q.V.: Autoaugment: learning augmentation strategies from data. In: Proceedings of the IEEE/CVF Conference on Computer Vision and Pattern Recognition, pp. 113–123 (2019)
11. Duchi, J.C., Hashimoto, T., Namkoong, H.: Distributionally robust losses against mixture covariate shifts. Under Rev. **2**(1) (2019)
12. Fang, A., et al.: Data determines distributional robustness in contrastive language image pre-training (clip). In: International Conference on Machine Learning, pp. 6216–6234. PMLR (2022)
13. Garg, S., Erickson, N., Sharpnack, J., Smola, A., Balakrishnan, S., Lipton, Z.C.: RLSbench: domain adaptation under relaxed label shift. In: International Conference on Machine Learning, pp. 10879–10928. PMLR (2023)

14. Geiger, A., Lenz, P., Urtasun, R.: Are we ready for autonomous driving? The KITTI vision benchmark suite. In: 2012 IEEE Conference on Computer Vision and Pattern Recognition, pp. 3354–3361. IEEE (2012)
15. Goodfellow, I.J., et al.: Challenges in representation learning: a report on three machine learning contests. In: Lee, M., Hirose, A., Hou, Z.-G., Kil, R.M. (eds.) ICONIP 2013. LNCS, vol. 8228, pp. 117–124. Springer, Heidelberg (2013). https://doi.org/10.1007/978-3-642-42051-1_16
16. Goodfellow, I.J., Shlens, J., Szegedy, C.: Explaining and harnessing adversarial examples. arXiv preprint arXiv:1412.6572 (2014)
17. Halabi, S.S., et al.: The RSNA pediatric bone age machine learning challenge. Radiology **290**(2), 498–503 (2019)
18. Han, Z., et al.: Alignclip: navigating the misalignments for robust vision-language generalization. Mach. Learn. **114**(3), 1–19 (2025). https://doi.org/10.1007/s10994-025-06742-z
19. Hendrycks, D., Dietterich, T.: Benchmarking neural network robustness to common corruptions and perturbations. In: Proceedings of the International Conference on Learning Representations (2019)
20. Jia, C., et al.: Scaling up visual and vision-language representation learning with noisy text supervision. In: International Conference on Machine Learning, pp. 4904–4916. PMLR (2021)
21. Joo, Y., Park, H.C., Lee, O.J., Yoon, C., Choi, M.H., Choi, C.: Classification of liver fibrosis from heterogeneous ultrasound image. IEEE Access **11**, 9920–9930 (2023)
22. Kar, O.F., Tonioni, A., Poklukar, P., Kulshrestha, A., Zamir, A., Tombari, F.: Brave: broadening the visual encoding of vision-language models. In: European Conference on Computer Vision, pp. 113–132. Springer (2024). https://doi.org/10.1007/978-3-031-72640-8_7
23. Koh, P.W., et al.: Wilds: a benchmark of in-the-wild distribution shifts. In: International conference on machine learning, pp. 5637–5664. PMLR (2021)
24. Lee, R.S., Gimenez, F., Hoogi, A., Miyake, K.K., Gorovoy, M., Rubin, D.L.: A curated mammography data set for use in computer-aided detection and diagnosis research. Sci. Data **4**(1), 1–9 (2017)
25. Li, K., Lu, J., Zuo, H., Zhang, G.: Source-free multi-domain adaptation with generally auxiliary model training. In: 2022 International Joint Conference on Neural Networks (IJCNN), pp. 1–8. IEEE (2022)
26. Lim, S., Kim, I., Kim, T., Kim, C., Kim, S.: Fast Autoaugment. In: Advances in Neural Information Processing Systems, vol. 32 (2019)
27. Madry, A., Makelov, A., Schmidt, L., Tsipras, D., Vladu, A.: Towards deep learning models resistant to adversarial attacks. In: International Conference on Learning Representations (2018)
28. Maharana, S.K., Zhang, B., Karlinsky, L., Feris, R., Guo, Y.: Enhancing robustness of clip to common corruptions through bimodal test-time adaptation. arXiv preprint arXiv:2412.02837 (2024)
29. OpenAI: GPT-4 technical report. https://openai.com/research/gpt-4 (2023). Accessed 9 Jan 2025
30. Radford, A., et al.: Learning transferable visual models from natural language supervision. In: International Conference on Machine Learning, pp. 8748–8763. PMLR (2021)
31. Sarhan, A.M., et al.: Knee osteoporosis diagnosis based on deep learning. Int. J. Comput. Intell. Syst. **17**(1), 241 (2024)

32. Taori, R., Dave, A., Shankar, V., Carlini, N., Recht, B., Schmidt, L.: Measuring robustness to natural distribution shifts in image classification. Adv. Neural Inf. Process. Syst. (NeurIPS) **33**, 18583–18599 (2020)
33. Vapnik, V.: The Nature of Statistical Learning Theory. Springer Science & Business Media (1999). https://doi.org/10.1007/978-1-4757-3264-1
34. Vapnik, V.N.: An overview of statistical learning theory. IEEE Trans. Neural Netw. **10**(5), 988–999 (1999)
35. Verma, P., Van, M.H., Wu, X.: Beyond human vision: the role of large vision language models in microscope image analysis. arXiv preprint arXiv:2405.00876 (2024)
36. Wang, Q., Lin, Y., Chen, Y., Schmidt, L., Han, B., Zhang, T.: A sober look at the robustness of clips to spurious features. arXiv preprint arXiv:2403.11497 (2024)
37. Wang, X., Peng, Y., Lu, L., Lu, Z., Bagheri, M., Summers, R.M.: Chestx-ray8: hospital-scale chest x-ray database and benchmarks on weakly-supervised classification and localization of common thorax diseases. In: Proceedings of the IEEE Conference on Computer Vision and Pattern Recognition, pp. 2097–2106 (2017)
38. Wolf, T., et al.: HuggingFace's transformers: state-of-the-art natural language processing. arXiv preprint arXiv:1910.03771 (2019)
39. Xia, G.S., et al.: AID: a benchmark data set for performance evaluation of aerial scene classification. IEEE Trans. Geosci. Remote Sens. **55**(7), 3965–3981 (2017)
40. Xu, H., et al.: Demystifying clip data. In: 12th International Conference on Learning Representations, ICLR 2024 (2024)
41. Yin, D., Gontijo Lopes, R., Shlens, J., Cubuk, E.D., Gilmer, J.: A fourier perspective on model robustness in computer vision. In: Advances in Neural Information Processing Systems, vol. 32 (2019)
42. Zhai, X., Mustafa, B., Kolesnikov, A., Beyer, L.: Sigmoid loss for language image pre-training. In: Proceedings of the IEEE/CVF International Conference on Computer Vision, pp. 11975–11986 (2023)
43. Zheng, Z., Ma, M., Wang, K., Qin, Z., Yue, X., You, Y.: Preventing zero-shot transfer degradation in continual learning of vision-language models. In: Proceedings of the IEEE/CVF International Conference on Computer Vision, pp. 19125–19136 (2023)

Numbers Don't Lie: Hybrid Extraction and Validation of Quantitative Statements in Arguments with Semi-structured Information

Mirko Lenz[1,2(✉)], Lorik Dumani[2], Ralf Schenkel[2], and Ralph Bergmann[1,2]

[1] German Research Center for Artificial Intelligence (DFKI), Behringstraße 21, 54296 Trier, Germany
{mirko.lenz,ralph.bergmann}@dfki.de

[2] Trier University, Universitätsring 15, 54296 Trier, Germany
info@mirko-lenz.de, {dumani,schenkel,bergmann}@uni-trier.de

Abstract. Evidence in arguments may be stated in various forms, including quantitative statements (i.e., numerical relations between entities). This measurable information can be validated against reliable sources like Wikipedia to combat the spread of misinformation. In this paper, we propose a four-step pipeline that combines rule-based techniques with prompting strategies for generative language models in a hybrid fashion. We use regular expressions to identify candidates in claim-premise structures, extract statements using GPT-4o, augment the data with tables from Wikipedia, and validate statements through retrieval-augmented generation (RAG). The pipeline is evaluated on two existing argumentation corpora and the generated dataset is manually annotated to assess the quality of our predictions, showing promising results for extraction and mixed results for validation. Our code and data are available to foster further research in this area.

Keywords: Argumentation · Quantitative Statements · Validation · Large Language Models · Retrieval-Augmented Generation

1 Introduction

Argumentation is ubiquitous in human communication—having the most convincing argument is a key factor in persuading others. In addition to its persuasiveness, the quality of an argument is influenced by a number of factors, including the evidence used to support it [21]. Such evidence may be stated in various forms—for instance, as a reference to a scientific study, an exemplary case, or a quantitative statement. The latter is particularly interesting, as it provides a precise and measurable piece of information between two entities. We use the widely spread definition of an argument that describes it as *claim* (i.e., a standpoint the arguer wants the reader to take or object) and a *premise* (i.e., evidence or a reason to support or oppose the claim). Take the following

T. Braun et al. (Eds.): KI 2025, LNAI 15956, pp. 77–90, 2026.
https://doi.org/10.1007/978-3-032-02813-6_6

argument as an example: "*Global overpopulation is a myth, because humans can now produce 16 times as much wheat per acre as we could in the 1500s*". Here, the claim "*Global overpopulation is a myth*" is supported by a premise that contains a quantitative statement, namely that the wheat yield per acre has increased by a factor of 16. More precisely, it contains the entities "1500s", "today" (not explicitly stated), the quantity "wheat yield per acre", and the factor "16". The provided evidence can be verified by consulting data in a reliable source such as Wikipedia tables, enabling the reader to assess the argument's credibility. A particular challenge in this purpose is the fact that information from natural language texts might need to be extracted and rephrased. For example, one or multiple tables would most likely contain (in the best case) numbers from today and the 1500s, where a system has to derive whether the data is sufficient to represent the entities today and the 1500s, as well as whether the difference is (approximately) 16 times higher.

In this work, we focus on the, to the best of our knowledge, new task of automatic extraction and validation of quantitative statements. To combat the spread of misinformation, we propose a hybrid approach that combines rule-based techniques with prompting strategies involving Large Language Models (LLMs) in a pipeline: (i) identify quantitative operators in the argument using Regex, (ii) send the matches to a generative language model to extract the quantitative statement, (iii) scrape relevant Wikipedia articles (especially tables) for validation, and (iv) apply Retrieval-Augmented Generation (RAG) [16] to validate the extracted statement against the articles. To evaluate our approach, we developed an open source application that can perform all steps of the pipeline and tested it on two different datasets. Thus, we seek to answer the following research question: *How reliable are LLM-based techniques for extracting and validating quantitative statements in arguments?* Compared to existing fact-checking techniques, our approach is unique in that it incorporates the claim-premise-structure of arguments—enabling the validation of reasoning structures rather than individual statements.

The remainder of this paper is structured as follows: After discussing related work concerning argumentation and validation in Sect. 2, we present our approach in Sect. 3. The evaluation is carried out in Sect. 4, and we conclude with a discussion and an outlook in Sect. 5.

2 Related Work

In this section, we examine work in the area of fact-checking. Guo et al. [5] provide a comprehensive overview of this field. Here, it is crucial to distinguish the term fact-checking from fake news detection. The latter involves assessing news articles and labeling items based on aspects not related to veracity (e.g., satire detection [17,22]). Actual fact-checking is the task of assessing whether the claims made are true, consisting of (i) claim detection, (ii) evidence retrieval, and (iii) claim verification (i.e., verdict prediction and justification production). Existing work typically covers individual components of the task, such as (i) identifying rumors on social media [9,23] (ii) detecting the stance of a given piece of evidence towards a claim [7,13], or (iii) producing explanations and justifications for fact checks [11]. In our work, we focus more on the premise than on the claim. However, in Computational Argumentation (CA) premises can also serve as claims for other premises [15]. We do not validate natural language statements as common, but rather quantities that are contained in them, specifically using tables from the Web. Naturally, not all sources on the Web are credible. Like most fact-checking approaches, we rely on credible sources, in our case Wikipedia [5].

As in Thorne et al. [20], most approaches use natural language sentences from Wikipedia to verify claims. To the best of our knowledge, we are only aware of the work of Gupta et al. [6] and Chen et al. [3], which utilize semi-structured data and use tables and info boxes from Wikipedia. The work of Gupta et al. [6] is only comparable to a limited extent to our work, because while they restrict their dataset InfoTabS to tables consisting of info boxes from Wikipedia—more precisely, 23,738 (premise, hypothesis) pairs from 2,540 info boxes—we scraped all tables from a Wikipedia page whose content is related to a quantitative statement. Furthermore, Chen et al. [3] present the dataset TabFact comprising 16k Wikipedia tables as evidence for 118k human-annotated natural language statements labeled as entailed or refuted. They also present the two methods Table-BERT and Latent Program Algorithm, which lag far behind human performance. Koleva et al. [10] investigate the entity linking task in tables, and together with a new dataset, they present an LLM prompting strategy for the recognition of named entities in table cells. Lastly, our approach has a certain similarity to subject-predicate-object triples, which can be double-checked in knowledge bases such as DBpedia. Guo et al. [5] criticize that it is difficult to extract these from texts, but with the advent of instruction-based LLMs, our work overcomes this limitation.

Our approach combines claim-premise arguments with quantitative statements, so we need corpora that contain such structures. The aforementioned TabFact or InfoTabS datasets do not contain this essential information, so we use two other datasets for this paper: (i) Args containing 382,545 arguments whose premises are divided into 6,146,646 sentences [1] and (ii) Kialo containing 208,969 arguments with 302,555 sentences [15]. The use of other corpora containing argument graphs or debates is also possible and an interesting direction for future work.

3 Pipeline Steps

In this section, we describe our approach consisting of four sequential phases: (i) *matching* potential quantitative statements in arguments through RegEx, (ii) *extracting* these statements using LLMs, (iii) *augmenting* the found statements with information from Wikipedia, and (iv) *validating* the extracted statements against the augmented data using RAG. Our approach is designed to be generic and applicable to any argumentation corpus. We use real example from our evaluation (see Fig. 4a) to illustrate each step in this section: a *pro-abortion* premise stating "it takes $245,000 to raise a kid, that's 490 times the amount of money it takes to have an abortion".

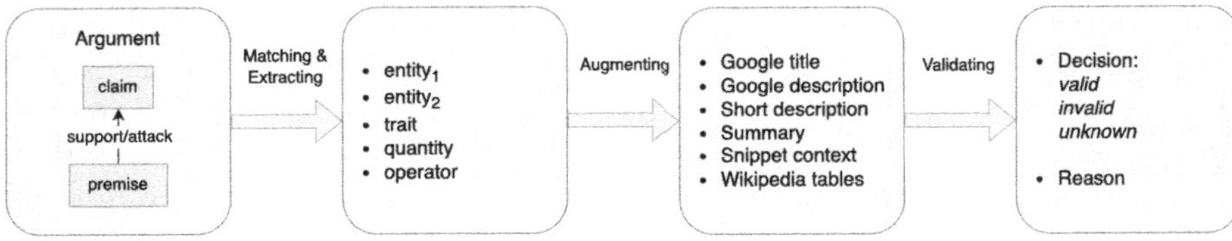

Fig. 1. Architectural overview of our proposed pipeline. Starting from claim-premise structures, quantitative statements are extracted and then augmented with Wikipedia data to validate their contents.

Table 1. The number of distinct claims matches for each RegEx and dataset.

pattern name	Args	Kialo	example
$\text{REGEX}_{\text{qualitative}}$	15,145	509	*is larger than*
$\text{REGEX}_{\text{as } \textit{attribute} \text{ as}}$	4,137	134	*as strong as*
$\text{REGEX}_{\text{times}}^{\text{decimal}}$	2,546	118	*4 times higher*
$\text{REGEX}_{\text{times}}^{\text{number as word}}$	1,760	106	*four times higher*
$\text{REGEX}_{\text{percentage}}$	963	94	*4% higher*
$\text{REGEX}_{\text{times as } \textit{attribute} \text{ as}}$	115	13	*four times as fast as*
total distinct claims	20,195	590	

3.1 Matching Regular Expressions

Since prompting on datasets is very costly, we employed regular expressions as pre-filters to find potential quantitative statements in arguments. For this purpose, we designed six regular expressions which are available in our GitHub repository and an overview shown in Table 1. They were developed as follows: (i) Two people developed them sequentially by testing them on our datasets, with the second taking up and refining the work of the first. (ii) We ran 103 simple regular expressions such as "`\w+ one of`", "`\w+ [(vs.)(vs)(versus)] \w+`", or "`\d+ of`" for the datasets and evaluated a maximum of 99 of each (depending on how many were found) on a scale of $0, 1, 2$ to determine whether the statements found are quantitative (0: not quantitative, 1: partially quantitative, 2: completely quantitative). They contained 7,479 matches with a mean of 72 and a median of 99 per expression. (iii) The most promising expressions were manually merged into the six general ones shown in Table 1. In our example, the pattern $\text{REGEX}_{\text{times}}^{\text{decimal}}$ matches "490 times" in the premise.

3.2 Extracting Quantitative Statements

Through the pattern matching results, we have a set of candidate premises that contain potentially quantitative statements. Next, we use a prompting strategy combined with an LLM to obtain the referenced *entities* (e.g., game console and computer), the *trait* that is compared (e.g., price), the *quantity* (e.g., 4), and the *operator* between the entities (e.g., greater). To make an informed prediction,

You are an assistant that extracts quantitative statements from arguments.

Task Description: An argument consists of a claim (a statement that is being argued) and a premise (a statement that supports or attacks the claim). The stance indicates whether the premise supports or attacks the claim. The goal is to extract quantity statements from the premise that are relevant to the claim. A quantity statement consists of two entities (e.g., "computers" and "consoles"), a trait between them (e.g., "cost"), an operator (e.g., "greater"), and a quantity (e.g., "2"). The operator indicates the relationship between the two entities, and the quantity specifies the amount of the trait that one entity has compared to the other.

Input: You will be provided with a claim, its premise, and the stance between them. As a starting point, a pattern-based approach has been used to identify sentences in the premise that contain some free-form operator. The operator indicates the relationship between two currently unknown entities in the sentence. As additional context, you are provided the entire regex pattern that matched the sentence together with the operator.

Output: You shall extract all relevant information to call the function PREDICTSTATEMENTS.

Constraints: If *quantity* == 1.0, the operator "equal" or "approx" must be used. If *quantity* is any other value, the operator must be one of the other four options. The premise id will later be used to match the extracted quantity statements with the provided premise, so make sure to keep it.

Fig. 2. System message for extracting quantitative statements.

we provide the following properties of the argument to the LLM: The claim, its stance, all connected premises, and the list of matched sentences (including the full regular expression that caused each match). With the extensive context length of recent LLMs, we are able to send *all* premises connected to the investigated claim to the model—this should help to deal with challenges like coreferences. Dealing with multiple matches at once brings certain benefits, such as improved context awareness, but due to the probabilistic nature of LLMs, makes it harder to enforce predictable output. We use *function calling* as supported by an increasing number LLMs, allowing the definition of the expected response format using JSON schemas. For models without this feature, libraries such as Instructor[1] can be used to enforce the response format. Consequently, a claim with multiple pattern matching results in different premises only needs one request to the model—significantly speeding up the process and decreasing the cost of inference. For our example, the LLM extracts: {'raising a kid' > 'abortion' : 490.0×cost}. It works by providing a JSON schema describing the response format—in our case, a list of objects with the following properties: ID of the premise, the names of the two entities to be compared, their trait, operator, and quantity. It is accompanied by the system message shown in Fig. 2.

3.3 Augmenting Statements with Wikipedia Data

The statements have been extracted from an argument and, as such, may contain incorrect data—for instance, the premise may contain exaggerations concerning the quantity to make a stronger point. Although LLMs are capable text generators, they are also prone to hallucination and are not directly usable for

[1] github.com/567-labs/instructor.

validation purposes. By grounding their predictions in some kind of background knowledge, we aim to mitigate this issue. We chose regular Wikipedia articles as our source of truth instead of more structured resources such as DBpedia because this eliminates the need of encoding knowledge graphs for use with LLMs. For specialized domains, this could be extended to include other sources as well (e.g., scientific articles). We use Google instead of Wikipedia's built-in search to gain access to valuable metadata such as snippets for each result. We employ the following search string: *"{entity_1} {trait} {quantity} times {operator} than {entity_2} site:en.wikipedia.org"*. Our example leads to: *"raising a kid cost 490.0 times greater than abortion site:en.wikipedia.org"*.

From the list of the ten most relevant results, we then extract several pieces of information: Starting from the Google search, we take the *Google snippet title*, which is identical to the Wikipedia title, and the *Google snippet description*, which contains text parts relevant to the search string. On the Wikipedia page itself, we extract several textual elements: (i) a *short description*, which is coded in the HTML code, (ii) the *summary* of the article, which is usually found in the text before the first heading, i.e., the text before the first h2 element in the HTML code, and (iii) the *snippet context*, i.e. the paragraph in the article that is most *similar* to the Google snippet description. In addition, we extract all *Wikipedia tables* in the article including their HTML code and ignore all attributes except *href* located in *a* tags. We use BM25 [18] to compute the similarity between the Google snippet and the paragraphs in the article to favor instances where the entity and trait are actually mentioned in the article. Compared to semantic-aware embedding approaches, BM25 is based on term frequency and inverse document frequency, which is more suitable for our use case.

Conflicting information may be extracted from the list of search results and the Wikipedia tables. We deliberately did not add a mechanism to resolve these conflicts as we wanted to evaluate the model's ability to handle such situations. In addition, it may even be the case that subjective elements are involved in the validation process, potentially mandating the inclusion of such conflicting information to make an informed decision.

3.4 Validating Quantitative Statements

After retrieving relevant Wikipedia articles and extracting tables, this information can be fed into the LLM to predict validations for the previously extracted quantitative statements. This RAG-based approach aims to reduce the risk of hallucinations and increase the reliability of the generated response [4]. With each table represented using HTML and each of the ten search results potentially containing multiple tables, passing all extracted information to the LLM has the potential of exceeding the model's maximum context size—for instance, the GPT-4o used in our evaluation is limited to 128,000 tokens. Requesting more tokens yields an error; thus, we prune the context by allocating a fixed number of tokens per article. We settled on a limit of 10,000 for GPT-4o so that even in the worst case scenario 28,000 tokens (or about 20% of the context size) are available for the remaining input data like the system prompt and the generated output.

You are an assistant that verifies quantitative statements via provided retrieval results.
Task Description: In the previous step, you extracted a quantitative statement from an argument. A quantity statement consists of two entities (e.g., "computers" and "consoles"), a trait between them (e.g., "cost"), an operator (e.g., "greater"), and a quantity (e.g., "2"). Via a web search, we identified relevant Wikipedia pages that contain additional context such as tables and summaries. The goal is to validate the extracted quantity statement based on the provided context.
Input: You will be provided with the extracted quantity statement, the claim, the premise, their stance, the web search string, and the Wikipedia search results. A quantity value of 1.0 acts as the reference point and the value 0.0 indicates that no meaningful quantity could be extracted. The tables have been extracted in their HTML representation. Only the given information shall be used to validate the quantity statement.
Output: You shall extract all relevant information to call the function PREDICTVALIDATION.
Constraints: Do not use any external information beyond the provided context. If no data is available for the queried validation, respond with UNKNOWN.

Fig. 3. System message for validating quantitative statements.

Each table is then sequentially added to the context if and only if the mentioned limit is not exceeded. In this process, we maintain the natural order of the tables as found on the page. Due to the generative nature of LLMs, the provided reason is not guaranteed to correspond to the predicted decision. Nevertheless, our tests showed a better decision quality when the reason was included in the prompt—similar to how chain-of-thought reasoning works better in certain cases. Similar to the statement extraction step, we use function calling—this time, to obtain a single object with the validation decision (valid/invalid/unknown) and the reasoning behind. The system message is shown in Fig. 3. In our example, the model returns "Unknown" as Wikipedia lacks specific abortion cost data to verify the quantitative statement.

There is a limitation to this validation technique—temporal elements in arguments may introduce ambiguity. Consider the following statement used in Sect. 1: "*Global overpopulation is a myth, because humans can now produce 16 times as much wheat per acre as we could in the 1500s.*" The "now" in this statement is not explicitly defined, which could lead to different interpretations (e.g., due to the knowledge cutoff of the model) with the consequence that multiple answers may be correct.

4 Evaluation

Having presented our hybrid pipeline for extracting and validating quantitative statements in arguments, we now evaluate its effectiveness on the two argument corpora ARGS and KIALO presented in Sect. 3. After presenting our methodology, we analyze the generated dataset and discuss the results obtained from our manual annotation.

4.1 Methodology

Our final pipeline is implemented in Python and is available as open-source software.[2] We use OpenAI's model gpt-4o-2024-05-13 for all experiments due to its favorable cost-performance ratio. The dataset KIALO contains 2,423 claim matches, while ARGS contains 36,542 matches, out of which we sampled 2,500 for our evaluation. These claims were then passed to GPT-4o to extract quantitative statements, resulting in 3,385 claim-premise pairs for ARGS and 2,671 for KIALO. To avoid rate limits imposed by Google when retrieving relevant Wikipedia articles, we created two variants of the datasets: (i) $\text{ARGS}_{\text{EVAL}}$ and $\text{KIALO}_{\text{EVAL}}$ containing the extracted quantitative statements and the validation decisions based on Wikipedia data for 50 randomly sampled claims (leading to 124 claim-premise pairs), and (ii) $\text{ARGS}_{\text{FULL}}$ and $\text{KIALO}_{\text{FULL}}$ containing the extracted quantitative statements for all 6,056 claim-premise pairs. In this section, we use variant (i) as it contains all the necessary information for our evaluation. The full set of extracted statements may still be a valuable resource for future research, so all four datasets are available on request via Zenodo.[3]

To assess the quality of the extracted quantitative statements and their validation, we performed a manual annotation study. We asked two student annotators that are familiar with argumentation to label the generated statements and validations as *correct*, *partly correct*, *incorrect*, or *unknown*. They were also tasked with manually fixing the quantitative statements if necessary and validating them against data from Wikipedia. The annotation task was performed using the tool *Label Studio*.[4] The code for setting up the annotation interface is included in our GitHub repository.

4.2 Analysis of the Dataset

In this section, we give an overview of the generated datasets. Pattern matching identified substantially more quantitative arguments in the ARGS dataset (ca. 7%) compared to the KIALO dataset (ca. 0.3%), with frequently occurring operators including phrases like "is better than" and "is more than". For the predicted statements, we found that entities appeared mostly once and concrete numerical comparisons were uncommon, whereas operators such as "greater", "equal", and "less" had overlap in both datasets. Data augmentation was largely successful, on average our approach found more than one table per extracted statement. The predicted validations were mostly "unknown". By analyzing the corresponding explanations, the main reason was insufficient context information.

4.3 Analysis of the Manual Annotation

Each annotator labeled the same 124 claim-premise pairs from the EVAL datasets as *correct*, *partly correct*, *incorrect*, or *unknown*, This allows us to calculate the

[2] github.com/recap-utr/quantigpt (MIT license).
[3] zenodo.org/records/15720817.
[4] github.com/HumanSignal/label-studio.

Table 2. Inter-annotator agreement with two perspectives of Krippendorff's α.

dataset	**task**	α_{all}	α_{known}	C
ARGS	statements	.453	.519	.565
	validations	.103	.132	.580
KIALO	statements	.265	.339	.510
	validations	−.044	.016	.491

Table 3. Distribution of the merged labels.

dataset	**task**	**fully correct**	**partly correct**	**incorrect**	**unknown**
ARGS	statements	.434	.341	.174	.051
	validations	.732	.094	.138	.036
KIALO	statements	.509	.318	.127	.046
	validations	.700	.100	.127	.073

Inter-Annotator Agreement (IAA) to measure the reliability between annotators. We use Krippendorff's α [12]—it is suitable for multiple annotators, accounts for missing values, and handles categorical data. Our custom distance function $d(x, y)$ is shown in Sect. 4.3—it penalizes large differences by assigning a distance of 3 for all pairs that do not match.

$$d(x, y) = \begin{cases} 0 & \text{if } x = y, \\ 1 & \text{if } x \in \{\text{correct}, \text{incorrect}\} \wedge y = \text{partly correct}, \\ 3 & \text{otherwise.} \end{cases}$$

Table 2 shows the IAA results. In addition to the overall α_{all}, we also calculate the α_{known} for known labels only by excluding items where at least one annotator chose *unknown*. The percentage of concordant labels C shows that annotators agreed on the same label in only 50–60% of cases. Assessing the extraction of quantitative statements yielded a moderate agreement for the ARGS dataset and a fair agreement for the KIALO dataset [14]. The validation task showed poor agreement for both datasets (even being negative for KIALO). Multiple reasons could be the cause for this, such as the complexity of the task, the subjectivity of argumentation, vague instructions, and ambiguity in the data. The annotators noted that the Wikipedia sources were off-topic in some cases, possibly contributing to the low agreement.

Next, we analyze the label distribution. For the results in Table 3, we merged the individual labels from the two annotators into a single dataset with 248 labels regardless of agreement or disagreement. For the quantitative statements, the annotators assigned the label *incorrect* in less than 20% of the cases and *unknown* in approximately 5%. At the same time, 30% of the statements were labeled as *partly correct* (i.e., one or multiple parts of the statement were wrongly extracted). For the validation task, the annotators agreed on the label *correct* in about 70% of the cases, while the label *incorrect* was assigned in about 15%. At the same time, the IAA for the validation task was lower than for the extraction task, limiting the significance of the otherwise promising results for the validation task.

Returning to the research question, we conclude that the LLM-based techniques for extracting quantitative statements in arguments are not fully reliable. About half of the extracted statements were deemed correct by the annotators, but the validation task showed poor agreement between the annotators—requiring at least some manual intervention for the remaining half. Given the relatively low number of incorrect labels, our approach may be suitable for interactive fact-checking systems that combine human and machine intelligence. For our evaluation dataset, the two annotators manually fixed the incorrect and partly correct statements, which could be used to fine-tune the LLM in future work. By making available the annotated datasets, we hope to foster further research in this area.

4.4 Qualitative Analysis

Having obtained an overview of the dataset and the results of the manual annotation, we now analyze one successful and one failed outcome (see Fig. 4) of our pipeline and investigate the annotators' decisions. The two examples are taken from Args—the largest of the two corpora. Together with the original argument, we show the predicted statement/validation and (if necessary) the corrected statement/validation from both annotators. After presenting the predictions and annotations, we continue with a discussion of the results. For the sake of brevity, we use the following notation for the quantitative statements:

$$\text{`entity\_1' operator `entity\_2'} : \text{quantity} \times \text{trait}$$

For the argument in Fig. 4a, our pipeline was able to correctly extract the quantitative statement. One of the annotators agreed with that predicted validation *unknown*, the other however found additional evidence to mark this statement as *invalid*. This indicates that the augmentation step should be improved to inject more relevant contextual information into the validation prompt.

For the argument in Fig. 4b, the pipeline failed to extract the correct quantitative statement. Notably, the two annotators extracted different statements from the argument, the first one focusing on the gravitational constant, the second one on Superman's speed. Consequently, the validation corrections differ as well—thus, both extraction and validation were found to be lacking. We also observe subjective elements w.r.t. the extraction of the quantitative statement. Validation suffers from a lack of contextual awareness of the LLM: It failed to deduce that the gravitational constant of the earth is compared to the one of Superman's origin planet and instead states that its value is constant anywhere on the earth. This may be caused by the incorrect extraction of the quantitative statement and the mentioned subjectivity—thus being a cascading error.

Claim: Abortion
Premise: On average, it takes $245,000 to raise a kid, that's 490 times the amount of money it takes to have an abortion.
Stance: Pro
Predicted Statement: 'raising a kid' > 'abortion' : 490.0 × cost
Predicted Validation: Unknown—The extracted data provides the average cost of raising a child and some information on abortion but does not provide a specific cost for obtaining an abortion. Without an actual value for the cost of abortion, it is impossible to verify the quantitative relationship that the cost of raising a child is 490 times the cost of an abortion.
Corrected Validation 1: Invalid—In 2022, a medication abortion cost was about $580 at Planned Parenthood, though it could be more, up to around $800 ($245000 / $580 < 490) (Abortion in the United States and Cost of raising a child).
Corrected Validation 2: Unknown—No comparison context available in premise, without knowing the country/region/... the values can't be compared.

(a) Successful example.

Claim: I believe that Superman, without a doubt, would beat Goku in a fight.
Premise: Let's assume the radius of his hand is 3 in, or 0.0762 m. Then assuming the gravitational constant is equal to ours.
Stance: Con
Predicted Statement: 'gravitational constant' = 'ours' : 1.0 × constant
Predicted Validation: Valid—The data confirms that the gravitational constant, often denoted by G, is a universal constant that applies both in Earth's context and in broader physical laws such as Newton's law of universal gravitation and Einstein's general relativity. It does not change depending on the context or location, affirming that the claim of the gravitational constant being equal to our own is valid.
Corrected Statement 1: 'krypton' = 'earth' : 1.0 × gravitational constant
Corrected Validation 1: Invalid—Krypton has a much higher gravity than Earth which is the reason for Kryptonians' super powers (Krypton (comics)).
Corrected Statement 2: 'superman' = 'light' : 0.14 × speed
Corrected Validation 2: Unknown—Non decidable unless more specific context is available, depending on the source/variant of Superman talked about, the speeds vary wildly (Superman).

(b) Failed example.

Fig. 4. Two exemplary outcomes of our pipeline.

4.5 Limitations

Our evaluation provides an overview of the performance of our pipeline, but it is not without limitations. Fact-checking is an immensely time-consuming task, which can take several days for a single statement, even if several trained and professional fact-checkers are working to find evidence for it. In our study, we evaluated a small dataset that still required substantial effort. We commissioned two annotators because four eyes see more than two. But, as usual when dealing with natural language, one actually needs more annotators. For reasons of efficiency, we initially identified the statements to be verified using regular expressions from the texts, which we refined over the period of more than a year by testing them on these datasets. The recall could be increased by adding more regular expressions. Naturally, other datasets may require other regular expressions. Still, we opted for a precision-oriented approach, meaning that even more regular expressions would increase the recall, but not necessarily decrease the precision.

The use of instruction-based LLMs such as ChatGPT to solve Natural Language Processing (NLP) problems is controversial due to the fact that their power and, above all, their limitations are not completely transparent. Results are rarely identical due to random seeds and are also highly dependent on the prompts sent to the system. However, we believe that the power of generative models far exceeds that of older fine-tuning-based approaches and have therefore opted for this approach. In our evaluation, we relied solely on a single LLM model, GPT-4o by OpenAI. We did not compare it to other models, especially open-weight models, and left this for future work.

5 Conclusion and Future Work

In this paper, we present a hybrid pipeline for extracting and validating quantitative statements in arguments. We used a combination of pattern matching, LLMs, and RAG to extract and validate the statements. Our approach is evaluated on two argument corpora, ARGS and KIALO, and the results are checked by two student annotators. We provide a working implementation and two datasets containing the extracted statements and their validation decisions for further research. The evaluation shows promising results for the extraction task with less than 20% being rated as incorrect, but mixed results for the validation task—meaning that our approach is mostly suitable for interactive scenarios and conclude that LLM-based techniques for extracting quantitative statements in arguments are not fully reliable.

In future work, we plan to improve the extraction of quantitative statements by refining the regular expressions in an attempt to improve the context provided to the LLM. In addition to solving the task via pattern matching—which could oversimplify the task—we plan to investigate more complex methods such as template filling [8]. We will also evaluate the use of additional sources for data augmentation, such as DBpedia [2] and ConceptNet [19]. With recent advances in multimodal LLMs, another possible approach is to extract images from the matched Wikipedia pages and use them as additional context for the validation task. We plan to test open-source LLMs where the training data is known s.t. we can make more informed decisions about potential sources of background knowledge. Lastly, we aim to increase the IAA for the validation task by providing more detailed instructions to the annotators and possibly using more annotators to reduce the impact of individual biases.

Acknowledgments. We thank Mircea-Luchian Pojoni, Martin Hoppe, and Kilian Bartz for their invaluable help in doing the annotations.

References

1. Ajjour, Y., Wachsmuth, H., Kiesel, J., Potthast, M., Hagen, M., Stein, B.: Data acquisition for argument search: the args.me corpus. In: Benzmüller, C., Stuckenschmidt, H. (eds.) KI 2019. LNCS (LNAI), vol. 11793, pp. 48–59. Springer, Cham (2019). https://doi.org/10.1007/978-3-030-30179-8_4

2. Bizer, C., et al.: DBpedia - a crystallization point for the web of data. J. Web Semant. 154–165 (2009)
3. Chen, W., et al.: Tabfact: a large-scale dataset for table-based fact verification. In: 8th International Conference on Learning Representations, ICLR 2020, Addis Ababa, Ethiopia, 26–30 April 2020. OpenReview.net (2020). https://openreview.net/forum?id=rkeJRhNYDH
4. Gao, Y., et al.: Retrieval-Augmented Generation for Large Language Models: A Survey (2024)
5. Guo, Z., Schlichtkrull, M.S., Vlachos, A.: A survey on automated fact-checking. Trans. Assoc. Comput. Linguist. **10**, 178–206 (2022). https://doi.org/10.1162/TACL_A_00454
6. Gupta, V., Mehta, M., Nokhiz, P., Srikumar, V.: INFOTABS: inference on tables as semi-structured data. In: Jurafsky, D., Chai, J., Schluter, N., Tetreault, J.R. (eds.) Proceedings of the 58th Annual Meeting of the Association for Computational Linguistics, ACL 2020, Online, 5–10 July 2020, pp. 2309–2324. Association for Computational Linguistics (2020). https://doi.org/10.18653/V1/2020.ACL-MAIN.210
7. Hardalov, M., Arora, A., Nakov, P., Augenstein, I.: A survey on stance detection for mis- and disinformation identification. In: Carpuat, M., de Marneffe, M., Ruíz, I.V.M. (eds.) Findings of the Association for Computational Linguistics: NAACL 2022, Seattle, WA, United States, 10–15 July 2022, pp. 1259–1277. Association for Computational Linguistics (2022). https://doi.org/10.18653/V1/2022.FINDINGS-NAACL.94
8. Huang, K.H., Tang, S., Peng, N.: Document-level entity-based extraction as template generation. In: Proceedings of the 2021 Conference on Empirical Methods in Natural Language Processing, pp. 5257–5269. Association for Computational Linguistics (2021)
9. Islam, M.R., Liu, S., Wang, X., Xu, G.: Deep learning for misinformation detection on online social networks: a survey and new perspectives. Soc. Netw. Anal. Min. **10**(1), 82 (2020). https://doi.org/10.1007/S13278-020-00696-X
10. Koleva, A., Ringsquandl, M., Hatem, A., Runkler, T., Tresp, V.: Wiki-TabNER: Advancing Table Interpretation Through Named Entity Recognition (2024)
11. Kotonya, N., Toni, F.: Explainable automated fact-checking: a survey. In: Scott, D., Bel, N., Zong, C. (eds.) Proceedings of the 28th International Conference on Computational Linguistics, COLING 2020, Barcelona, Spain (Online), 8–13 December 2020, pp. 5430–5443. International Committee on Computational Linguistics (2020). https://doi.org/10.18653/V1/2020.COLING-MAIN.474
12. Krippendorff, K.: Reliability in content analysis: some common misconceptions and recommendations. Hum. Commun. Res. 411–433 (2004)
13. Küçük, D., Can, F.: Stance detection: a survey. ACM Comput. Surv. **53**(1), 12:1–12:37 (2021). https://doi.org/10.1145/3369026
14. Landis, J.R., Koch, G.G.: The measurement of observer agreement for categorical data. Biometrics 159–174 (1977)
15. Lenz, M., et al.: Towards an argument mining pipeline transforming texts to argument graphs. In: Computational Models of Argument, pp. 263–270. IOS Press (2020)
16. Lewis, P., et al.: Retrieval-augmented generation for knowledge-intensive NLP tasks. In: Advances in Neural Information Processing Systems, pp. 9459–9474. Curran Associates, Inc. (2020)
17. Oshikawa, R., Qian, J., Wang, W.Y.: A survey on natural language processing for fake news detection. In: Calzolari, N., et al. (eds.) Proceedings of the 12th Language

Resources and Evaluation Conference, LREC 2020, Marseille, France, 11–16 May 2020, pp. 6086–6093. European Language Resources Association (2020). https://aclanthology.org/2020.lrec-1.747/
18. Robertson, S.E.: Okapi at TREC-3, pp. 109–126 (1994)
19. Speer, R., Chin, J., Havasi, C.: ConceptNet 5.5: an open multilingual graph of general knowledge. In: Proceedings of the Thirty-First AAAI Conference on Artificial Intelligence, pp. 4444–4451. AAAI Press (2017)
20. Thorne, J., Vlachos, A., Christodoulopoulos, C., Mittal, A.: FEVER: a large-scale dataset for fact extraction and verification. In: Walker, M.A., Ji, H., Stent, A. (eds.) Proceedings of the 2018 Conference of the North American Chapter of the Association for Computational Linguistics: Human Language Technologies, NAACL-HLT 2018, New Orleans, Louisiana, USA, 1–6 June 2018, Volume 1 (Long Papers), pp. 809–819. Association for Computational Linguistics (2018). https://doi.org/10.18653/V1/N18-1074
21. Wachsmuth, H., et al.: Computational argumentation quality assessment in natural language. In: Proceedings of the 15th Conference of the European Chapter of the Association for Computational Linguistics: Volume 1, Long Papers, pp. 176–187. Association for Computational Linguistics (2017)
22. Zhou, X., Zafarani, R.: A survey of fake news: fundamental theories, detection methods, and opportunities. ACM Comput. Surv. **53**(5), 109:1–109:40 (2021). https://doi.org/10.1145/3395046
23. Zubiaga, A., Aker, A., Bontcheva, K., Liakata, M., Procter, R.: Detection and resolution of rumours in social media: a survey. ACM Comput. Surv. **51**(2), 32:1–32:36 (2018). https://doi.org/10.1145/3161603

A Hybrid Constraint-Based, Greedy, and Local Search Approach for the Transshipment Problem

Sven Löffler(✉), Viktoria Abbenhaus, George Assaf, and Petra Hofstedt

Brandenburg University of Technology Cottbus - Senftenberg,
Konrad-Wachsmann-Allee 5, 03046 Cottbus, Germany
{Sven.Loeffler,Viktoria.Abbenhaus,George.Assaf,Hofstedt}@b-tu.de

Abstract. The efficient resolution of logistics problems, particularly those aimed at minimizing costs and reducing environmental impact, represents a critical challenge in our globalized world. A prominent example of such problems is the Transshipment Problem, which seeks to determine the most cost-effective paths from sources (e.g., producers) through transshipment points to sinks (e.g., customers). Approaches to addressing this problem range from greedy algorithms, which may rapidly yield locally optimal solutions, to constraint-based methods that, given sufficient resources and computation time, can identify globally optimal solutions. In this study, we propose a hybrid approach that integrates greedy strategies into the solution process of constraint modeling for the Transshipment Problem. This integration aims to expedite the discovery of high-quality initial solutions while preserving the global optimization capabilities inherent in constraint-based search methods. To validate the effectiveness of this new hybrid approach, we conducted an extensive series of experiments, which demonstrate its significant advantages in solving the Transshipment Problem compared to both a conventional constraint model and pure greedy methods.

Keywords: Constraint Satisfaction and Optimization · Transshipment Problem · Planning, Routing and Scheduling · Search and Optimization

1 Introduction

Over the years, supply chains and producer-consumer relationships have been evolved into complex, globally distributed networks. This globalization has led to a multitude of logistics problems, the resolution of which can significantly reduce both costs and CO_2 emissions. One notable example of such a logistics problem is the Transshipment Problem (TP).

The TP is a generalization of the classical transportation problem [1]. While the classical transportation problem considers only sources (e.g., factories) and sinks (e.g., markets), the transshipment problem introduces intermediate nodes,

T. Braun et al. (Eds.): KI 2025, LNAI 15956, pp. 91–103, 2026.
https://doi.org/10.1007/978-3-032-02813-6_7

referred to as transshipment nodes. These nodes represent points where goods can be transferred, stored, or rerouted during the logistics process.

The objective of the transshipment problem is to minimize the total cost of transporting goods from sources through transshipment nodes to sinks while satisfying capacity and demand constraints. Conceptually, the TP is a network flow problem [3] where material flows are optimized on a directed graph.

Various greedy algorithms have been developed to address the TP [3,7,8, 10,12]. These algorithms are advantageous due to their rapid execution times. However, their primary limitation is that they typically find only locally optimal solutions and, in some cases, fail to find any feasible solution. On the other hand, global optimization approaches, such as constraint programming, can guarantee globally optimal solutions given sufficient computational resources and time. The drawback of these approaches lies in their often-prohibitive computational requirements, which may result in unacceptably long execution times even for suboptimal solutions. In some scenarios, constraint-based methods fail to deliver a solution within a reasonable time frame, whereas simpler greedy algorithms can provide a feasible solution within seconds.

To address these challenges, this paper proposes a hybrid approach that integrates greedy algorithms (problem-specific knowledge) into the global constraint-solving framework (a general-purpose solution method) to enhance its solving speed. Our approach combines the fast discovery of high-quality initial solutions with the global optimization capabilities of constraint programming, ensuring that a globally optimal solution is found given sufficient computational time.

Additionally, neighborhood-based local search methods, which improve existing solutions by exploring similar, potentially better alternatives, are incorporated into our hybrid approach. This aims to further accelerate the discovery of improved solutions after an initial feasible solution has been identified.

The remainder of the paper is structured as follows: Sect. 2 discusses the fundamentals of the Transshipment Problem, related work, and of constraint programming. Section 3 presents our four novel constraint-based greedy search approaches for the TP and their combination with neighborhood-based local search approaches in constraint programming. Section 4 outlines our experimental setup and analyzes the results. Finally, Sect. 5 summarizes the paper, presents conclusions, and discusses future work.

2 Preliminaries

This section begins with an introduction of the Transshipment Problem and presents greedy algorithms capable of solving it. Subsequently, a brief overview of the fundamentals of constraint programming is provided.

2.1 The Transshipment Problem and Greedy Solution Algorithms

The Transshipment Problem (TP) is a network flow problem involving three types of nodes: warehouses, transshipment points, and customers. Warehouses act as source nodes, while customers serve as sink nodes. Products are transported from warehouses to customers through transshipment points. In more complex real-world applications, transshipment points enable additional handling operations (e.g., packaging, splitting, consolidation) and/or a change of transportation modes [4].

We formally define the TP as follows. The problem involves w warehouses (source nodes), t transshipment points, and c customer nodes. Each warehouse $i \in \{1, 2, \ldots, w\}$ has a supply s_i, and each customer $j \in \{1, 2, \ldots, c\}$ has a demand d_j. Warehouses are connected to all transshipment points, and each transshipment point is connected to all customers. Each such connection (i, j), where $i \in \{1, 2, \ldots, w\}$ and $j \in \{1, 2, \ldots, t\}$ or $i \in \{1, 2, \ldots, t\}$ and $j \in \{1, 2, \ldots, c\}$, has a given maximum capacity $cap_{i,j}$ and a given cost $c_{i,j}$ per unit transported along the route from i to j. Tansshipment points serve solely as transfer nodes where goods cannot be stored. This means that the total inflow at a transshipment point must equal the total outflow.

The total transportation cost of the TP can be expressed as: $total\_cost = \Sigma_{(\forall i,j)} x_{i,j} * c_{i,j}$. Here, $x_{i,j}$ represents the number of goods transported along the connection (i, j), with $x_{i,j}$ constrained to $0 \leq x_{i,j} \leq cap_{i,j}$. The objective is to find an assignment for all $x_{i,j}$ that minimizes $total\_cost$ of transporting goods from warehouses, through transshipment points, to customers, while ensuring that no capacity constraints are violated and all customer demands are met [1].

In this paper, we assume resource balance, meaning that the total supply of all warehouses equals the total demand of all customers. The TP is an important computationally NP-hard combinatorial network flow problem. In the described variant, there are potentially $2^{w*t+t*c}$ possible edge selection combinations, from which the most cost-efficient one needs to be identified.

Greedy Solution Algorithms for the Transshipment Problem. There are various approaches for solving TPs. In this section, we focus on four greedy algorithms. Greedy search methods iteratively select a locally optimal option at each decision point to incrementally improve the current solution. Although greedy search approaches do not guarantee globally optimal solutions, they are computationally efficient and can often provide quick solutions [2,15].

Greedy Cost Minimization Heuristic [3]. This method iteratively selects the cheapest available transport route (i, j) and assigns the maximum possible flow $cap_{i,j}$ until the supply s_i of the sources (warehouses) and the demand d_j of the customers are satisfied.

In the first step, the connections (i, j) between warehouses and transshipment points are sorted in ascending order based on their transport costs $c_{i,j}$. Each connection $x_{i,j}$ is then assigned as much flow as possible without exceeding capacity or supply limits. This process is repeated until all warehouses have sent their supply to the transshipment points.

In the second step, this procedure is applied analogously to the connections between transshipment points and customers. However, since the assignment of warehouses to transshipment points is performed first without considering the connections and costs from transshipment points to customers, this can lead to significantly higher costs in the second stage. In some cases, it may even result in an infeasible solution where capacity constraints of certain connections are violated.

Demand-Satisfaction Greedy Algorithm [8]. This method focuses on fulfilling customer demand as quickly as possible by prioritizing customers with the highest unmet demand. The approach is divided into two phases, starting with the customers. In the first step, iteratively select the customer with the highest unsatisfied demand. Identify the transshipment point whose connection to the customer has the lowest transportation cost. Assign the maximum possible flow along this path. Then, update the remaining capacities and the customer's demand. Continue assigning connections to the customer until their demand is fully satisfied, and then proceed to the next customer. This process is repeated until all customer demands are fulfilled.

In the second step, the procedure is repeated starting from the transshipment points until all demands of the transshipment points are satisfied by the warehouses.

Since the transshipment points are not limited by capacity, the first step (connecting customers and transshipment points) can be executed exactly as described. However, due to connection capacities and limited supply, the second step (connecting transshipment points to warehouses) may result in higher costs and potential capacity violations.

Supply-Driven Greedy Algorithm [10]. This method follows a procedure similar to the Demand-Satisfaction Greedy Algorithm, with the difference being that it focuses on the supply from the warehouses. Initially, warehouses are sorted in descending order of their supply, and each warehouse is connected to the transshipment points at the lowest possible cost.

Next, the available supply at the transshipment points is sorted in descending order, and the transshipment points are connected to customers, again minimizing transportation costs. This approach aims to optimize the flow from warehouses to customers through transshipment points, ensuring efficient distribution while minimizing overall costs.

Route Enumeration and Greedy Assignment [7,12]. This algorithm calculates all potential routes from warehouses through transshipment points to customers and sorts them by cost in ascending order. Goods are transported along the least costly routes, subject to the constraints of warehouse supply, connection capacities, and customer demand, until all customer demands are satisfied. After each route assignment, the relevant capacities, supplies, and demands are updated.

Unlike previous algorithms that focus solely on flows to or from transshipment points, this method considers complete end-to-end routes. It is particularly effective for networks with a manageable number of transshipment points and customers, where the number of possible routes remains computationally feasible.

However, calculating all potential routes can be computationally intensive in larger networks, leading to a combinatorial explosion. Additionally, greedy algorithms as heuristic approaches do not guarantee optimality.

2.2 Constraint Programming

Constraint Programming (CP) is a declarative method for modeling and solving complex problems, particularly those classified as NP-complete and NP-hard. CP has been successfully applied to various domains, including scheduling, graph coloring, optimization, and satisfiability problems (SAT) [13].

The process of constraint programming typically consists of two main phases: a) declarative modeling the problem using constraints and b) automated solving the model with an independently operating solver.

The primary responsibility of the CP user is to create the application-specific problem model using constraints and then initialize and run the solver. The solver autonomously explores potential solutions, leveraging advanced techniques to find either an optimal or satisfactory solution. This separation of responsibilities simplifies the problem-solving process and allows domain experts to focus on the abstract representation of their real-world challenges without delving into the computational intricacies of the underlying methods. In the following, we briefly introduce the key terminologies of constraint programming.

A *Constraint Satisfaction Problem (CSP,* [5]*)* is formally defined as a 3-tuple $P = (X, D, C)$, consisting of the following components: $X = \{x_1, x_2, \dots, x_n\}$ is a set of variables, $D = \{D_1, D_2, \dots, D_n\}$ is a set of finite domains, where D_i represents the domain of the variable x_i, and $C = \{c_1, c_2, \dots, c_m\}$ is a set of constraints, with each constraint c_j defined over a subset of variables from X.

Additionally, a *Constraint Optimization Problem (COP)* extends the CSP by adding an objective function. A COP is defined as $P = (X, D, C, f)$, where $f \in X \rightarrow \mathbb{R}$ is an explicitly identified optimization function that allows the evaluation and comparison of solution quality. The goal of a COP is to either minimize or maximize the objective function to achieve an optimal solution [5].

A *constraint*, represented by the tuple (X', R), consists of a relation R and an ordered set of variables X', which is a subset of X (the variables of the corresponding CSP or COP). The relation R is defined over these variables [9].

A *solution* to a CSP or COP consists of an assignment of values d_i from their respective domains D_i to all variables x_i, such that all constraints C are satisfied [5,9].

3 A Hybrid Local Search Constraint Approach

This section names the constraint-based modeling approach adopted for the TP. Additionally, we discuss how the four greedy algorithms and a local search method were seamlessly integrated into the constraint solution process to enhance efficiency and accelerate convergence. We adopted a Choco-based implementation of the COP as described in [4] which is mainly based on weighted sum constraints. Choco [14], a Java library, serves as an efficient and versatile constraint solver, offering robust support for constraint propagation, search heuristics, global constraints, and optimization capabilities.

The integration of greedy search methods did not require modifications to the constraint model but involved adjustments to the solver's search strategy. In finite domain constraint solvers, the search strategy is determined by two components: a variable selection heuristic H^{var} that decides the next variable of a set of candidates X to assign and a value selection heuristic H^{val} that specifies the value to assign.

Choco allows these heuristics to be flexibly configured for different subsets of variables within the CSP/COP, enabling the composition of a tailored overall search strategy (see Eq. 1). Initially, the search strategy is applied over the variable set X_1 using variable selection heuristic H_1^{var} and value selection heuristic H_1^{val}. Subsequently, the process continues with variable selection heuristic H_2^{var} and value selection heuristic H_2^{val} over the variable set X_2, and so forth. This capability was leveraged in [11] to effectively combine the advantages of local neighborhood search methods with the global nature of constraint solving. This approach will later be employed to combine our newly developed greedy constraint-based methods with a local neighborhood search strategy, enabling the refinement of initial solutions and further exploration of the solution space around the best solution found so far.

$$\begin{aligned} \textit{Total Search Heuristic} &= (H_1^{var}, H_1^{val}, X_1), (H_2^{var}, H_2^{val}, X_2)... \\ &\text{where } X_1, X_2, ... \subseteq X \text{ of the CSP/COP.} \end{aligned} \tag{1}$$

3.1 Greedy Cost Minimization Heuristic

The realization of the Greedy Cost Minimization Heuristic (see Sect. 2.1) as a constraint search strategy is achieved through an appropriate combination of variable and value selection heuristic, H^{var} resp. H^{val}. The variable selection heuristic consists of a static component, executed once during initialization, and a dynamic component, invoked repeatedly during the search for the next variable.

Static Component: Sort all routes from warehouses to transshipment points (*routeWT*) and all routes from transshipment points to customers (*routeTC*) in ascending order based on their transportation costs.

Dynamic Component: Iterate through the routes (w, t) in list *routeWT*, sorted from least to most expensive. If the variable $x_{w,t}^{W,T}$ representing the flow on the route between warehouse w and transshipment point t has not yet been instantiated, select it. If $x_{w,t}^{W,T}$ is not instantiated, it indicates that the capacity for this connection has not been exhausted. Otherwise, constraint propagation would have set $x_{w,t}^{W,T}$ to 0. Therefore, $x_{w,t}^{W,T}$ always corresponds to the most cost-effective route among the remaining feasible options.

After assigning all values to the variables $x_{w,t}^{W,T}$ the algorithm continues with the same behavior for the routes (t, c) in list *routeTC*. Thus, the variable $x_{t,c}^{T,C}$ with the lowest costs is instantiated first.

Value Selection: Once a variable x is selected, the $IntDomainMax$ value selection heuristic can be applied. This heuristic always assigns the maximum value d_{max} from the domain D_x, effectively allocating the maximum possible flow to the selected route. At this stage, constraint propagation has already excluded any values exceeding the route's capacity, ensuring the assignment remains feasible.

The combination of a static initialization step to prioritize cost-efficient routes and a dynamic heuristic for iterative variable selection ensures that the most economical flows are prioritized early in the search. Using $IntDomainMax$ for value selection further guarantees efficient resource allocation while adhering to capacity constraints, making this approach both robust and computationally effective for the TP.

3.2 Demand-Satisfaction Greedy Algorithm

Similar to the previous approach, this method again employs a combination of variable selection and value selection heuristic. The variable selection heuristic is still divided into a static component, executed once during initialization, and a dynamic component, applied iteratively during the search process.

Static Component: Sort all customers in descending order of their demand. For each customer and each transshipment point, sort all edges in ascending order based on transportation costs.

Dynamic Component: Iterate through the list of customers sorted by demand (starting with the highest demand). For each customer c, traverse the sorted list of incoming routes from transshipment points. Select the variable $x_{t,c}^{T,C}$ corresponding to the lowest-cost route (t, c) that has not yet been instantiated. This ensures that the cheapest available edges for a given customer are selected first. If all edges for a customer have already been assigned, the process continues with the customer with the next highest demand.

Once all customers have been fully supplied, the corresponding route assignments from transshipment points to customers are finalized. The next step is to determine the demand at each transshipment point based on the assigned flows (from transshipments points to customers). The transshipment points are then processed analogously as the customer nodes before. They are sorted in descending order of their total demand, and routes from warehouses to transshipment points are assigned iteratively, beginning with the transshipment point with the highest demand.

Once a variable is selected, the value is assigned using the $IntDomainMax$ heuristic, which always selects the maximum feasible value from the variable's domain. At this stage, constraint propagation has already eliminated excessive capacities, ensuring that only valid assignments remain.

3.3 Supply-Driven Greedy Algorithm

This algorithm follows an approach analogous to the Demand-Satisfaction Greedy Algorithm, with the primary difference being that, in the static phase, warehouses are sorted by their supply instead of customers by their demand.

In the dynamic phase, the selection process is adjusted accordingly: The algorithm first identifies the most cost-effective routes from the warehouse with the highest remaining supply to the transshipment points. Subsequently, once the transshipment points have accumulated supplies, the most cost-effective routes from transshipment points to customers are determined.

The value selection heuristic remains unchanged and continues to use the *IntDomainMax* strategy, ensuring that the maximum allowable flow is assigned to each selected route.

3.4 Route Enumeration and Greedy Assignment

In this approach, the traditional two-step process of selecting a variable first and then assigning a value could not be applied rigidly. Instead, a search strategy was developed that, at each step, directly returns a variable along with a concrete value. This results in an atypical traversal of the search tree. Unlike conventional search strategies, where all possible values for a given variable are systematically explored, this approach allows for a more flexible progression. If a selected value $d_1 \in D_x$ for variable x does not lead to a feasible solution, the strategy does not necessarily attempt alternative values $d_2, d_3, ... \in D_x$ for x next. Instead, another variable y may receive a value assignment, bypassing the typically structured order of variable and value selection.

This unstructured exploration sacrifices completeness, as it does not systematically store all partial assignments (which would be infeasible due to the problem size). Therefore, to ensure completeness, this newly developed search strategy should be complemented by a subsequent exhaustive search phase that iterates through all (remaining unconsidered) assignments.

Although our developed Route Enumeration and Greedy Assignment search strategy (S^{REGA}) does not follow the classical variable and value selection heuristics, it still consists of a static and a dynamic component.

Static component: All possible routes (w, t, c) from warehouses via transshipment points to customers are precomputed and sorted in ascending order based on their transport costs. Using the Route Enumeration and Greedy Assignment method described in Sect. 2.1, an initial solution is determined. From this solution, the corresponding values $v_{w,t}^{W,T}$ and $v_{t,c}^{T,C}$ for the variables in $X^{W,T}$ resp. $X^{T,C}$ are derived. For example, if two routes $(3, 4, 2)$ with flow 4 and $(4, 4, 2)$ with flow 3 belong to the solution, then: $x_{3,4}^{W,T} = 4$, $x_{4,4}^{W,T} = 3$, and $x_{4,2}^{T,C} = 4 + 3 = 7$.

Dynamic component: Starting with the sheapest route (w, t, c), the algorithm attempts to assign the precomputed variable value $v_{w,t}^{W,T}$ to $x_{w,t}^{W,T}$, which represents the sub-route (w, t). If this is not feasible, the largest possible value smaller than $v_{w,t}^{W,T}$ is assigned instead.

Unlike previous methods, S^{REGA} completes entire routes sequentially, meaning that after setting $x_{w,t}^{W,T}$, it immediately assigns $x_{t,c}^{T,C}$ with its corresponding value. This differs from approaches where all sub-routes from warehouses to transshipment points (or from transshipment points to customers) are set first.

Since this method is not inherently complete, a standard search strategy as *DomOverWDeg* [6] is subsequently applied to exhaustively explore all remaining variable-value combinations. This results in the overall search strategy:

$$TotalSearchHeuristic = S^{REGA}, (H_{default}^{var}, H_{default}^{val}, X). \tag{2}$$

3.5 The Combination of Constraint-Based Greedy and Local Search

In [11] a method to combine local search with the global nature of constraint programming is introduced. Initially, pure greedy algorithms as presented in Sect. 2.1 are used to compute solutions and the best one is used as initial solution s_0. Subsequently, the search explores the neighborhood of this solution to identify improved solutions.

In order to link the local neighborhood-based approach from [11] to our COP for the TP the overall search strategy is extended. In [11] exists a variable n which is associated with the exact number of deviations from the inital solution s_0. The variable n is first assigned values from its domain $D_n = \{0, 1, ..., w * t + t * c\}$ in ascending order. This means that the search process begins with the exact initial solution s_0, where there are no deviations ($n = 0$) in the choice of routes or flow assignments. Gradually, the search is allowed to introduce increasing variations ($n = 1$, $n = 2$, ...), either by employing different routes or by altering the flow, compared to the initial solution. This controlled expansion of the neighborhood enables the exploration of alternative solutions while still remaining close to the inital solution, thereby balancing intensification around promising areas with diversification to potentially improve the overall objective.

For example, the variable selection heuristic *InputOrder* can be applied exclusively to the variable n. In combination with the value selection heuristic *IntDomainMin*, n is initially assigned the smallest value in its domain, and then progressively larger values are tried.

Afterwards, the search strategies described in Sect. 3.1 to Sect. 3.4 are applied to the remaining variables. Additional constraints (see [11]) ensure that the constraint solver prioritizes solutions that differ from the initial solution s_0 by exactly n modifications. Despite being guided by the initial solution, this approach remains complete and can find a globally optimal solution given sufficient computational resources. After enforcing the n-bounded search, the solver continues with another search strategy (as presented in former sections):

$$\begin{aligned} &TotalSearchHeuristic = \\ &\qquad (H_{InputOrder}^{Var}, H_{IntDomainMin}^{Val}, \{n\}), (H_1^{Var}, H_1^{Val}, X_1), ... \end{aligned} \tag{3}$$

4 Experiments and Results

This section begins by detailing the experimental setup, followed by an evaluation and discussion of the results obtained by the conducted experiments.

4.1 The Experimental Setup

All experiments were carried out on a Dell computer featuring an 4th Gen Intel(R) Core(TM) i7-4770 quad-core processor running at a clock speed of 3.40 GHz and 32 GB DDR3 RAM, operating at 3401 MHz. The operating system used was Microsoft Windows 10 Enterprise. The Java programming language with JDK version 17.0.5 and the constraint solver Choco-Solver version 4.10.7 [14] was utilized. Unless otherwise specified during greedy or local search experiments, the default search strategy *DomOverWDeg* [6] was employed.

We explored a wide range of solution approaches, which can be broadly categorized into four main areas: pure greedy search methods G, greedy-based constraint approaches C, local search greedy-based constraint approaches L, and local search greedy-based constraint approaches with selected neighborhood S. The pure greedy approaches G are those described in Sect. 2.1: G^{GC}: Greedy Cost Minimization Heuristic, G^{DS}: Demand-Satisfaction Greedy Algorithm, G^{SD}: Supply-Driven Greedy Algorithm, and G^{RE}: Route Enumeration and Greedy Assignment.

The constraint-based approaches C consist of a model using the default search strategy of the Choco Solver C^D, as well as the implementations of various greedy search strategies presented in Sect. 3: C^{GC}, C^{DS}, C^{SD}, and C^{RE}.

L and S are variants of the previously described greedy algorithms, adapted with local search techniques from [11]. Initially, the pure greedy algorithm G^{GC}, G^{DS}, G^{SD}, or G^{RE} are used to compute solutions and the best one is used as initial solution s_0. Subsequently, the search explores the neighborhood of this solution to identify improved solutions. After enforcing the n-bounded search of the neighborhood of s_0, the solver continues with Choco's default search strategy L^D. In contrast, the methods L^{GC}, L^{DS}, L^{SD}, and L^{RE} replace the default search with the respective greedy search heuristics from Sect. 3.

To reduce computational complexity and the number of constraints required to enforce n, the S variants (S^D, S^{GC}, S^{DS}, S^{SD}, and S^{RE}) only bind n to variables in $X^{W,T}$ and $X^{T,C}$ that are non-zero in the initial solution s_0. This adjustment expands the neighborhood for a given n, allowing for a wider search space while reducing the computational overhead of enforcing the n-bounded constraints.

Our TP instances are randomly generated with the following parameters: number of warehouses $w \in \{10, 50, 100, 500, 1000\}$, number of transshipment points $t \in \{10, 50, 100, 200\}$, and number of customers $c \in \{10, 50, 100, 500, 1000\}$. This yields $5 \times 4 \times 5 = 100$ distinct TP instances, each of which is to be solved using the 19 different solution approaches within a 5-minute time limit. The locations of warehouses, transshipment points, and customers are randomly positioned on a 2000×2000 2D-grid, and the transportation costs are computed using the Euclidean distance between these points.

Furthermore, we always ensured that the total demand equaled the total supply. These values were then randomly distributed across the nodes in the first or last layer of the transshipment problem, respectively.

4.2 Results and Evaluation

Table 1 presents the results of our experimental run. Each row in the table summarizes the outcomes of one of the 19 previously described approaches. In the columns, the number of instances for which at least one solution was found if solvable or unsatisfiability is detected (*#Sol*), the number of instances for which a global optimum was identified and proven (*#Global*), the number of instances where the obtained solution was optimal compared to the other 18 approaches (*#Best*), the average cost of the best solution (*∅Cost*), and the number of cases in which the respective approach provided better (+), equal (=), or worse (−) solutions compared to the next simpler version. For instance, the number of improvements, unchanged results, and deteriorations from S^{RE} to L^{RE} is recorded in the row corresponding to S^{RE}. Similarly, the differences between L^{RE} and C^{RE} are documented in the row for L^{RE}, while the changes from C^{RE} to G^{RE} are listed in the row for C^{RE}.

Table 1. The obtained results of the various solution approaches for the 100 TP instances.

Approach	#Sol	#Global	#Best	∅Cost	Imps. (+, =, −)
G^{GC}	35	0	0	1,739,714	(-,-,-)
G^{DS}	44	0	0	2,068,681	(-,-,-)
G^{SD}	44	0	0	1,974,139	(-,-,-)
G^{RE}	85	0	33	1,214,101	(-,-,-)
C^{D}	68	14	14	2,267,796	(-,-,-)
C^{GC}	90	14	15	1,470,481	(74, 26, 0)
C^{DS}	**100**	14	14	1,392,599	(67, 33, 0)
C^{SD}	99	14	14	1,396,502	(79, 21, 0)
C^{RE}	**100**	**15**	55	1,214,031	(18, 82, 0)
L^{D}	**100**	14	47	1,213,474	(86, 14, 0)
L^{GC}	**100**	14	47	1,213,474	(85, 14, 1)
L^{DS}	**100**	14	47	1,213,440	(86, 14, 0)
L^{SD}	**100**	14	48	1,213,459	(86, 14, 0)
L^{RE}	**100**	14	47	1,213,474	(2, 81, 17)
S^{D}	**100**	14	48	1,213,436	(3, 96, 1)
S^{GC}	**100**	14	50	1,213,410	(10, 90, 0)
S^{DS}	**100**	14	49	1,213,456	(7, 90, 3)
S^{SD}	**100**	14	48	1,213,444	(7, 91, 2)
S^{RE}	**100**	14	**86**	**1,213,080**	(44, 55, 1)

Since the approaches in L and S each use the best solution among the four greedy algorithms G as their initial solution, they cannot perform worse than the pure greedy methods.

Some problem instances (14 of 100) were unsolvable, which the greedy approaches G were unable to detect. None of the greedy algorithms consistently found a solution, even when one existed (see row #Sol). In contrast, all constraint-based approaches C, L, and S could quickly determine when a TP instance had no solution (in less than 10 s).

In summary, the following observations can be made from the results: The constraint-based greedy approaches C clearly outperform both the pure greedy methods G (with 18 to 79 improvements and no deteriorations) and the general constraint approach C^D. Although the general constraint approach C^D finds a solution in at least 68% of the cases in the time limit of 5 min, its solutions are significantly inferior compared to those of the other methods (see column ⌀Cost), particularly the greedy approaches G.

With the exception of G^{RE}, which performed relatively well, the pure greedy algorithms G produced significantly worse objective values ⌀Cost compared to the new developed constraint approaches (the greedy approaches without G^{RE}: 1,739,714 to 2,068,681, the new greedy constraint-based approaches C: 1,214,031 to 1,470,481). For the calculation of the average best cost ⌀Cost, note that an algorithm that failed to find a solution was assigned the worst solution found by the other approaches.

The integration of the greedy constraint approaches L with the local search method from [11] guarantees that a solution is always found and further improves the objective function values. Moreover, modifying the local search procedure so that only those variables that are non-zero in the initial solution are considered in the neighborhood S leads to additional improvements (see the Imps. (+,=,-) column).

Ultimately, the S^{RE} approach stands out as the best method: among the 19 methods it finds the best solution in 86% of the 100 test instances and achieves the best optimization value with an average cost of 1,213,080 (see column ⌀Cost).

5 Conclusion

In this work, we introduced and developed new variants for integrating greedy search algorithms into the global context of constraint programming and combined it with local search methods for COPs. Overall, the newly developed approaches effectively prune suboptimal search spaces within the general branch-and-bound framework, leading to faster discovery of improved solutions and potentially accelerating the completion of the global search process. The proposed methods were demonstrated and tested using the Transshipment Problem. It was shown that incorporating greedy and local search strategies into the global constraint-based approach allows solutions to be found for large instances while surpassing the quality of purely local methods.

Future research will focus on enhancing the interaction between different methods, such as greedy and local search, and methods of constraint programming to better leverage their respective strengths. In upcoming studies, further variations of greedy search will be integrated into our constraint approaches, aiming to improve solution performance even further. Furthermore, the greedy constraint approaches should also be investigated and developed for problems other than the Transshipment Problem.

References

1. Agadaga, G.O., Akpan, N.P.: Transshipment problem and its variants: a review. Math. Theory Model. **7**(2), 19–32 (2017)
2. Aggarwal, C.C.: Artificial Intelligence - A Textbook. Springer (2021). https://doi.org/10.1007/978-3-030-72357-6
3. Ahuja, R.K., Magnanti, T.L., Orlin, J.B.: Network Flows - Theory, Algorithms and Applications. Prentice Hall (1993)
4. Özgür Akgün: CSPLib problem 083: Transshipment problem. http://www.csplib.org/Problems/prob083
5. Apt, K.: Constraint Satisfaction Problems: Examples. Cambridge University Press, New York (2003). Principles of Constraint Programming: chapter 2
6. Boussemart, F., Hemery, F., Lecoutre, C., Sais, L.: Boosting systematic search by weighting constraints. In: Proceedings of the 16th European Conference on Artificial Intelligence, ECAI'2004, including Prestigious Applicants of Intelligent Systems, PAIS 2004, Valencia, Spain, 22–27 August 2004, pp. 146–150 (2004)
7. Dantzig, G.B., Ramser, J.H.: The truck dispatching problem. Manage. Sci. **6**(1), 80–91 (1959)
8. Dantzig, G.B., Thapa, M.N.: Linear Programming: Theory and Extensions, vol. 2. Springer (2003)
9. Dechter, R.: Constraint Networks, pp. 25–49. Elsevier Morgan Kaufmann (2003). Constraint Processing: chapter 2
10. Hillier, F.S., Lieberman, G.J.: Introduction to Operations Research. McGraw-Hill (2015)
11. Löffler, S., Hofstedt, P.: A constraint-based greedy-local-global search for the warehouse location problem. In: Maglogiannis, I., Iliadis, L.S., MacIntyre, J., Avlonitis, M., Papaleonidas, A. (eds.) Artificial Intelligence Applications and Innovations - AIAI 2024, Part III. IFIP Advances in Information and Communication Technology, vol. 713, pp. 291–304. Springer, Cham (2024). https://doi.org/10.1007/978-3-031-63219-8_22
12. Magnanti, T.L., Wong, R.T.: Network design and transportation planning: models and algorithms. Transp. Sci. **18**(1), 1–55 (1984)
13. Marriott, K., Stuckey, P.J.: Programming with Constraints - An Introduction. MIT Press, Cambridge (1998)
14. Prud'homme, C., Fages, J.G., Lorca, X.: Choco documentation (2017). http://www.choco-solver.org. Accessed 20 Mar 2025
15. Russell, S., Norvig, P.: Artificial Intelligence: A Modern Approach, 3 edn. Prentice Hall (2010)

Re-examining Learning Linear Functions in Context

Omar Naim[1,2(✉)], Guilhem Fouilhé[1,2], and Nicholas Asher[1,3]

[1] Institut de Recherche en Informatique de Toulouse, IRIT, Toulouse, France
omar.naim.docs@gmail.com, {guilhem.fouilhe-lafforgue,nicholas.asher}@irit.fr
[2] Université de Toulouse, Toulouse, France
[3] CNRS, Toulouse, France

Abstract. We explore in-context learning (ICL), a popular paradigm for inference with Large Language Models (LLMs), in a controlled experimental setup using synthetic training data. Using a range of small transformer models trained from scratch, we focus on a mathematical task with simple yet precise prompts: learning a linear function f from a sequence of inputs x_i and their corresponding function values $f(x_i)$. Our findings challenge the prevailing narrative that transformers adopt algorithmic approaches like linear regression to in-context learn (ICL) a linear function. We observe that all models have "boundary values" that limit generalizability. While we can extend boundary values with training distributions over a wider range, we lose the precision of models trained on distributions with more restricted ranges. Thus, we see a dilemma for ICL at least in some tasks: either models will lack generalizability or precision.

Keywords: Transformers · LLMs · In-context learning

1 Introduction

Large language models (LLMs) exhibit a remarkable capability to learn by analogy, known as in-context learning (ICL) [6]. This approach is especially useful for adapting a model to a task with very few resources at inference time. Despite several theoretical reconstructions of ICL, there have been few studies on exactly how ICL works in practice. Are models actually learning to reason or are they simply memorizing training data, as [13] have suggested for other tasks? We address this gap by focusing on a simple mathematical task, the in-context learning of one-dimensional linear functions, where the training regime and data are straightforward and clean. This allows us to have greater control over the data, enabling a more effective analysis of the model's ICL behavior.

We distinguish between two notions of "learning a function". The first, assumed by the literature since [9], says that a model has learned a function if, when sampling values and coefficients of linear functions at test time from the same distribution used in training the model yields close to 0 expected error. Our models can perform ICL for linear functions in this sense, which we call ICL_1. On the other hand, another notion, call it ICL_2, might say that learning a linear function requires specifying the parameters

T. Braun et al. (Eds.): KI 2025, LNAI 15956, pp. 104–117, 2026.
https://doi.org/10.1007/978-3-032-02813-6_8

a, b, where $f(x) = ax + b$, from data points in the prompt, something which is clearly possible using a variety of techniques like linear regression. [1,2,12,22] claimed that transformers use such techniques and so exhibit ICL_2. However, we give strong empirical evidence that none of our transformer models use such techniques or are performing ICL_2 of linear functions.

The reason our models cannot ICL_2 is that, across all training distributions, they have "boundary values" B^-, B^+ that constrain their predictions. When the inputs x_i and the function f parameters are chosen such that $f(x_i) >> B^+$ or $<< B^-$, model performance degrades substantially. We investigate the conditions under which these boundary values arise. These specific values, frequently near the maximum and minimum observed during training, indicate that the model has likely memorized them. But while boundary values are defined by training, the models' degraded predictions do not come from the training data but are rather a sort of reasoning-based hallucination.

We studied ICL in over 30 models, ranging from transformer architectures with a single layer and one attention head (AH) to models with 12 layers and 8 AHs, as well as small attention-only models [17]. To train our models, we use sequences generated by functions $f(x) = ax + b$ where a, b are sampled from a distribution over function parameters and x is drawn from a distribution over inputs. The model learns an algorithm on that training data, and we then evaluate its predictions on a test distribution at inference time. Since ICL depends on a model's pretraining, such an investigation requires training from scratch, and that has constrained our study to relatively small transformer architectures.

2 Related Work

[6] introduced ICL as a paradigm where the model learn at inference from the prompt by analogy, with no changes to the parameters set during training. Despite its promise and popularity, this phenomenon is still not fully explored. [7] surveys the successes and challenges for ICL and argue that so far, research has only studied simple tasks like learning linear or Boolean functions.

[9] showed that a transformer trained from scratch (GPT-2 [20] with an embedding size of 256) performed in-context learning of linear functions given identical train and test Gaussian distributions.

Further research has offered several reconstructions for how ICL might work for the class of linear functions $\mathcal{L}$ in transformers. [1,2,12,22] provided a construction to show that transformers could learn via gradient descent to do linear regression in ICL, which would solve the task. [8] showed that transformers could perform ICL in virtue of using higher-order optimization techniques. [18,24,25,27] argued that ICL follows from Bayesian principles. [4] show that transformers can under certain assumptions implement many algorithms with near-optimal predictive power on various in-context data distributions. Given [19]'s result that full transformers with linear attention are Turing complete, however, such theoretical demonstrations are not surprising. Whereas most prior research has concentrated on what transformer models *can* or *could* do on this task, we examine how ICL works in practice under different training and testing distributions in order to establish what transformers *actually* do.

[25,26] show that when the training and inference distributions are shifted, ICL performance degrades. This work is closely aligned with our approach, as is the work of [10]. However, [10,26] make important modifications to transformer architectures by working with linear attention, whereas we look at attention layers as they are more commonly implemented with softmax. In addition, [26] uses a new kind of optimization or training with gradients and a special fixed initial point. This means that their architecture and training are quite different from what normally happens with transformers; they are interested in getting a revised transformer-like model to learn linear functions, while we want to find out whether existing transformers learn linear functions or something else. As we detail below, the results for the architectures of [10,26] are quite different from those we have for standard transformers. Finally, our proposed mathematical model for ICL differs from any of the proposed methods.

[5] trained small GPT-2-like models from scratch to show that transformers can ICL simple boolean functions, while their performance deteriorates on more complex tasks. [21] investigated how ICL in models evolves as the number of pretraining examples increases, in a setting where the training and test distributions are identical.

[17] offer an in-depth analysis of ICL across tasks using a general evaluation measure on prompt length. They propose that ICL is enabled by a learned copying and comparison mechanism for ICL. In Sect. 6, we show their proposal does not fit our task.

3 Experimental Setup

We investigate transformer models that are trained directly on an in-context learning objective. We trained several small decoder-only transformer models from scratch to perform ICL of linear functions.[1] Our models feature from 1 to 18 layers (L) and from 1 to 8 attention heads (AH) with an embedding size of 64 to 256.

The model's task is to predict the next value for $f(x_i)$ through a prompt of type $(x_1, f(x_1), ..., x_i)$. We refer to that prediction as $\hat{f}(x_i)$. To train the model to perform ICL $\mathcal{L}$, we looked for a θ^* that optimizes the following auto-regressive objective:

$$\theta^* = \arg\min_{\theta} \mathbb{E}_{x_i \in D_I, f \in D_F} \left[\sum_{i=0}^{k} \ell \left(f\left(x_{i+1}\right), \mathcal{L}_\theta \left((x_1, f(x_1), \ldots, f(x_i), x_{i+1}) \right) \right) \right]$$

where $\mathcal{L}_\theta$ is a "learner", ℓ is squared error and $f : x \rightarrow ax + b$ is a linear function with a, b sampled from some training distribution D_F for functions. Samples x_i are sampled from a training distribution for points D_I. We note that $f \in D_F$ and $x \in D_I$. At each training step, we choose at random, a function $f \in D_F$ and then a sequence of points $x_i \in D_I$ from a distribution D_I. We use a batch size of 64 and train for 500k steps. The models saw over 1.3 billion training examples for each distribution we studied. For D_F and D_I we used several distributions: the normal distribution $\mathcal{N}(0, 1)$, uniform ("rectangle") distributions over specified intervals, and bimodal distributions.

We evaluate the models across various test distributions for both function classes D_F^t and data points D_I^t. During training, we always set $D_F = D_I$. For testing, however,

[1] Our code can be found in https://github.com/omyokun/re-examining-LF-ICL.

we sometimes consider scenarios where $D_F^t \neq D_I^t$. To evaluate ICL performance of a model on (D_I^t, D_F^t), we generate a set of $N = 100$ functions in D_F^t. Our data samples for testing are composed of $N_b = 64$ batches, each containing $N_p = 41$ points in D_I^t. For each function f_j, in each batch b, for all points, we predict for each x_k^b, $k \geq 2$, $f_j(x_k^b)$ given the prompt $(x_1^b, f_j(x_1^b), ..., x_{k-1}^b, f_j(x_{k-1}^b), x_k^b)$. We then calculate for each function the mean average over all the points N_p of all batches N_b, then do a mean average over all functions. Formally this gives us:

$$\epsilon_\sigma = \frac{1}{N}\sum_{j=1}^{N}\sum_{b=1}^{N_b}\frac{1}{N_b}\left(\frac{1}{N_p}\sum_{k=3}^{N_p}(\text{pred}_k^{b,j} - f_j(x_k^b))^2\right)$$

We also have to define an *error rate* for our problem. The error rate we chose is: $r_\epsilon = \frac{\epsilon_\sigma}{|\epsilon_* - \epsilon_0|}$ where ϵ_* is the best ϵ_σ error for a model M with $\hat{f}(x)$ calculated with Least Squares, and ϵ_0 is the worst ϵ_σ error for a model M such that $\hat{f}_M(x) = 0, \forall x$. In all error calculations, we exclude the first two predictions of each batch from the squared error calculation. We need at least two points to be able to find a linear function and the first two predictions by the model are hence almost always wrong.

4 Models ICL$_1$ but do not ICL$_2$ $\mathcal{L}$

When trained on $D_F = D_I = \mathcal{N}(0,1)$ and tested on $D_F^t = D_I^t = \mathcal{N}(0,1)$, even many small models were able to converge to near-zero average error. Error rates fluctuated across batches and were not consistently zero, and so similar to [11]'s findings on MSE estimation by transformers.

These results hold for full transformer architectures models, as well as for models without feedforward networks. In fact the 12-layers, 8-attention heads model without feedforward networks performed equivalently to the full transformer model (see Fig. 1). Together these were our best models. However, models with only MLP layers, were unable to do the task.

Observation 1. *Attention layers are necessary and sufficient to ICL$_1$ the class of linear functions. Models with only MLP were unable to ICL$_1$.*

Varying the test distributions, however, altered performance considerably. All the models had systematic and non 0 average error once we chose $f \in D_F^t = \mathcal{N}(0,\sigma)$ for $\sigma > 2$. While [26] claim shifting the distribution sampled at inference of the functions does not affect performance, Fig. 1 clearly shows that for transformer models with soft attention, this task shift at inference reduces performance dramatically. Figure 1 shows that the error rate increases substantially and non-linearly as $D_F^t = \mathcal{N}(0,\sigma)$ and σ increases. To ensure that comparisons between models are meaningful, for each $\mathcal{N}(0,\sigma)$, we set a seed when generating the 100 random linear functions, ensuring that each model sees the same randomly chosen functions and the same set of prompting points x_i. The Table 1 in the Appendix contains the full scores for average error.

The heatmaps in Fig. 2 show how our models generalized outside of three different sets of training distributions. Figure 2 shows that the generalization ability of models trained on $D_F = D_I = \mathcal{N}(0,1)$ rapidly degrades once we sample elements far

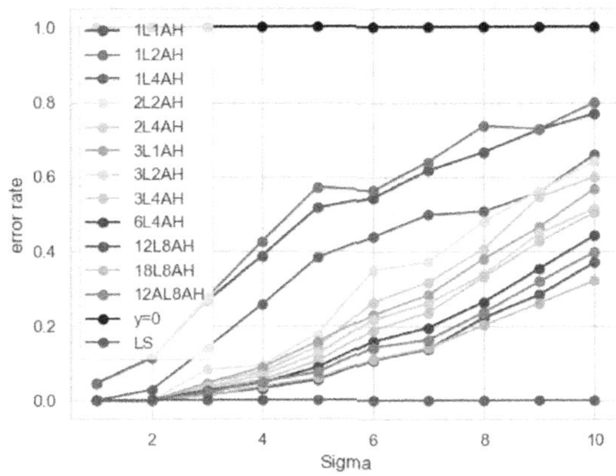

Fig. 1. Evolution of error rates for various models with $D_F = D_I = D_I^t = \mathcal{N}(0, 1)$ and D_F^t for various $\mathcal{N}(0, \sigma)$. $y = 0$ illustrates a model that predicts $f(x_n) = 0, \forall f$ and $\forall x_n$. And LS represents linear or ridge regression, which is trivially a perfect estimator given our totally clean input data.

away from the training distribution, such as when we set D_F^t and D_I^t to $\mathcal{N}(0, \sigma)$ for $\sigma > 1$. On the other hand, while the errors in the performance of models trained on $D_F = D_I = \mathcal{N}(0, 100)$ are large when tested on $\mathcal{N}(0, 1)$, performance improves as we move toward testing regimes that take elements $x_i, f(x_i)$ within intervals [a,b] that contain a higher proportion of examples seen in training (Fig. 2). This trend is even clearer in the heatmap for the model trained on $\mathcal{N}(0, 10)$ in the same figure. The model shows relatively good performance on $D_F^t = D_I^t = \mathcal{N}(0, 1)$ though not as good as the model trained on $\mathcal{N}(0, 1)$. However, the model trained on $\mathcal{N}(0, 10)$'s performance improves as the testing regime comes to include a higher proportion of functions seen in training. During testing, we also observed that models trained on distributions with a larger variance[2] provided better generalization performance in the sense that such models had lower squared errors than models trained on distributions with a smaller variance. Thus, a model trained on $\mathcal{N}(0, 10)$ has a better generalization ability and more robust performance than the model trained on $\mathcal{N}(0, 1)$.

To further test our hypothesis we evaluated the performance of our models using bimodal and uniform distributions for training data. The bimodal distribution $0.5\mathcal{N}(-1, 1) + 0.5\mathcal{N}(1, 1)$ expanded the range of values $f(x)$ the model will likely see during training. Models trained on this distribution for D_F or on uniform distributions with a larger span like $\mathcal{U}(-5, 5)$ had more robust performance than models trained with $D_F = \mathcal{N}(0, 1)$ at least with $D_F^t = D_I^t = \mathcal{N}(0, \sigma)$ and $n \geq 6$. The best models had superior performance with $D_F^t = \mathcal{N}(0, \sigma)$ for $\sigma \geq 3$ (see Table 1, in which $D_I^t = \mathcal{N}(0, 1)$). But they also had limited generalization capability. For error values on various distributions and models see Table 1.

[2] A larger variance means that a larger interval includes a large majority if not the totality of the examples seen in training.

Table 1. Comparison to show the evolution of squared ϵ type error depending on the distribution according to which we take the parameters, without taking into account the error of the prediction of the first and second prompts. Embedding size for all models is 64 except for 12L and 18L models (256K). For models N, $D_F = \mathcal{N}(0,1)$; $D_i = D_i^t = \mathcal{N}(0,1)$. For B, $D_F = 0.5\mathcal{N}(-1,1)+0.5\mathcal{N}(1,1)$ and U, $D_F = \mathcal{U}(-5,5)$. For B and U testing, $D_i^t = \mathcal{U}(-1,1)$ and $D_F^t = \mathcal{N}(0,\sigma)$. We show error rates for models prompted without and with the natural ordering on the prompts [sorted], for the large model size. 3NN refers to [17]'s method which generates values through the average of the 3 nearest neighbors.

models \ σ	1	2	3	4	5	6	7	8	9	10
$1L1AH_N$	0.1	0.8	5.1	13.1	26.9	39.7	53.0	84.8	120.0	153.2
$1L4AH_N$	0.0	0.2	2.7	8.7	19.9	32.0	42.8	64.5	92.3	131.2
$2L1AH_N$	0.0	0.1	2.0	4.9	13.7	27.0	36.1	64.9	99.0	134.0
$2L4AH_N$	0.0	0.0	0.9	2.6	7.5	19.3	27.3	51.8	90.2	119.4
$3L1AH_N$	0.0	0.0	0.9	3.0	8.2	16.8	24.4	48.4	76.7	113.2
$3L4AH_N$	0.0	0.0	0.6	1.9	5.5	13.8	20.4	42.2	70.3	100.4
$6L4AH_N$	0.0	0.0	0.5	1.6	4.6	11.6	16.8	33.7	58.3	87.9
$12L8AH_N$	0.0	0.0	0.3	1.1	2.9	7.9	11.9	28.3	46.9	73.5
$18L8AH_N$	0.0	0.0	0.2	1.1	2.8	7.1	10.3	22.9	40.3	64.6
$6L4AH_B$,	0.01	0.04	0.23	0.44	1.19	2.15	3.08	4.8	9.98	18.01
$6L4AH_U$,	0.02	0.04	0.11	0.24	0.57	1.36	1.82	4.62	10.23	15.07
$12L8AH_N$,	0.0	0.0	0.32	1.34	3.14	8.8	12.13	30.14	49.37	73.93
sorted $12L8AH_N$	0.0	0.01	0.32	1.63	3.69	8.39	10.06	27.11	43.23	58.56
$12L8AH_B$	0.0	0.01	0.08	0.29	0.78	2.23	3.66	9.04	18.68	30.23
sorted $12L8AH_B$	0.01	0.03	0.18	0.25	0.74	2.27	2.62	6.87	13.73	20.8
$12L8AH_U$	0.0	0.01	0.13	0.71	1.92	6.78	10.92	27.91	38.75	64.39
sorted $12L8AH_U$	0.01	0.01	0.13	0.75	2.12	6.18	10.5	26.8	36.3	53.48
REF$_N$**: y = 0**	2.19	7.05	19.22	33.94	52.23	73.08	86.02	127.43	165.27	199.31
REF_U**: y = 0**	1.52	4.43	13.55	19.94	30.81	44.75	52.71	76.11	105.43	128.52
3NN	0.03	0.14	0.27	0.66	1.09	1.32	1.75	2.45	2.95	4.01

To sum up our findings,

Observation 2. *(i) Models have better performance over intervals that contain a larger proportion of examples in the training distribution. (ii) Models trained on a distribution with larger variance (up to a certain point) had better generalization ability but less accuracy than models trained on distributions with smaller variance.*

We also tested an alternative hypothesis to that in Observation 2 on which the total number of points observed in the distribution, rather than the proportion, is the critical factor. We trained one model on $\mathcal{U}(-1,1)$ for $100k$ steps and another on $\mathcal{U}(-5,5)$ for $500k$ steps, ensuring that both models were exposed to a statistically equivalent number

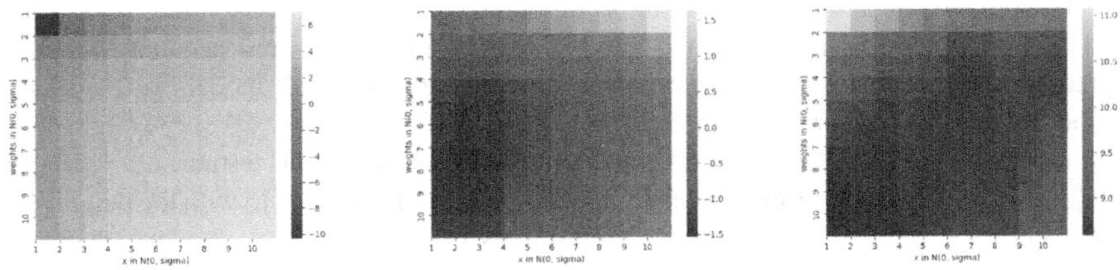

Fig. 2. Heatmaps showing the evolution of log error rates for a model trained on $D_I = D_F = \mathcal{N}(0,1)$ (left), one trained on $D_I = D_F = \mathcal{N}(0,10)$ (middle) and one trained on $D_I = D_F = \mathcal{N}(0,100)$ (right). During testing, inputs and weights are drawn independently from $D_I^t \sim \mathcal{N}(0,\sigma_1)$ and $D_F^t \sim \mathcal{N}(0,\sigma_2)$, with $\sigma_1, \sigma_2 \in \{1,\ldots,10\}$. Note that while darker shades are always better, the shades have different values in the figures above to help with interpretability.

of points within the interval $[-1,1]$. The results remained similar, showing that the total number of points observed is not the critical factor.[3]

Model performance on out-of-distribution examples also improved somewhat when the sequence of x_i in the training prompts (p_1) was sorted, to follow the natural order on $\mathbb{R}$, especially for larger models. Error rates with sorting are lower, by up to a third—compared to error rates without sorting for small values of σ, with $D_F^t = \mathcal{N}(0,\sigma)$. We also tested prompts ($p2$) on a 12L8AH model in a kind of "curriculum learning"; models with $p2 < p1$ did less well than models with $p1 = p2$.

We did a heatmap analysis of the weights of the elements in the attention matrices of the last layer of our model. The results are in Fig. 3. As can be seen, the model learns to pretty much ignore the first element and its prediction on that (the first two elements in its context) after the 6th prediction. Attention weights are largely uniform over the rest of the elements in the sequence. This suggests that it uses most if not all contextually given pairs of input function values, $x_i, y_i = f(x_i)$, to predict a value y_n for input x_n.

All this leads to the following observation.

Observation 3. *A model takes into account the whole sequence to predict $\hat{f}(x_n)$, not just some small, fixed subsequence.*

Contrary to what [2,22] have suggested, our observations show that our models did not use linear regression to ICL the target function. If they had, we would not observe the error patterns or variations in performance, nor would the models disregard the class form of the target function in making their predictions; otherwise, the use of all elements in the sequence would have been superfluous. In conclusion, all models tested failed to ICL_2 $\mathcal{L}$. Moreover, we have seen a trade off between accuracy and generalizability from Observation 2; given current transformer models, we cannot have both accuracy and generalizability.

---

[3] A checkpoint analysis further confirmed this indicating that optimal performance is typically reached at a fixed training step number, beyond which model performance fluctuates and does not really improve.

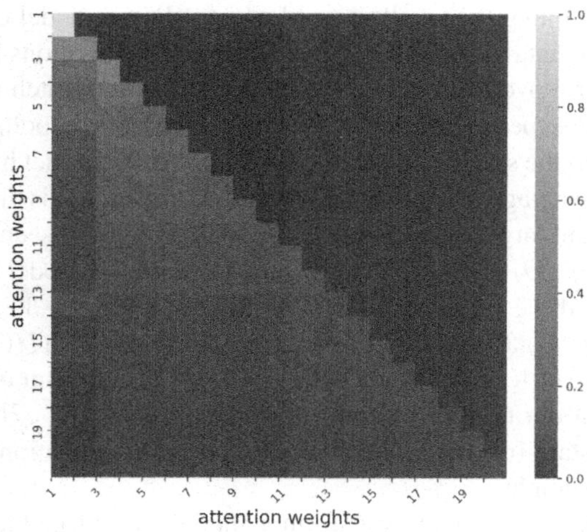

Fig. 3. Heatmap showing the evolution of attention weights in an attention head of the last layer of 12AL8AH model

5 Error Analysis and Boundary Values

During all our experiments, we noticed that all our models presented what we called *boundary values*, which are values that the model cannot exceed during inference predictions (see Fig. 4). These *boundary values* act as a filtering mechanism, restricting the model to generate outputs only within a specific range. This constraint prevents the generation of values beyond these limits, effectively preventing the model from generalizing its good performance on the task over a small interval to larger values outside that interval. This behavior can be seen also in [10]'s figures, despite their use of a transformer based on linear attention.

Since these models are trained from scratch, the only factor that can determine the boundary values is the training process itself. We conjecture that the model memorizes some bounds that are the largest and smallest values encountered during training, and then sets them as boundary limits. This effectively functions as a bandpass filter.

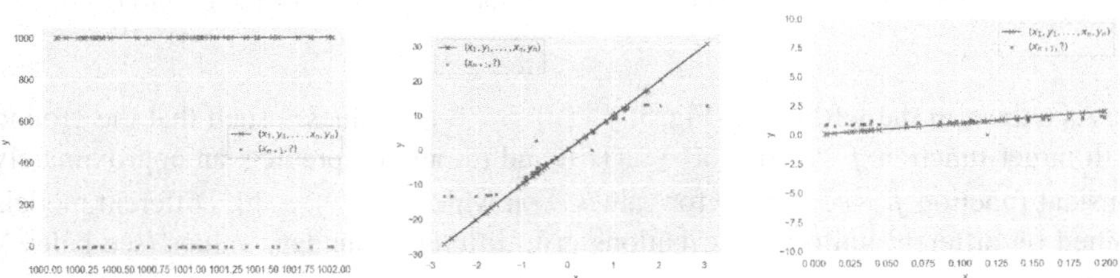

Fig. 4. Plots for model 12L8AH, trained on $D_I = D_F = \mathcal{N}(0, 1)$ for $f(x) = x$ for high values (left) of x and $f(x) = 10x$ for normal (middle) then for low values of x (right)

However, determining the smallest and largest values a model could have encountered by training for data and coefficients in Gaussian distributions is impossible. As a result, to have control over this parameter, we trained from scratch models on uniform distributions. These experiments enabled us to confirm our hypothesis that boundary values correspond to the smallest and largest values that the model has encountered and memorized. One advantage of training on a uniform distribution is that we know exactly what the smallest and largest values that the model could have seen in its training. For example, setting D_F, D_I to $\mathcal{U}(-5,5)$, the largest value the model could have seen is $30 = 5 \times 5 + 5$ and the smallest value it could have seen is -30. Most likely it saw values significantly > -30 and < 30. All our models trained on $\mathcal{U}(-5,5)$ estimate the target function more or less well for x with $f(x) \in [-30, 30]$. But once we took inputs x_i or coefficients of our target function f to force $f(x_i) \notin [B^-, B^+]$, the estimations become either constant functions or chaotic. Figure 5 with equation $f(x) = 40x + 40$ provides another example.

Boundary values hold for all transformer models tested including attention only models. Linear transformer architectures also show boundary values [10]. For example with $D_F = D_I = \mathcal{U}(-5,5)$, consider as an illustrative example the target function, $f(x) = 9x$, with our largest trained model. The model approximates $f(x)$ well within a certain range $[B^-, B^+] = [-30, 30]$ that corresponds to the maximal and minimal values it could have seen. On the other hand, it predicts $\hat{f}(x)$ as a constant function for x such that $\hat{f}(x) \notin [B^-, B^+]$ within a certain range (See Fig. 5). Indeed, model errors depended on how many functions and values sampled forced the result outside $[B^-, B^+]$, as well as on the proportion of a model's training sequences in the test distribution (see Observation 2).

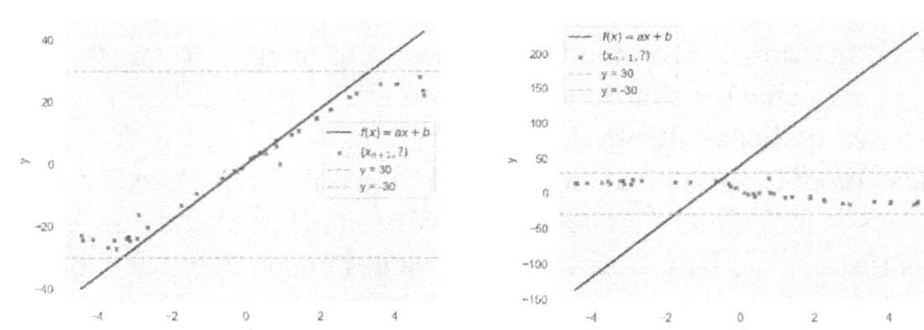

Fig. 5. Plots for $f(x) = 9x$ and $f(x) = 40x + 40$ for a 12L8ah model trained on $\mathcal{U}(-5,5)$

As shown in the left plot of Fig. 5, $\hat{f}^+(v) \approx 30$ for values v such that the ground truth target function f satisfies $30 \leq f(v)$, and the model predicts an approximately constant function $\hat{f}^-(v) \approx -30$ for values v on which $f(v) \leq -30$. Different models trained on different uniform distributions give different boundary values (see below). To summarize:

Observation 4. *(i) Boundary values $B^-, B+$ are determined by maximal and minimal elements seen during training. (ii) For all models M and for values v such that $B^+ <$*

$f(v) < B^+ + \alpha$, where α is a constant determined by M, $\hat{f}_M(v) \approx B^+$, and for $B^- - \alpha < f(v) < B^-$, $\hat{f}_M(v) \approx B^-$. (iii) For functions and data samples when the values of $f(v)$ in the prompt sequence are such that $f(v) > B^+ + \alpha$ or $< B^- - \alpha$, the model assigns $\hat{f}(v)$ a random value within $[B^-, B^+]$.

The parameter α in Observation 4 accounts for the fact that larger models trained on the same distribution and the same number of data will ICL $\mathcal{L}$ functions over a slightly larger number of intermediate values than smaller models, as Fig. 1 suggests. Figure 7 in the Appendix shows plots of the predictions of two models (12L8AH, and 6L4AH) for $D_F, D_I = \mathcal{N}(0,1)$ for target $f(x) = 10x$. The larger model has boundary values $\approx -13.7, 13.7$, the smaller one boundary values $\approx -12, 12$.

Since a model's predictions do not go beyond the boundary values defined by training, a consequence of Observation 4 is that error rates depend on the distance of the target function's values from the boundary values and hence the majority of the data points in the model's training. Figures 1 and 2 verify this progressive increase in error rate as the testing distribution gets larger and larger.

Training with distributions over much larger intervals, for example $\mathcal{N}(0,10)$ or $\mathcal{N}(0,100)$, can extend boundary values. Models trained on $\mathcal{N}(0,10)$ show much better generalization ability than the models trained on $\mathcal{N}(0,1)$. On the other hand, they are less accurate on small values as Fig. 2 shows. Models trained with $D_F = \mathcal{U}(-10,10) = D_I$ generalized even better, but their performance on $\mathcal{N}(0,1)$ was significantly worse than models trained on $\mathcal{N}(0,1)$. The $\mathcal{N}(0,100)$ performed poorly in our testing scenarios but improved as values got larger. Our observations thus present a dilemma: training on restricted intervals is needed for high accuracy, but it inevitably hampers generalizability. Generalizability requires training over larger intervals but that reduces accuracy.

As evident from Sect. 2, we are not the first to investigate out of distribution behavior. [10] investigated out of distribution behavior but only on $D_I \neq D_I^t$ (not shifts in D_F). Unlike [10], we found that larger models had better performance than smaller ones, when $f(x), x$ were within boundary values. Our results also differed from [26]'s shifting from D_I. They shift the prompt distribution but not that of the query. We shift both prompt and query distributions. With our shifts, we found that the choice of points is important and model performance degrades considerably when the values of the functions on the chosen points lie beyond the boundary values, which [26] do not. As far as we know, we are the first to take boundary values and their dependence on model parameters as key indications of what is actually going on in ICL.

6 What and how are the Models Learning?

In Sect. 4, we gave our reasons why the hypotheses of [2,22] about models being able to ICL linear functions via linear regression are incorrect. The boundary values B^-, B^+ of Sect. 5 provide further evidence that models are not performing linear regression to define a linear function; otherwise, the models' predictions within $[B^-, B^+]$ should carry over to intervals larger than those defined by the majority of their training samples.

A lack of full generalizability is perhaps expected[4] but the limitations on learnability we have seen are quite strong.

The presence of boundary values suggests that our models have memorized at least a portion of their training. Do our models simply overfit the data? We argue no. The pretraining data has no noise, and it is too large to be memorized by our models (our largest models with 256 size embeddings have 32M parameters; each parameter would have to encode on average sequences for over 100 different functions). Moreover, our models performed similarly on several different training distributions for D_F and D_I and tested on $\mathcal{N}(0, \sigma)$ for $\sigma \in \{1, 2\}$. To test overfitting, we added noise to our input, but that did not improve error rates. This implies that the models did not overfit to their training regimes. Our models seem to reason as well. Given 100 samplings with $D_F^t = \mathcal{N}(0, 1)$ nets on average 20 functions with coefficients the model with $D_F = D_I = \mathcal{U}(-1, 1)$ has not seen in training, we would expect the model's performance to degrade more substantially than it did if only memory were involved. Moreover, models trained on $\mathcal{N}(0, 10)$ had substantially improved generalization ability over $\mathcal{N}(0, \sigma)$ for all σ we tested, as can be seen in Fig. 2, even though the odds of having a sequence from training repeated on the test are very low.

To go back to the question of what models are learning, [17] conjectures that an attention layer only model generates values by averaging the three nearest neighbors. We tested [16]'s conjecture by using it to calculate the output. The results were too far from the values we got experimentally (Table 1). Using [17]'s suggestion also gives results that make the wrong predictions for points at or around *boundary values* and other points (see Fig. 4 of $f(x) = x$ for high values in the Appendix).

Instead given a prompt sequence $(x_1, y_1(= f(x_1)), x_2, y_2, \cdots, x_n, ?)$, we hypothesize that the model optimizes a projection π based on stored projections $\pi_n : Z \mapsto Y$, where Z are inputs it has seen and Y are the corresponding outputs y_i it has seen paired with x_i in X, to predict a value for $y_n \in [B^-, B+]$. Given Observation 1, this projection can be fully determined by the attention layers in the model. Our observations show that more attention heads, but even more importantly more attention layers, improve model performance. This is, we conjecture, because they provide more memory for storing nearby sequences from which the model can draw to optimize π. To sketch a possible projection, we mimic the autoregressive reasoning. We start just with a target input x_1^t. The attention matrices have encoded a projection in their weights that takes $g_0 : x_1^t \mapsto z_i^m$, where $z_i^m = x_1^t \pm e$ for which there is also a stored projection $p_0 : x_i^m \mapsto y_i^m$. It then sets π_0 such that $\pi_0(x_1^t) = p_0(g_0(x_i^m)) = y_i^m \pm d$, where d is close to e. This is usually wrong, since there are many possible choices for g_0 and p_0 from training. Now suppose the model has received x_1^t, y_1^t, x_2^t. It now tries to predict $\pi_1(x_2^t)$. To do this it must find projections g_1, p_1, and π_1 such that $\pi_1(x_2^t) = p_1 \circ g_1(x_2^t)$ and $\pi_1(x_1^t) = y_1^t$. Several choices for p_1, g_1, π_1 may obey these constraints and so the model may interpolate between them. π_1 at level l is defined via the context matrix given by attention [14].

Now, what happens when the model has to predict a value for x_i^t such that either x_i or the projection lies beyond the boundary values it has learned in training? The

[4] A linear function is a Π_1^0 set in the Borel hierarchy, which [3] show is not learnable using ordinary LLM assumptions.

boundary values and the parameter α_M from Observation 4 define the limits of what a "close" element is. If $\pi(x_i) = y_i$ is not too far from $[B^-, B^+]$, then the closest element the model can find is the boundary value or something near it. Far away elements are $Ker(\pi)$ and so sent to 0 or some nearby value. Observation 4 establishes that the projection is not linear over intervals larger than $[B^-, B+]$, and that's what our algorithm also predicts.

So, even though memory is a crucial component of ICL_1, it's not everything. The autoregressive component of our algorithm above resembles basic causal language modeling, but it incorporates a basic reasoning element of closeness. The closer a sequence stored in memory of a model M z is to the target sequence, the higher the probability that $\pi_M(x_n) = ay_n^z$, for $a < \alpha_M$.

7 Conclusion

While transformers are theoretically capable of learning linear regression to perform ICL $\mathcal{L}$, we have demonstrated they do not do this in practice. Our models achieve strong statistical results comparable to state-of-the-art algorithms when training and test distributions are the same. However, we aimed to assess whether our models truly learn the underlying class representation, and they do not. All tested models failed to robustly learn the class form $f(x) = ax+b$ of linear functions on both noisy and non-noisy data. Had models used linear regression on our clean data, they would have exactly inferred the class form of linear functions and correctly estimated the parameters a, b.

We also found that attention layers were both necessary and sometimes sufficient for good ICL performance. More layers and more heads improved performance. Generalization ability improved as perhaps expected when models were trained on distributions with larger variance.

We also analyzed model errors and emphasized the importance of boundary values that define intervals within which the vast majority of all of values of the training sequences occur. End users, who typically lack knowledge about the training data distribution for the models they use, need to know about model sensitivity to training and testing regimes. Otherwise, they may unwittingly employ ICL on tasks significantly beyond the original training domain with a consequent substantial loss of accuracy. This limitation will affect larger models working on different tasks.

Finally, we have offered a hypothesis about how models learn. Our hypothesis H explains our observations, but we have not fully shown that transformer models actually implement this hypothesis. In future work, we will attempt to isolate various stages of the algorithm by probing what models do at various layers, following [15,23].

Acknowledgments. We gratefully acknowledge the support of the EU grant TUPLES, the grants SARER, Summ-RE (ANR-20-CE23-0017), and the AI Cluster ANITI (ANR-19-PI3A-0004). This work used the HPC resources from CALMIP (Grant 2016-P23060).

References

1. Ahn, K., Cheng, X., Daneshmand, H., Sra, S.: Transformers learn to implement preconditioned gradient descent for in-context learning. In: Advances in Neural Information Processing Systems, vol. 36, pp. 45614–45650 (2023)
2. Akyürek, E., Schuurmans, D., Andreas, J., Ma, T., Zhou, D.: What learning algorithm is in-context learning? Investigations with linear models. arXiv preprint arXiv:2211.15661 (2022)
3. Asher, N., Bhar, S., Chaturvedi, A., Hunter, J., Paul, S.: Limits for learning with large language models. In: 12th Joint Conference on Lexical and Computational Semantics (*Sem). Association for Computational Linguistics (2023)
4. Bai, Y., Chen, F., Wang, H., Xiong, C., Mei, S.: Transformers as statisticians: provable in-context learning with in-context algorithm selection. In: Advances in Neural Information Processing Systems, vol. 36 (2024)
5. Bhattamishra, S., Patel, A., Blunsom, P., Kanade, V.: Understanding in-context learning in transformers and LLMs by learning to learn discrete functions. arXiv preprint arXiv:2310.03016 (2023)
6. Brown, T., et al.: Language models are few-shot learners. In: Advances in Neural Information Processing Systems, vol. 33, pp. 1877–1901 (2020)
7. Dong, Q., et al.: A survey on in-context learning. arXiv preprint arXiv:2301.00234 (2022)
8. Fu, D., Chen, T.Q., Jia, R., Sharan, V.: Transformers learn higher-order optimization methods for in-context learning: a study with linear models. arXiv preprint arXiv:2310.17086 (2023)
9. Garg, S., Tsipras, D., Liang, P.S., Valiant, G.: What can transformers learn in-context? A case study of simple function classes. In: Advances in Neural Information Processing Systems, vol. 35, pp. 30583–30598 (2022)
10. Giannou, A., Yang, L., Wang, T., Papailiopoulos, D., Lee, J.D.: How well can transformers emulate in-context newton's method? arXiv preprint arXiv:2403.03183 (2024)
11. Liu, J.W., Grogan, J., Dugan, O.M., Arora, S., Rudra, A., Re, C.: Can transformers solve least squares to high precision? In: ICML 2024 Workshop on In-Context Learning (2024)
12. Mahankali, A., Hashimoto, T.B., Ma, T.: One step of gradient descent is provably the optimal in-context learner with one layer of linear self-attention. arXiv preprint arXiv:2307.03576 (2023)
13. Mirzadeh, I., Alizadeh, K., Shahrokhi, H., Tuzel, O., Bengio, S., Farajtabar, M.: GSM-symbolic: understanding the limitations of mathematical reasoning in large language models. arXiv preprint arXiv:2410.05229 (2024)
14. Naim, O., Asher, N.: On explaining with attention matrices. In: ECAI 2024, pp. 1035–1042. IOS Press (2024)
15. Nanda, N., Chan, L., Lieberum, T., Smith, J., Steinhardt, J.: Progress measures for grokking via mechanistic interpretability. arXiv preprint arXiv:2301.05217 (2023)
16. Olausson, T.X., et al.: Linc: a neurosymbolic approach for logical reasoning by combining language models with first-order logic provers. In: The 2023 Conference on Empirical Methods in Natural Language Processing (2023)
17. Olsson, C., et al.: In-context learning and induction heads. arXiv preprint arXiv:2209.11895 (2022)
18. Panwar, M., Ahuja, K., Goyal, N.: In-context learning through the Bayesian prism. arXiv preprint arXiv:2306.04891 (2023)
19. Pérez, J., Barceló, P., Marinkovic, J.: Attention is turing-complete. J. Mach. Learn. Res. **22**(75), 1–35 (2021)
20. Radford, A., Wu, J., Child, R., Luan, D., Amodei, D., Sutskever, I., et al.: Language models are unsupervised multitask learners. OpenAI Blog **1**(8), 9 (2019)
21. Raventós, A., Paul, M., Chen, F., Ganguli, S.: Pretraining task diversity and the emergence of non-Bayesian in-context learning for regression. In: Advances in Neural Information Processing Systems, vol. 36 (2024)

22. Von Oswald, J., et al.: Transformers learn in-context by gradient descent. In: International Conference on Machine Learning, pp. 35151–35174. PMLR (2023)
23. Wang, K., Variengien, A., Conmy, A., Shlegeris, B., Steinhardt, J.: Interpretability in the wild: a circuit for indirect object identification in GPT-2 small. arXiv preprint arXiv:2211.00593 (2022)
24. Wu, J., Zou, D., Chen, Z., Braverman, V., Gu, Q., Bartlett, P.L.: How many pretraining tasks are needed for in-context learning of linear regression? arXiv preprint arXiv:2310.08391 (2023)
25. Xie, S.M., Raghunathan, A., Liang, P., Ma, T.: An explanation of in-context learning as implicit Bayesian inference. arXiv preprint arXiv:2111.02080 (2021)
26. Zhang, R., Frei, S., Bartlett, P.L.: Trained transformers learn linear models in-context. J. Mach. Learn. Res. **25**(49), 1–55 (2024)
27. Zhang, Y., Zhang, F., Yang, Z., Wang, Z.: What and how does in-context learning learn? Bayesian model averaging, parameterization, and generalization. arXiv preprint arXiv:2305.19420 (2023)

ODExAI: A Comprehensive Object Detection Explainable AI Evaluation

Loc Phuc Truong Nguyen[1(✉)], Hung Truong Thanh Nguyen[2], and Hung Cao[2]

[1] Friedrich-Alexander-Universität Erlangen-Nürnberg, Erlangen 91054, Germany
loc.pt.nguyen@fau.de
[2] University of New Brunswick, Fredericton E3B 5A3, Canada
{hung.ntt,hcao3}@unb.ca

Abstract. Explainable Artificial Intelligence (XAI) techniques for interpreting object detection models remain in an early stage, with no established standards for systematic evaluation. This absence of consensus hinders both the comparative analysis of methods and the informed selection of suitable approaches. To address this gap, we introduce the *Object Detection Explainable AI Evaluation* (ODExAI), a comprehensive framework designed to assess XAI methods in object detection based on three core dimensions: localization accuracy, faithfulness to model behavior, and computational complexity. We benchmark a set of XAI methods across two widely used object detectors (YOLOX and Faster R-CNN) and standard datasets (MS-COCO and PASCAL VOC). Empirical results demonstrate that region-based methods (e.g., D-CLOSE) achieve strong localization (PG = 88.49%) and high model faithfulness (OA = 0.863), though with substantial computational overhead (Time = 71.42 s). On the other hand, CAM-based methods (e.g., G-CAME) achieve superior localization (PG = 96.13%) and significantly lower runtime (Time = 0.54 s), but at the expense of reduced faithfulness (OA = 0.549). These findings demonstrate critical trade-offs among existing XAI approaches and reinforce the need for task-specific evaluation when deploying them in object detection pipelines. Our implementation and evaluation benchmarks are publicly available at: https://github.com/Analytics-Everywhere-Lab/odexai

Keywords: Explainable AI · Object detection · Evaluation framework

1 Introduction

Deep neural networks (DNNs) have greatly advanced object detection, but their opaque decision-making limits interpretability [4,36,38]. This reduces trust and restricts deployment in sensitive domains such as healthcare [11,37,41,58] and security [6,33,40], where models must offer not only high accuracy but also transparent, fair, and domain-aligned reasoning [10,54,61]. Explainable AI (XAI) addresses this need by making detection decisions more interpretable.

T. Braun et al. (Eds.): KI 2025, LNAI 15956, pp. 118–133, 2026.
https://doi.org/10.1007/978-3-032-02813-6_9

XAI has emerged as a distinct research area, driven by advances in deep learning, DARPA-funded initiatives [17], and the recognition of a right to explanation in the GDPR [20], EU AI Act [35]. In recent years, a wide range of XAI methods for object detectors has been proposed. While this reflects growing interest in the field, it also presents several challenges [24]. The number of available methods is increasing rapidly, and many offer only minor variations, such as Spatial Sensitive Grad-CAM [60] and Spatial Sensitive Grad-CAM++ [59], or produce similar outputs despite differing implementations, as seen with D-RISE [44] and ODAM [64]. As discussed by several works [4,34,36,38,40], explanation quality must be assessed across several dimensions, such as faithfulness to the model's reasoning, conciseness, fairness, and alignment with domain knowledge. Yet two critical questions remain open: *how to select an appropriate XAI method*, and *how to benchmark it in a standardized manner* that ensures objective evaluation and supports generalizable findings.

To address these challenges, we introduce the *Object Detection Explainable AI Evaluation* (ODExAI), a structured approach for evaluating and comparing XAI methods for object detection. Recent work has focused on developing metrics to benchmark interpretability techniques across settings. These metrics enable quantitative assessment of localization, faithfulness, and complexity, offering insight into each method's strengths, limitations, and suitability. The main contributions of this work are:

1. We propose ODExAI Framework grounded in three key interpretability dimensions: faithfulness, localization, and complexity. Each dimension is supported by a set of quantitative metrics, enabling systematic and task-specific assessment.
2. We conduct a toy experiment to illustrate the use of ODExAI in evaluating three novel XAI methods: D-CLOSE [56], G-CAME [39], and D-RISE [44], across two widely used detectors, YOLOX [14] and Faster R-CNN [47], using the MS COCO [25] and PASCAL VOC [12] datasets.
3. We conduct a comprehensive analysis of the trade-offs between XAI methods, highlighting their performance across different interpretability dimensions. Based on the evaluation outcomes, we provide practical guidelines for selecting suitable XAI methods according to task-specific requirements.

2 Background

This section reviews object detection models and the XAI methods applied to interpret their outputs. It also outlines recent progress and key challenges, particularly the lack of a unified and standardized evaluation framework for explanation techniques.

2.1 Object Detection Models

DNNs have significantly advanced object detection through various techniques, each offering distinct advantages. Single-stage detectors, such as the YOLO [45] family (e.g., YOLOv3 [46], YOLOv4 [3], and YOLOX [14]) and SSD [26], are optimized for real-time detection, offering a balance between speed and accuracy. Two-stage detectors, known for their high precision, include Faster R-CNN [47], Mask R-CNN [18] (which extends detection to instance segmentation), and Cascade R-CNN [7], which refines results through multi-stage processes. Anchor-free detectors simplify detection by directly predicting object centers or corners, with notable examples such as CornerNet [23], CenterNet [9], and FCOS [55] (Fully Convolutional One-Stage Object Detection). Transformer-based detectors, including DETR [55] (DEtection TRansformer), Deformable DETR [8] (which accelerates convergence), and Swin Transformer [27] (known for hierarchical feature extraction), leverage attention mechanisms to improve performance.

2.2 Explainable AI for Object Detection

While XAI techniques are widely used in classification and segmentation, their application to object detection remains limited due to constraints in flexibility, efficiency, and adaptability [16]. As noted in [39], these methods fall into two main categories: region-based and CAM-based saliency approaches. Region-based methods assess the importance of image regions by masking inputs and observing changes in output. Examples include LIME [48], RISE [43], and object detection variants such as SODEx [49], D-RISE [44], and D-CLOSE [56], which enhance precision through task-specific adaptations. These approaches are model-agnostic and intuitive. In contrast, CAM-based methods like Grad-CAM [50] use feature maps and partial derivatives to generate saliency maps, offering efficiency but often yielding less informative results for detection tasks. Designed primarily for classification, they lack dedicated adaptations for detection. The recently proposed G-CAME [39] addresses these gaps by providing stable, efficient explanations across both one-stage and two-stage detectors.

2.3 Development and Gaps in XAI for Object Detection

XAI is essential for object detection for understanding how models assign class labels and bounding boxes from visual features. It helps verify whether predictions rely on relevant regions, supports error analysis, and enhances transparency. In healthcare, XAI has been applied to thyroid nodule detection [41], tumor localization [11], and abnormality identification [58]. In autonomous driving, it aids in diagnosing detection failures [52] and assessing robustness under varying weather conditions [53]. In environmental monitoring, it explains outputs in deforestation tracking [5] and ecological change analysis [51]. In security, it promotes fairness in detecting individuals with disabilities [40], while in agriculture, it supports interpretation in crop health monitoring [33] and pest detection [6].

Although XAI methods are broadly applicable, their evaluation remains inconsistent and lacks standardization [2,24]. This stems from subjective assessment criteria and methodological diversity. In many cases, formal procedures are absent, leading to ad hoc strategies tailored to specific models, datasets, or use cases [28]. As a result, findings are often difficult to compare or reproduce, limiting progress toward reliable and generalizable explanations. Several initiatives have attempted to address this gap. Quantus [19], an open-source toolkit, offers computational metrics for evaluating explanation properties. Other studies seek to clarify evaluation processes [24,34] or provide practical guidelines for designing evaluation protocols [2,29]. However, current approaches lack consensus, and no unified evaluation framework exists for XAI for object detection [21,31,41].

3 The Proposed ODExAI Framework

The absence of a standardized evaluation framework in XAI has led to the use of ad hoc metrics, often tailored to specific methods. Although these metrics cover diverse evaluation criteria, their proliferation introduces uncertainty about task relevance, hindering reproducibility and generalization. To address this, ODExAI defines three core evaluation properties for object detection, each illustrated by schematic diagrams (Fig. 2, 3 and 4). Metrics are grouped by property based on shared objectives, enabling a more structured, consistent, and interpretable evaluation process.

3.1 Framework Overview

Our framework consists of two main components: Saliency Map Extraction with XAI and Explanation Evaluation. The overall procedure is illustrated in Fig. 1. A detailed description of each component is provided below:

1. **Saliency Map Extraction with XAI**: This part focuses on generating saliency maps using XAI methods. Users begin by uploading an image and selecting the task, with object detection models assigned, e.g., Faster R-CNN [47] and YOLOX [14]. After prediction, the user selects the target object and an explanation method, e.g., D-CLOSE [56], G-CAME [39], and D-RISE [44], to generate the saliency map.
2. **Explanation Evaluation**: This part provides an overview of each explanation method across three dimensions: localization, faithfulness, and complexity. Once the explanation is generated, users can initiate the evaluation, which is visualized as a spider plot. Each dimension includes multiple metrics, with corresponding numerical values that quantify the method's performance.

3.2 Localization Metrics

Localization quality is measured by how well an explanation aligns with a user-defined region of interest (ROI) [4], as shown in Fig. 2. This involves comparing the most relevant pixels in the explanation map to labeled regions such as bounding boxes or segmentation masks. Since the ROI is assumed to be critical to the model's decision [1,4,62], localization is considered high when relevance is concentrated within it. We quantify this using the Pointing Game (PG) [62] and Energy-based Pointing Game (EBPG) [57] metrics.

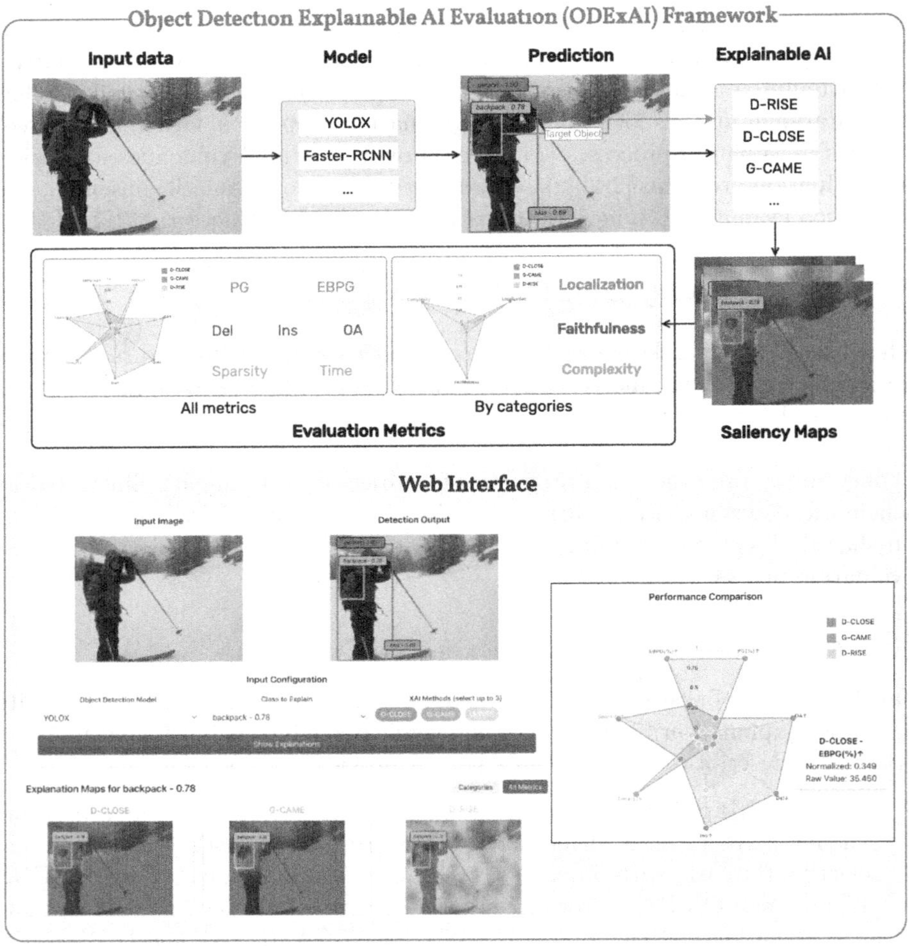

Fig. 1. Overview of the architecture and web interface of our proposed Object Detection Explainable AI Evaluation (ODExAI).

Pointing Game (PG). [62] evaluates the spatial precision of saliency maps by measuring their ability to localize target object categories. For each saliency map, the metric identifies the point of maximum saliency and checks if it falls within the ground truth annotation of the target object. A hit is recorded if the point lies within the annotated region; otherwise, it is counted as a miss. The localization accuracy is defined as:

$$\text{PG} = \frac{\#\ \text{Hits}}{\#\ \text{Hits} + \#\ \text{Misses}} \tag{1}$$

where # Hits and # Misses denote the total number of correct and incorrect localizations, respectively. The overall performance is calculated by averaging

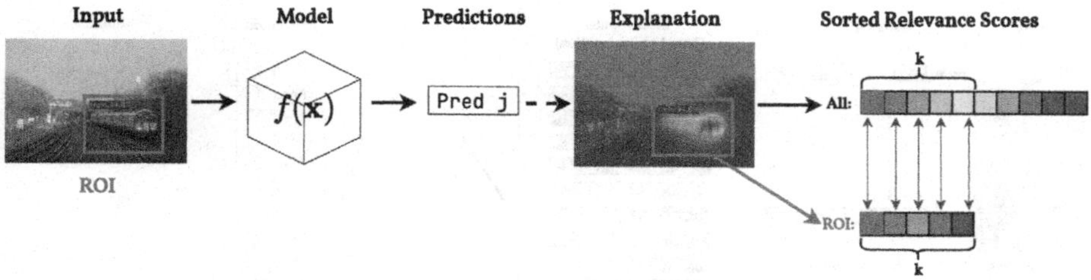

Fig. 2. Demonstration of the localization property using region-specific relevance comparisons. Localization measures how well an explanation aligns with the spatial region that is most responsible for the model's prediction. To evaluate this, a region of interest (ROI) is predefined in the input image (red box), marking the area where the most relevant evidence is expected to reside. In this example, the target is the object "train". After model inference, an explanation map is generated, and pixel-wise relevance scores are computed. The k relevance scores within the ROI are then compared with the top-k scores from the full explanation map. A method with strong localization is expected to assign the highest relevance to pixels inside the ROI.

the accuracy across all categories. A high PG score for an XAI method indicates its ability to generate accurate and interpretable saliency maps.

Energy-Based Pointing Game (EBPG). [57] evaluates the spatial accuracy of saliency maps by analyzing how much of the map's energy (activation) falls within the annotated bounding box of the target object. Unlike the traditional PG, which focuses on a single maximum point, this metric considers the overall distribution of energy within the relevant regions. The localization accuracy is defined as:

$$\text{EBPG} = \frac{\sum L^c_{(i,j)\in \text{bbox}}}{\sum L^c_{(i,j)\in \text{bbox}} + \sum L^c_{(i,j)\notin \text{bbox}}} \tag{2}$$

where $L^c_{(i,j)}$ represents the saliency map intensity at pixel (i, j), and bbox refers to the bounding box region of the target object. The overall performance is calculated by averaging the accuracy across all categories. A high EBPG value indicates that the XAI-generated saliency maps can effectively localize the target region and suppress noise outside the bounding box.

3.3 Faithfulness Metrics

Faithfulness measures how changes to highly relevant input features affect the model's prediction [56], as shown in Fig. 3. It is evaluated by progressively inserting or deleting pixels with high relevance scores and comparing the resulting predictions to the original. Since relevance reflects each feature's contribution,

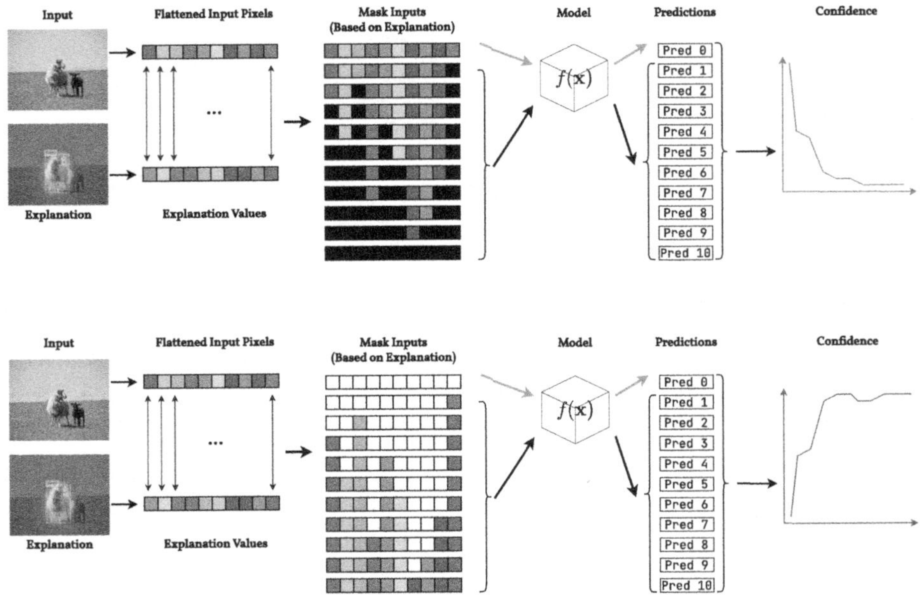

Fig. 3. Illustration of the faithfulness property using progressive perturbation. Faithfulness evaluates how the most relevant pixels affect the model's prediction. Explanation scores are first ranked to identify the most relevant features, highlighted in red. Their corresponding positions are then mapped onto the flattened input, as indicated by the arrows. In the top panel, pixels are progressively removed (masked in black), while in the bottom panel, they are progressively added to a blank input. At each step, the modified input is passed through the model, and the prediction confidence is recorded. A faithful explanation produces a steep confidence drop when highly relevant pixels are removed and a sharp increase when they are added, whereas perturbing less relevant regions yields smaller changes.

significant prediction shifts are expected when important features are modified. We quantify faithfulness using the Deletion and Insertion [43] and Over-All (OA) [63] metrics.

Deletion (Del) and Insertion (Ins). The Deletion [43] metric evaluates the effect of removing key features by tracking the decline in predicted class probability as important pixels, identified by the saliency map, are progressively removed. A steep probability drop, quantified by a low area under the curve (AUC), denotes a strong explanation. In contrast, the Insertion [43] metric measures the probability increase when pixels are gradually introduced, with a higher AUC indicating better explanations.

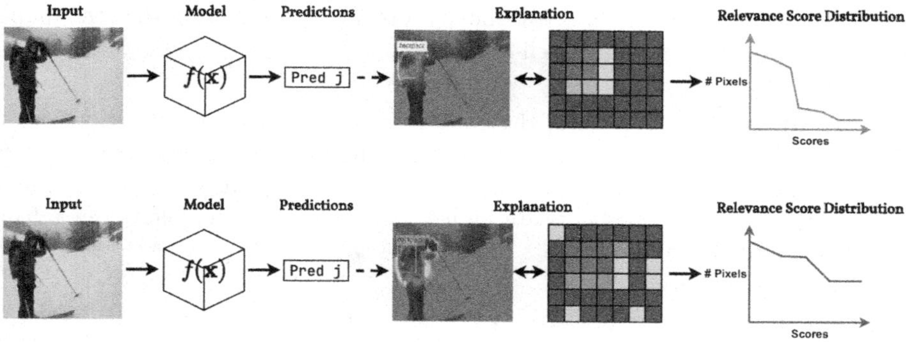

Fig. 4. Visualization of the complexity property via relevance score distributions. Complexity measures how concentrated or dispersed an explanation is over the input space. After inference, relevance scores are assigned to all input pixels and sorted to generate a distribution curve. In the top panel, low complexity is illustrated by a steep curve where few pixels carry most of the relevance. In contrast, the bottom panel shows high complexity, with scores distributed more uniformly, resulting in a flatter curve. Low-complexity explanations are desirable because they highlight key regions, reduce the effective input space, and focus the analysis on essential model components, leading to more efficient and interpretable outputs.

Over-All (OA). [63] metric is defined as AUC(Insertion) − AUC(Deletion), provides a comprehensive evaluation of saliency map quality by considering both its positive and negative effects on the model's predictions. A higher score indicates the saliency map effectively highlights critical regions (high Insertion) while minimizing the influence of irrelevant ones (low Deletion).

3.4 Complexity Metrics

Complexity reflects the conciseness of an explanation, ideally highlighting a small set of key features [4]. Concise explanations are easier to interpret and tend to convey more useful information with less noise [4,15]. We also incorporate computation time into the complexity measure, as high runtime can limit the practicality of XAI methods in real-time or large-scale systems [22]. Complexity is quantified using the Sparsity metric [15] and by recording the runtime of each method. The overall concept is illustrated in Fig. 4.

Sparsity. [15] metric measures how concentrated or dispersed the scores in a saliency map are, reflecting the focus of attention on specific regions. It is defined as the ratio of the maximum score (S_{max}) to the mean score (S_{mean}):

$$\text{Sparsity} = \frac{S_{\text{max}}}{S_{\text{mean}}} \tag{3}$$

It should be noted that all saliency map scores are min–max normalized beforehand to ensure consistency across methods. A high sparsity value corresponds to a high $S_{\text{max}}/S_{\text{mean}}$ ratio, i.e., a low mean score S_{mean}, indicating that the map's focused areas are narrow and well-defined.

Computation Time. records the time taken to compute explanations for a given model input. It is typically reported as the average runtime per instance and depends on factors such as model complexity, input size, and explanation method. Lower values indicate higher efficiency, while higher values reflect greater computational cost.

4 Demonstrative Example

This section presents a toy experiment to demonstrate the utility of the ODExAI framework. We first introduce the evaluated XAI methods, benchmark datasets, and object detection architectures, followed by the experiment and a detailed report of the results.

4.1 Benchmark Datasets and Object Detection Models

We evaluate XAI methods on two standard object detection benchmarks: **MS COCO 2017** [25], containing 123,287 images across 80 object categories, and **PASCAL VOC 2012** [12], with 11,530 images spanning 20 classes. These datasets are selected for their diversity in scene complexity and annotation quality. To ensure architectural variety, we employ **Faster R-CNN** [47], a two-stage detector known for high accuracy, and **YOLOX** [14], a single-stage, anchor-free model optimized for real-time performance.

4.2 XAI Methods for Object Detection

In this work, we focus on XAI methods from the region-based and CAM-based categories, as defined in [39] for object detection models. A brief description of the selected methods is given as follows:

1. **D-CLOSE** [56] is a region-based XAI method that generates saliency maps by segmenting the input image at multiple levels and combining their features to enhance explanation quality. Random masks are used to perturb the input, and the similarity scores between the masked outputs and the target prediction are calculated to weigh the importance of each region. The saliency maps are normalized using a density map to reduce noise, and multi-scale feature fusion aggregates fine-grained to coarse-grained features.
2. **D-RISE** [44] is a region-based XAI method designed to generate saliency maps for object detectors by jointly addressing localization and classification. It applies randomized binary masks to the input and evaluates their impact on detection outcomes. The importance of each masked region is estimated using a similarity metric that combines Intersection over Union (IoU) for spatial alignment, cosine similarity for classification consistency, and objectness scores. The final saliency map is obtained by aggregating the masks weighted by their contributions.

3. **G-CAME** [39] improves CAM-based explanation methods for object detection by integrating a Gaussian kernel for saliency map refinement. Specifically, it combines gradient-based localization with Gaussian weighting, effectively suppressing background noise and enhancing the detection of small objects. The method consists of four steps: selection of target layers, gradient-based object localization, feature map weighting, and Gaussian masking.

4.3 Results

We evaluate the selected XAI methods for object detection using YOLOX and Faster R-CNN on the MS COCO and PASCAL VOC datasets across three core criteria: faithfulness, localization, and complexity, each measured using dedicated quantitative metrics. All experiments are conducted under identical settings, with 2000 perturbed masks for both D-CLOSE and D-RISE. The evaluation is run on a system with an Nvidia Tesla T4 GPU and 24GB of RAM. Results in Table 1 show that ODExAI consistently measures explanation quality and informs method selection for specific detection tasks.

Faithfulness Results. Across both datasets and models, D-CLOSE consistently outperformed in faithfulness metrics. For example, on MS-COCO with YOLOX, it achieved the highest Insertion (0.908) and OA (0.881), with the lowest Deletion (0.027). This trend held across all settings, with D-CLOSE achieving the highest OA on three out of four model–dataset combinations. These results indicate that region-based explanations generated by D-CLOSE are tightly aligned with the model's true decision process, highlighting its high faithfulness.

Localization Results. In contrast, G-CAME excelled in the localization dimension, achieving the highest EBPG and PG scores across all configurations. For instance, on YOLOX with MS COCO, it reached PG = 94.31% and EBPG = 67.16%, significantly outperforming D-CLOSE and D-RISE. This consistent strength in localization suggests that CAM-based methods are highly effective at identifying relevant object regions, even when their explanations do not fully capture the model's internal reasoning.

Table 1. Evaluation results from ODExAI

Dataset	**MS-COCO**						**PASCAL VOC**					
Model	YOLOX			Faster R-CNN			YOLOX			Faster R-CNN		
Method	D-CLOSE	G-CAME	D-RISE	D-CLOSE	G-CAME	D-RISE	D-CLOSE	G-CAME	D-RISE	D-CLOSE	G-CAME	D-RISE
Ins↑	**0.908**	0.703	0.812	**0.912**	0.718	0.867	**0.804**	0.512	0.775	**0.826**	0.534	0.783
Del↓	**0.027**	0.059	0.043	**0.049**	0.169	0.152	0.128	0.183	**0.103**	**0.171**	0.284	0.206
OA↑	**0.881**	0.644	0.769	**0.863**	0.549	0.715	**0.676**	0.329	0.672	**0.655**	0.250	0.577
PG(%)↑	87.86	**94.31**	86.55	88.49	**96.13**	84.12	81.48	**93.96**	79.01	82.94	**95.13**	80.11
EBPG(%)↑	35.45	**67.16**	18.47	36.13	**70.11**	20.74	29.23	**58.39**	28.11	31.84	**57.84**	29.94
Sparsity↑	**25.02**	2.84	4.43	**26.13**	4.95	5.91	**26.84**	3.81	12.45	**27.12**	4.91	6.21
Time(s)↓	70.12	**0.43**	98.67	71.42	**0.54**	99.11	70.94	**0.48**	99.19	72.94	**0.63**	99.98

Complexity Results. G-CAME was also the most computationally efficient, with runtimes under 1 s across all settings. For instance, on MS-COCO with YOLOX, it required only 0.43 s, compared to 70.12 s for D-CLOSE and 98.67 s for D-RISE. However, its Sparsity was the lowest, indicating more diffuse explanations. In contrast, D-CLOSE produced the most concise maps but at a significantly higher cost.

5 Discussion

This section discusses the trade-offs between XAI methods revealed by ODExAI, offers guidance on selecting suitable approaches, acknowledges current limitations, and outlines directions for future development.

5.1 Trade-Off Analysis and Suggestions for Method Selection

The evaluation reveals clear trade-offs among XAI methods for object detection, showing that no single approach satisfies all interpretability requirements [30,42]. Region-based methods like D-CLOSE demonstrate strong faithfulness and produce sparse saliency maps due to their perturbation-based design: they systematically mask input regions and measure changes in model predictions, directly attributing importance to causally influential areas [39,56]. This yields explanations closely aligned with the model's decision process, as confirmed by faithfulness metrics, and results in concise, object-focused maps with minimal noise [56]. However, the repeated inference across multiple segmentation levels imposes significant computational overhead, limiting practicality for real-time or resource-constrained applications [39,56].

In contrast, CAM-based methods such as G-CAME achieve superior localization and low runtime by using internal feature maps and gradient signals from specific layers to generate saliency maps [39]. On the one hand, selecting appropriate layers for explanation requires specialized knowledge from users [13,38]. On the other hand, this design enables rapid computation and produces highly localized explanations, as reflected in strong PG and EBPG scores. However, because they rely on aggregated activations rather than direct input perturbations, these methods may miss the exact features responsible for decisions, leading to lower faithfulness [32,39].

D-RISE, while not leading in any single metric, maintains consistently moderate performance across all evaluation dimensions, making it a robust option for ensemble-based detection or sensitivity analysis, where explanation diversity is valuable [39,44,56]. These results demonstrate ODExAI's effectiveness in highlighting method-specific trade-offs and guiding the choice of XAI techniques based on task-specific interpretability goals.

5.2 Limitations and Future Development

Despite promising results, the current framework presents several limitations. First, the number of included XAI methods and evaluation metrics remains limited, which may restrict generalizability. Future work will expand the framework to incorporate additional methods and metrics, aiming for broader coverage and the development of more comprehensive evaluation tools. Second, some dimensions mix higher-better and lower-better metrics, which may introduce inconsistency. A possible solution is to assign weights to account for these directional differences. Third, the framework lacks a proper normalization scheme, which can be misleading when metrics differ in scale or unit. A promising approach is the skill score proposed in [4], which compares an explanation method's performance relative to a reference baseline designed to violate the assumptions of each evaluation property. The skill score quantifies how much an explanation improves over this worst-case scenario.

6 Conclusion

In this work, we introduce the *Object Detection Explainable AI Evaluation* (ODExAI), a structured system for assessing XAI methods for object detection. ODExAI supports the selection of explanation techniques based on three core dimensions: localization accuracy, faithfulness to model predictions, and computational complexity. To validate the framework, we evaluate several XAI methods on two widely used object detectors and standard benchmark datasets. Results show that region-based methods perform well in both localization and faithfulness but incur higher computational cost. CAM-based methods, by contrast, are more efficient and provide good localization but are less faithful. These findings reveal key trade-offs and underscore the need for task-specific evaluation when selecting XAI methods. Future work will expand the framework with additional methods and metrics, and establish consistent, interpretable evaluation procedures for each dimension.

Acknowledgment. This work is supported by the UNB-FCS Startup Fund (22–23 START UP/H CAO).

References

1. Arras, L., Osman, A., Samek, W.: CLEVR-XAI: a benchmark dataset for the ground truth evaluation of neural network explanations. Inf. Fusion **81**, 14–40 (2022). https://doi.org/10.1016/j.inffus.2021.11.008
2. Arya, V., et al.: One explanation does not fit all: a toolkit and taxonomy of AI explainability techniques. arXiv preprint: arXiv:1909.03012 (2019)
3. Bochkovskiy, A., Wang, C.Y., Liao, H.Y.M.: YOLOv4: optimal speed and accuracy of object detection. arXiv preprint: arXiv:2004.10934 (2020)
4. Bommer, P.L., Kretschmer, M., Hedström, A., Bareeva, D., Höhne, M.M.C.: Finding the right XAI method–a guide for the evaluation and ranking of explainable AI methods in climate science. Artif. Intell. Earth Syst. **3**(3), e230074 (2024). https://doi.org/10.1175/AIES-D-23-0074.1
5. Buchelt, A., et al.: Exploring artificial intelligence for applications of drones in forest ecology and management. For. Ecol. Manage. **551**, 121530 (2024). https://doi.org/10.1016/j.foreco.2023.121530

6. C, B., R, S., A, A., Moorthy, R.S.: A novel approach for jute pest detection using improved VGG-19 and XAI. In: 2024 15th International Conference on Computing Communication and Networking Technologies (ICCCNT), pp. 1–5 (2024). https://doi.org/10.1109/ICCCNT61001.2024.10723939
7. Cai, Z., Vasconcelos, N.: Cascade R-CNN: delving into high quality object detection. In: 2018 IEEE/CVF Conference on Computer Vision and Pattern Recognition (CVPR), pp. 6154–6162. IEEE Computer Society, Los Alamitos (2018). https://doi.org/10.1109/CVPR.2018.00644
8. Carion, N., Massa, F., Synnaeve, G., Usunier, N., Kirillov, A., Zagoruyko, S.: End-to-end object detection with transformers. In: Computer Vision – ECCV 2020: 16th European Conference, Glasgow, UK, August 23–28, 2020, Proceedings, Part I, pp. 213–229. Springer-Verlag, Berlin, Heidelberg (2020). https://doi.org/10.1007/978-3-030-58452-8_13
9. Duan, K., Bai, S., Xie, L., Qi, H., Huang, Q., Tian, Q.: CenterNet: keypoint triplets for object detection. In: 2019 IEEE/CVF International Conference on Computer Vision (ICCV), pp. 6568–6577 (2019). https://doi.org/10.1109/ICCV.2019.00667
10. Ennab, M., Mcheick, H.: Enhancing interpretability and accuracy of AI models in healthcare: a comprehensive review on challenges and future directions. Front. Robot. AI **11**, 1444763 (2024). https://doi.org/10.3389/frobt.2024.1444763
11. Esmaeili, M., Vettukattil, R., Banitalebi, H., Krogh, N.R., Geitung, J.T.: Explainable artificial intelligence for human-machine interaction in brain tumor localization. J. Personalized Med. **11**(11) (2021). https://doi.org/10.3390/jpm11111213
12. Everingham, M., Van Gool, L., Williams, C.K.I., Winn, J., Zisserman, A.: The pascal visual object classes (VOC) challenge. Int. J. Comput. Vision **88**(2), 303–338 (2010). https://doi.org/10.1007/s11263-009-0275-4
13. Famiglini, L., Campagner, A., Barandas, M., La Maida, G.A., Gallazzi, E., Cabitza, F.: Evidence-based XAI: an empirical approach to design more effective and explainable decision support systems. Comput. Biol. Med. **170**, 108042 (2024). https://doi.org/10.1016/j.compbiomed.2024.108042
14. Ge, Z., Liu, S., Wang, F., Li, Z., Sun, J.: YOLOX: exceeding YOLO series in 2021. arXiv preprint: arXiv:2107.08430 (2021)
15. Gomez, T., Fréour, T., Mouchère, H.: Metrics for saliency map evaluation of deep learning explanation methods. In: El Yacoubi, M., Granger, E., Yuen, P.C., Pal, U., Vincent, N. (eds.) Pattern Recognition and Artificial Intelligence, pp. 84–95. Springer International Publishing, Cham (2022). https://doi.org/10.1007/978-3-031-09037-0_8
16. Grami, A.: The Gaussian Distribution, pp. 201–238. John Wiley & Sons (2019). https://doi.org/10.1002/9781119300847.ch7
17. Gunning, D.: DARPA's explainable artificial intelligence (XAI) program. In: Proceedings of the 24th International Conference on Intelligent User Interfaces, IUI '19, p. ii. Association for Computing Machinery, New York (2019). https://doi.org/10.1609/aimag.v40i2.2850
18. He, K., Gkioxari, G., Dollár, P., Girshick, R.: Mask R-CNN. In: 2017 IEEE International Conference on Computer Vision (ICCV), pp. 2980–2988 (2017). https://doi.org/10.1109/ICCV.2017.322
19. Hedström, A., et al.: Quantus: an explainable AI toolkit for responsible evaluation of neural network explanations and beyond. J. Mach. Learn. Res. **24**(34), 1–11 (2023)
20. Hoofnagle, C.J., van der Sloot, B., and, F.Z.B.: The European Union general data protection regulation: what it is and what it means*. Inf. Commun. Technol. Law **28**(1), 65–98 (2019). https://doi.org/10.1080/13600834.2019.1573501

21. Kadir, M.A., Mosavi, A., Sonntag, D.: Evaluation metrics for XAI: a review, taxonomy, and practical applications. In: 2023 IEEE 27th International Conference on Intelligent Engineering Systems (INES), pp. 000111–000124 (2023). https://doi.org/10.1109/INES59282.2023.10297629
22. Kakogeorgiou, I., Karantzalos, K.: Evaluating explainable artificial intelligence methods for multi-label deep learning classification tasks in remote sensing. Int. J. Appl. Earth Obs. Geoinf. **103**, 102520 (2021). https://doi.org/10.1016/j.jag.2021.102520
23. Law, H., Deng, J.: CornerNet: detecting objects as paired keypoints. Int. J. Comput. Vision **128**(3), 642–656 (2019). https://doi.org/10.1007/s11263-019-01204-1
24. Le, P.Q., Nauta, M., Nguyen, V.B., Pathak, S., Schlötterer, J., Seifert, C.: Benchmarking eXplainable AI - a survey on available toolkits and open challenges. In: Elkind, E. (ed.) Proceedings of the Thirty-Second International Joint Conference on Artificial Intelligence, IJCAI-23, pp. 6665–6673. International Joint Conferences on Artificial Intelligence Organization (2023). https://doi.org/10.24963/ijcai.2023/747, survey Track
25. Lin, T.Y., et al.: Microsoft COCO: common objects in context. In: Fleet, D., Pajdla, T., Schiele, B., Tuytelaars, T. (eds.) Computer Vision – ECCV 2014, pp. 740–755. Springer International Publishing, Cham (2014). https://doi.org/10.1007/978-3-319-10602-1_48
26. Liu, W., et al.: SSD: single shot multibox detector. In: Leibe, B., Matas, J., Sebe, N., Welling, M. (eds.) Computer Vision – ECCV 2016, pp. 21–37. Springer International Publishing, Cham (2016). https://doi.org/10.1007/978-3-319-46448-0_2
27. Liu, Z., et al.: Swin transformer: hierarchical vision transformer using shifted windows . In: 2021 IEEE/CVF International Conference on Computer Vision (ICCV), pp. 9992–10002. IEEE Computer Society, Los Alamitos (2021). https://doi.org/10.1109/ICCV48922.2021.00986
28. Lopes, P., Silva, E., Braga, C., Oliveira, T., Rosado, L.: XAI Systems Evaluation: A Review of Human and Computer-Centred Methods. Applied Sciences **12**(19) (2022). https://doi.org/10.3390/app12199423
29. Mohseni, S., Zarei, N., Ragan, E.D.: A multidisciplinary survey and framework for design and evaluation of explainable AI systems. ACM Trans. Interact. Intell. Syst. **11**(3–4) (2021). https://doi.org/10.1145/3387166
30. Molnar, C., König, G., et al.: General pitfalls of model-agnostic interpretation methods for machine learning models, pp. 39–68. Springer International Publishing, Cham (2022). https://doi.org/10.1007/978-3-031-04083-2_4
31. Moradi, M., et al.: Model-agnostic explainable artificial intelligence for object detection in image data. Eng. Appl. Artif. Intell. **137**, 109183 (2024). https://doi.org/10.1016/j.engappai.2024.109183
32. Mumuni, F., Mumuni, A.: Explainable artificial intelligence (XAI): from inherent explainability to large language models. arXiv preprint: arXiv:2501.09967 (2025)
33. Natarajan, S., Chakrabarti, P., Margala, M.: Robust diagnosis and meta visualizations of plant diseases through deep neural architecture with explainable AI. Sci. Rep. **14**(1), 13695 (2024). https://doi.org/10.1038/s41598-024-64601-8
34. Nauta, M., et al.: From anecdotal evidence to quantitative evaluation methods: a systematic review on evaluating explainable AI. ACM Comput. Surv. **55**(13s) (2023). https://doi.org/10.1145/3583558
35. Neuwirth, R.: The EU Artificial Intelligence Act: Regulating Subliminal AI Systems. Routledge Research in the Law of Emerging Technologies, Taylor & Francis (2022)

36. Nguyen, H., et al.: LangXAI: integrating large vision models for generating textual explanations to enhance explainability in visual perception tasks. In: Larson, K. (ed.) Proceedings of the Thirty-Third International Joint Conference on Artificial Intelligence, IJCAI-24, pp. 8754–8758. International Joint Conferences on Artificial Intelligence Organization (2024). https://doi.org/10.24963/ijcai.2024/1025, demo Track
37. Nguyen, H., et al.: Heart2Mind: human-centered contestable psychiatric disorder diagnosis system using wearable ECG monitors. arXiv preprint: arXiv:2505.11612 (2025)
38. Nguyen, H.T.T., Nguyen, L.P.T., Cao, H.: XEdgeAI: a human-centered industrial inspection framework with data-centric explainable Edge AI approach. Inf. Fusion **116**, 102782 (2025). https://doi.org/10.1016/j.inffus.2024.102782
39. Nguyen, K., Nguyen, H., Nguyen, K., Truong, B., Phan, T., Cao, H.: Efficient and concise explanations for object detection with gaussian-class activation mapping explainer. In: Proceedings of the Canadian Conference on Artificial Intelligence (2024)
40. Nguyen, T.T.H., Nguyen, V.T.K., Cao, Q.H., Truong, V.B., Nguyen, Q.K., Cao, H.: Enhancing the Fairness and Performance of Edge Cameras with Explainable AI. In: 2024 IEEE International Conference on Consumer Electronics (ICCE), pp. 1–4 (2024)
41. Nguyen, T.T.H., Truong, V.B., Nguyen, V.T.K., Cao, Q.H., Nguyen, Q.K.: Towards trust of explainable AI in thyroid nodule diagnosis. In: International Workshop on Health Intelligence, pp. 11–26. Springer (2023)
42. Padmanabhan, D.C., Plöger, P.G., Arriaga, O., Valdenegro-Toro, M.: DExT: detector explanation toolkit. In: Longo, L. (ed.) Explainable Artificial Intelligence, pp. 433–456. Springer Nature Switzerland, Cham (2023). https://doi.org/10.1007/978-3-031-44067-0_23
43. Petsiuk, V., Das, A., Saenko, K.: RISE: randomized input sampling for explanation of black-box models. In: British Machine Vision Conference (BMVC) (2018)
44. Petsiuk, V., Jain, R., Manjunatha, V., Morariu, V.I., Mehra, A., Ordonez, V., Saenko, K.: Black-box explanation of object detectors via saliency maps. In: 2021 IEEE/CVF Conference on Computer Vision and Pattern Recognition (CVPR), pp. 11438–11447 (2021). https://doi.org/10.1109/CVPR46437.2021.01128
45. Redmon, J., Divvala, S., Girshick, R., Farhadi, A.: You only look once: unified, real-time object detection. In: 2016 IEEE Conference on Computer Vision and Pattern Recognition (CVPR), pp. 779–788 (2016). https://doi.org/10.1109/CVPR.2016.91
46. Redmon, J., Farhadi, A.: YOLOv3: an incremental improvement. arXiv preprint: arXiv:1804.02767 (2018)
47. Ren, S., He, K., Girshick, R., Sun, J.: Faster R-CNN: towards real-time object detection with region proposal networks. In: Cortes, C., Lawrence, N., Lee, D., Sugiyama, M., Garnett, R. (eds.) Advances in Neural Information Processing Systems, vol. 28. Curran Associates, Inc. (2015)
48. Ribeiro, M.T., Singh, S., Guestrin, C.: "Why Should I Trust You?": explaining the predictions of any classifier. In: Proceedings of the 22nd ACM SIGKDD International Conference on Knowledge Discovery and Data Mining, KDD '16, pp. 1135–1144. Association for Computing Machinery, New York (2016). https://doi.org/10.1145/2939672.2939778
49. Sejr, J.H., Schneider-Kamp, P., Ayoub, N.: Surrogate object detection explainer (SODEx) with YOLOv4 and LIME. Mach. Learn. Knowl. Extr. **3**(3), 662–671 (2021). https://doi.org/10.3390/make3030033

50. Selvaraju, R.R., Cogswell, M., Das, A., Vedantam, R., Parikh, D., Batra, D.: Grad-CAM: visual explanations from deep networks via gradient-based localization. In: 2017 IEEE International Conference on Computer Vision (ICCV), pp. 618–626 (2017). https://doi.org/10.1109/ICCV.2017.74
51. Song, L., et al.: Explainable artificial intelligence to interpret spatially-explicit impacts of future climate change on species distribution (2024)
52. Tahir, H.A., Alayed, W., Hassan, W.U., Haider, A.: A novel hybrid XAI solution for autonomous vehicles: real-time interpretability through LIME–SHAP integration. Sensors **24**(21) (2024). https://doi.org/10.3390/s24216776
53. Tarmissi, K., Mengash, H.A., Negm, N., Said, Y., Al-Sharafi, A.M.: Explainable artificial intelligence with fusion-based transfer learning on adverse weather conditions detection using complex data for autonomous vehicles. AIMS Math. **9**(12), 35678–35701 (2024). https://doi.org/10.3934/math.20241693
54. Teng, Q., Liu, Z., Song, Y., Han, K., Lu, Y.: A survey on the interpretability of deep learning in medical diagnosis. Multimedia Syst. **28**(6), 2335–2355 (2022). https://doi.org/10.1007/s00530-022-00960-4
55. Tian, Z., Shen, C., Chen, H., He, T.: FCOS: fully convolutional one-stage object detection. In: 2019 IEEE/CVF International Conference on Computer Vision (ICCV), pp. 9626–9635 (2019). https://doi.org/10.1109/ICCV.2019.00972
56. Truong, V.B., Nguyen, T.T.H., Nguyen, V.T.K., Nguyen, Q.K., Cao, Q.H.: Towards better explanations for object detection. In: Yanıkoğlu, B., Buntine, W. (eds.) Proceedings of the 15th Asian Conference on Machine Learning. Proceedings of Machine Learning Research, PMLR, vol. 222, pp. 1385–1400 (2024)
57. Wang, H., et al.: Score-CAM: score-weighted visual explanations for convolutional neural networks . In: 2020 IEEE/CVF Conference on Computer Vision and Pattern Recognition Workshops (CVPRW), pp. 111–119. IEEE Computer Society, Los Alamitos (2020). https://doi.org/10.1109/CVPRW50498.2020.00020
58. Wood, D.A., et al.: Deep learning models for triaging hospital head MRI examinations. Med. Image Anal. **78**, 102391 (2022). https://doi.org/10.1016/j.media.2022.102391
59. Yamauchi, T.: Spatial sensitive grad-CAM++: improved visual explanation for object detectors via weighted combination of gradient map. In: Proceedings of the IEEE/CVF Conference on Computer Vision and Pattern Recognition (CVPR) Workshops, pp. 8164–8168 (2024)
60. Yamauchi, T., Ishikawa, M.: Spatial sensitive GRAD-CAM: visual explanations for object detection by incorporating spatial sensitivity. In: 2022 IEEE International Conference on Image Processing (ICIP), pp. 256–260 (2022). https://doi.org/10.1109/ICIP46576.2022.9897350
61. Yang, G., Ye, Q., Xia, J.: Unbox the black-box for the medical explainable AI via multi-modal and multi-centre data fusion: a mini-review, two showcases and beyond. Inf. Fusion **77**, 29–52 (2022). https://doi.org/10.1016/j.inffus.2021.07.016
62. Zhang, J., Bargal, S.A., Lin, Z., Brandt, J., Shen, X., Sclaroff, S.: Top-down neural attention by excitation backprop. Int. J. Comput. Vision **126**(10), 1084–1102 (2017). https://doi.org/10.1007/s11263-017-1059-x
63. Zhang, Q., Rao, L., Yang, Y.: Group-CAM: group score-weighted visual explanations for deep convolutional networks. arXiv preprint: arXiv:2103.13859 (2021)
64. Zhao, C., Hsiao, J.H., Chan, A.B.: Gradient-based instance-specific visual explanations for object specification and object discrimination. IEEE Trans. Pattern Anal. Mach. Intell. **46**(9), 5967–5985 (2024). https://doi.org/10.1109/TPAMI.2024.3380604

Enhancing Semi-supervised Learning with a Meta-feature Based Safeguard System

Martin Schumann(✉)

Technical University of Munich, Arcisstraße 21, 80333 Munich, Germany
martin.schumann@tum.de

Abstract. The selection of Meta-Features can strongly influence the performance and accuracy of semi-supervised learning methods. In this paper, we analyze the role of different meta-features in enhancing the performance of semi-supervised learning models. We present experiments with benchmark data sets to determine which meta-features contribute most significantly to model accuracy and robustness.

We propose an enhanced safeguard system for semi-supervised learning that leverages meta-features to predict the potential benefits of pseudo-labeling, with a focus on simultaneously reducing computational resource consumption and improving the overall performance of semi-supervised learning models. By determining when to train a second predictor, our system optimizes computational efficiency.

Keywords: AutoML · ML · semi-supervised learning · Meta-Features

1 Introduction

One of the biggest problems in machine learning (ML) and artificial intelligence (AI) research is the scarcity of labeled data, which are required to train many models. However, most useful data models do not have labels, which requires expensive [16,22], time-consuming, labor-intensive, and not easily scalable human labeling, which is prone to errors [22,31].

For ML, a larger amount of correctly labeled data almost always leads to higher accuracy results of the model [16]. This is not a problem when ML is done with unsupervised learning, because it does not require labeled data, and instead identifies patterns directly [4,22]. However, it lacks "ground truth" for evaluation and is sensitive to how data is input and used (e.g., introducing biases) [7,27]. Supervised learning, which relies on labeled data, and is used e.g., for prediction, does not have some of the limitations that unsupervised learning has, but has others, including sensitivity to the quality of labels. As a result, ML research often relies on small, well-labeled datasets like the IMDB dataset [21] or MNIST [19]. Canned benchmark datasets, however, can contain incorrect [30], biased [6], or incomplete information [30]. Furthermore, older data might not correspond to changes in the real world [9].

T. Braun et al. (Eds.): KI 2025, LNAI 15956, pp. 134–147, 2026.
https://doi.org/10.1007/978-3-032-02813-6_10

To deal with the problems arising from this lack of up-to-date correctly labeled training data, the concept of self-training was introduced [17,28]. One approach works by having one supervised learning model create "pseudo-labels" for unlabeled data, which is then used, together with existing "human" labels, to train a second supervised learning model. This reduces the need for massive amounts of labeled data. To streamline this process, we utilize AutoML (Automated Machine Learning)-based libraries, which are able to automate many parts of an ML workflow [13].

However, pseudo-labeling introduces risks, as inaccurate labels can degrade performance. We therefore ask our main research question: can an AutoML-trained ML model achieve better performance by utilizing previously unlabeled data, which a separately (AutoML) trained model has labeled?

In order to address this question, we have developed an approach that uses meta-features to decide if the newly generated pseudo-label is likely to lead to an improvement in performance. Our safeguard system is designed to evaluate the behavior of the previously trained model using our meta-feature approach. It will decide if this model will be further used for labeling or not. In the latter case, the original model will be kept as the newly trained one is likely to result in worse performance.

To study our approach in detail, we split up our initial question into two research questions: RQ1: Does our approach perform favorably to AutoGluon's built-in support for semi-supervised learning? and RQ2: Which models and settings influence the performance of the safeguard system?

The remainder of the paper is structured as follows: In Sect. 2, we briefly describe AutoML. Section 3 discusses related work. In Sect. 4, we describe our approach, including our improved AutoML pipeline and its safeguard system. In Sect. 5, we answer the 2 research questions and present experimental results. Section 6 concludes the paper and outlines potential directions for future work.

2 Background

AutoML (Automated Machine Learning) as a concept describes the process of automating and simplifying ML tasks and workflows. There is a wide field of libraries, which accomplish simplification by automating a lot of the time-consuming and tedious process of setting up an ML pipeline [13]. Among other things, they automate the data preprocessing and feature engineering steps, the model selection steps, and the hyperparameter optimization [13,32,33]. With the help of feature extraction and feature selection, the data is preprocessed and the relevant features are selected. For model selection, AutoML libraries try out different models to find the best one for a given dataset. Hyperparameter optimization is another important task, which involves using different algorithms to select the optimal hyperparameters for a given model.

One AutoML library is AutoGluon, which promises to be very easy to use while having competitive performance. It can achieve this by "ensembling multiple models and stacking them in multiple layers" [11], e.g., by combining the output of multiple models for one prediction. We chose it as it is one of the best performing AutoML libraries on current benchmarks [15], high quality of models, and ease of use. Semi-supervised learning is one way to deal with the problem of large amounts of unlabeled data existing. The basic concept is to have a subset of the data labeled and then employ specific strategies to utilize the remaining unlabeled data. One of these strategies is known as "pseudo-labeling", which involves an algorithm labeling some or all of the unlabeled data. These "pseudo-labeled" data points are then used like normal data points. The combination of labeled and pseudo-labeled data can then be used to train with supervised learning algorithms [34].

3 Related Work

There are two main papers that focus on unifying AutoML and semi-supervised learning. The first, by Van Engelen et al. introduces the term co-ensembling [10], where K classifiers are trained on labeled data, and $K-1$ of them generate pseudo-labels for unlabeled data. This combined data, selected based on a minimum probability threshold, is then used to train the K-th original classifier further. Although this process can be multi-step, their findings conclude that a single step is better, as multiple steps can lead to errors being multiplied. They evaluate their approach on the OpenML100 benchmark suite [10], establishing baseline performance on the same datasets used in our experiments.

The second, by Li et al., introduces the term AutoSSL (for auto semi-supervised learning) to describe unifying the two approaches (AutoML and SSL) from the view of SSL [20]. The authors use a clustering approach to characterize the data, and use meta-features of the data to inform the "automated learning" [20] process. This process utilizes well-known SSL algorithms and combines them with AutoML techniques to tune the hyperparameters of these algorithms. This approach is compared to a normal auto-sklearn approach, which the authors claim has "highly competitive" [20] performance compared to classic SSL techniques. Li et al.'s paper is also referenced by [10] as having a different approach, although both papers focus on how to combine AutoML and semi-supervised learning, with the differences lying in the angle from which the authors tackle the problem.

AutoGluon also has experimental support for "unlabeled_data" via the experimental "FT_TRANSFORMER" [3, Line 504] model, which "does not scale well to >100 features." [3, Line 504], limiting the usability compared to our approach. There is also some support for "pseudo-labeling" [3, Line 1904], which labels data above a certain probability threshold and uses it as extra training data to refit the original model.

4 Our Approach

4.1 The Improved AutoML Pipeline

In this section, we introduce our linear-ensembling pipeline (Fig. 1), consisting of the two models M1 and M2 to be trained. Each model is a classifier trained using AutoGluon ("AutoML" in Fig. 1). The models are trained one after the other, and the pipeline returns M2 as the final model. The pipeline utilizes three subsets of one dataset.

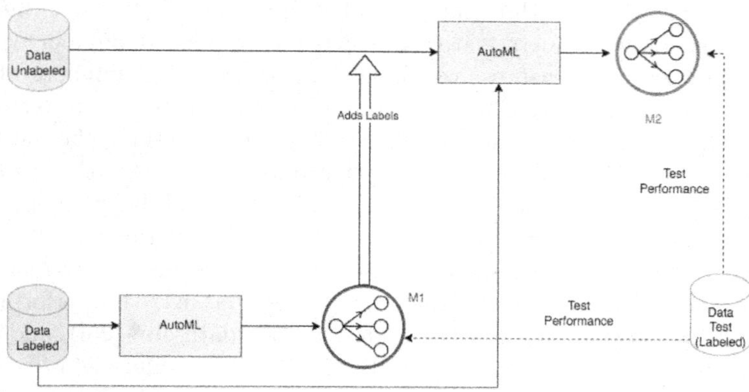

Fig. 1. Graphical representation of the linear-ensembling pipeline. M1 is the first model (labeler), M2 is the second model.

M1 is trained on labeled data and then pseudo-labels all the unlabeled data. It only keeps the data points which exceed a certain confidence threshold. M1 predicts a probability distribution for the possible labels for each data point, and if no label meets the threshold, the data point is not used for further training. This pseudo-labeling step is the "Adds labels" process in Fig. 1 and is shown in more detail in Fig. 2. All the newly pseudo-labeled data and labeled data are combined into a new dataset used to train the model M2, which is the final model. As it is possible that M2 sometimes performs worse than M1, we introduce a safeguard system (Sect. 4.3) to overcome this problem.

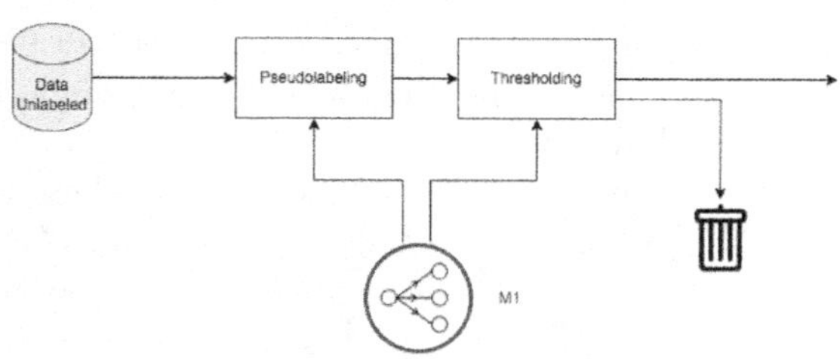

Fig. 2. The "Adds labels" process from Fig. 1. Data with pseudo-labels that do not meet a certain threshold get "thrown away".

4.2 Meta-Features

Meta-features, as defined by Rivolli et al., are functions that take a dataset and return "values that characterize the dataset" [26]. These calculated measures describe various aspects of datasets [1], and can be used to predict the performance of ML algorithms, among other applications [18]. It is important to distinguish between meta-features, which return a fixed number of values, and characterization measures, which may return varying numbers of values depending on the dataset. Characterization measures can be "converted" into meta-features through summarization functions, such as taking the mean [26].

Alcobaça et al. have developed a Python implementation of meta-features (pymfe [1]), that provides a comprehensive collection of different meta-features to choose from. These meta-features are split by Alcobaça et al. into multiple groups, including, but not limited to: model based ("measures extracted from a [simple] model" [25]), landmarking ("measures that use the performance of simple and fast learning algorithms to characterize datasets" [25]), and clustering (characteristics of dataset clusters) [1,25]. Which specific meta-features are included in these various groups can be found in the pymfe documentation [1].

In this work, we utilize meta-features for our safeguard system (Subsect. 4.3) to be able to predict if training with "pseudo-labeled" data will lead to an improvement in performance. We have decided to use the meta-feature groupings as defined in pymfe to find the group or groups of meta-features which provide the best predictive qualities. The theory behind our approach is that a dataset's meta-features reveal meaningful information, enabling us to predict the behavior of the pipeline (better/worse performance) more accurately.

4.3 Safeguard System

Adding "pseudo-labeled" data can hurt model performance, due to confirmation biases [2], where incorrect labels from M1 lead to M2 learning from poor predictions. This means that M2 might perform worse than M1, which was trained only on correctly labeled data. We have decided to alleviate the problem of deteriorated performance and the high cost of training by introducing a safeguard system. This system predicts whether a given M2 model will perform better or worse than the M1 model. Then, depending on settings, pseudo-labeling of data and the training of M2 will not occur.

The safeguard system operates as a classifier that predicts the relative performance of M1 and M2 models based on dataset characteristics. The system's design follows a two-phased approach: an offline training phase and an online prediction phase. During the **offline training phase**, data to train the safeguard system is gathered by running the complete linear-ensembling pipeline on a diverse collection of datasets. For each dataset, both M1 and M2 are trained and their accuracies compared. This information is saved as a single binary value indicating whether the second classifier outperformed the first ("plus" if it outperformed, "minus" otherwise). Simultaneously, meta-features are extracted from each dataset using the pymfe library. These meta-features serve as the input features, while the binary value serves as the target variable for training the safeguard classifier. During the **online prediction phase**, when a new dataset is encountered in the linear-ensembling pipeline, the safeguard system extracts the same set of meta-features from the initially labeled data. These meta-features are then fed into the trained safeguard classifier, which outputs a prediction of whether training M2 would be beneficial.

As can be seen in Fig. 3, if M2 is predicted to perform better ("plus"), the pipeline proceeds with pseudo-labeling and training of M2. Otherwise, if the classifier predicts that M2 will perform worse ("minus"), the pipeline will end and return M1 as the final model.

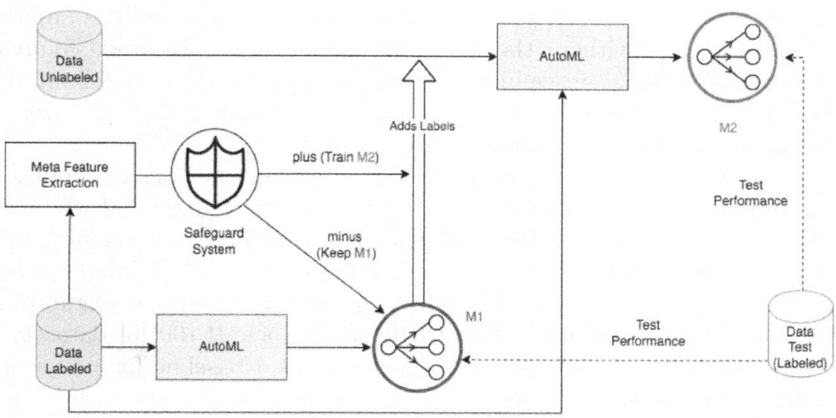

Fig. 3. Pipeline with safeguard system included. Pseudo-labeling and training of M2 only occurs if the safeguard system allows it. Otherwise M1 is kept as the main model.

The safeguard system is implemented both using scikit-learn's RandomForestClassifier [23] and AutoGluon. This is to make evaluating various performance differences easier, as is described in Subsect. 5.2. Both systems are trained using the same procedure, where the binary performance outcomes and meta-features are used as training data.

This system provides two major benefits. First, it offers insight into when and how an ML model performs better or worse after being trained with pseudo-labeled data, and for which datasets this occurs. Second, it saves time and compute resources by avoiding unnecessary training of M2, when the safeguard system predicts poor performance.

5 Experiments and Results

We run tests to compare whether utilizing pseudo-labeled data can improve the performance of a classifier, and we train and benchmark our safeguard system[1].

[1] The source code of our implementation can be found under https://github.com/martinlschumann/safeguard-system.

We run both of these tasks on the LRZ's interactive and serial Linux clusters [24]. We do not utilize GPUs for our work, as they bring no huge performance benefit for our datasets (tabular data). We use python 3.10 from anaconda3/2022.10 (part of spack/23.1.0).

For AutoGluon's fit method we keep most of the default settings [12], except for increasing the time limit to 600 seconds and changing the quality parameter. We primarily use AutoGluon's medium quality setting (which defines the speed, quality and size of models) for its balance of speed and performance, with high quality tests to validate our findings [12]. We decided not to utilize best quality as it consumes a lot of resources. The results of datasets trained using the medium quality setting are used when training the safeguard system. Our research has shown that the different quality settings have little impact on the results with and without the safeguard system, especially when comparing performance after linear-ensembling. For pymfe, we also add a time limit of 600 seconds. If the time limit is reached, our safeguard system method errors out and linear-ensembling continues with training M2.

To train the safeguard system, we used results from previous runs of the linear ensembling approach. The data was split into 90% training and 10% validation. For the scikit-learn RandomForestClassifier version of the classifier, certain hyperparameters were modified from their default values. These were optimized using scikit-learn's GridSearchCV[2] to find the best-performing configuration.

For the datasets, we utilize the widely used OpenML100 [5] datasets and their successors, the OpenML-CC18 [5] datasets as a baseline for performance comparison. The reason for choosing both is that a lot of research utilizes the original OpenML100, thus allowing for comparison to other research. For more variety and to see how the implementation functions on larger and more diverse datasets, we utilize the datasets that were compiled by Feurer et al. [14]. We put them in 3 dataset groups and named them CD1, CD2, and CD3.

5.1 Linear-Ensembling Versus Autogluon's Built In Semi-Supervised Learning Support (RQ1)

In this section, we address the question: Does our linear-ensembling approach perform favorably to AutoGluon's built-in support for semi-supervised learning?

We compare three different approaches using the "accuracy" measurement:

1. **Our Approach (*S1*)**: Linear-ensembling with two models. *M1*: Initial model, trained on labeled data only. *M2*: Final model which uses both labeled and pseudo-labeled data (labeled with M1).
2. **AutoGluon's built-in approach (*S2*)**: Using AutoGluon's native semi-supervised learning functionality
3. **Hybrid approach (*S3*)**. *M1*: Trained using AutoGluon's semi-supervised functionality. *M2*: Trained the same as our approach, i.e. labeled and pseudo-labeled data (labeled with M1).

[2] https://scikit-learn.org/stable/modules/generated/sklearn.model_selection.GridSearchCV.html.

Table 1. Comparison results of *S1* with *S2* and *S3* across various datasets and model configurations.

Datasets	Comparison	Better Performance			Equal Perf.
		S1	*S2*	*S3*	
`OpenML100`	S1 (M1) vs S2	15	**21**	–	43
	S1 (M1) vs S3 (M2)	**30**	–	24	25
	S1 (M2) vs S3 (M2)	**19**	–	16	44
	S1 (M2) vs S2	**31**	27	–	21
`OpenML-CC18`	S1 (M1) vs S2	12	**14**	–	28
	S1 (M1) vs S3 (M2)	**20**	–	18	16
	S1 (M2) vs S3 (M2)	9	–	**16**	29
	S1 (M2) vs S2	**25**	18	–	11
`CD1`	S1 (M1) vs S2	**11**	2	–	12
	S1 (M1) vs S3 (M2)	**17**	–	6	2
	S1 (M2) vs S3 (M2)	**11**	–	5	9
	S1 (M2) vs S2	9	**14**	–	2
`CD2`	S1 (M1) vs S2	**39**	28	–	24
	S1 (M1) vs S3 (M2)	**36**	–	28	27
	S1 (M2) vs S3 (M2)	**35**	–	28	28
	S1 (M2) vs S2	33	**37**	–	21
`CD3`	S1 (M1) vs S2	6	**8**	–	8
	S1 (M1) vs S3 (M2)	**7**	–	3	12
	S1 (M2) vs S3 (M2)	**6**	–	2	14
	S1 (M2) vs S2	7	7	–	8

We also compare *S1*'s M1 model to *S2*'s model, as the safeguard system is not in effect, so sometimes *S1*'s M1 model might have better results than *S1*'s M2 model.

Table 1 shows the performance of the various approaches across multiple datasets. Each row compares two specific approaches, with columns indicating how many times one approach outperformed the other in terms of accuracy. For the "Equal Perf." column, the accuracy of both approaches is identical.

The results show that our approach for linear-ensembling is almost always better than the hybrid approach (comparing "*S1* (M2) vs *S3* (M2)"). Our initial model also always outperforms the hybrid model (comparing "*S1* (M1) vs *S3* (M2)"). When comparing our initial model to AutoGluon's approach ("*S1* (M1) vs *S2*"), results are mixed across datasets. Similarly, when comparing our final linear-ensembled model to AutoGluon's approach ("*S1* (M2) vs *S2*"), neither consistently outperforms the other. These results indicate that while AutoGluon's semi-supervised support is occasionally better than our approach without the safeguard system, overall our linear-ensembling pipeline delivers better performance. Furthermore, integrating AutoGluon's semi-supervised support into our pipeline for training M1 provides only marginal benefits.

5.2 Safeguard System (RQ2)

In this section, we address the research question: which models and settings influence the performance of the safeguard system?

We compare the performance along 2 directions. The first is which classifier we use for the safeguard system, scikit-learn's RandomForestClassifier, or AutoGluon's TabularPredictor. For the AutoGluon classifier we also look at different metrics, including a custom metric which weighs false negatives or false positives heavily and is based on [29].

The second direction is looking at meta-feature performance and relevance. To measure the benefit of the safeguard system, we measure the mean accuracy with and without utilizing the safeguard system. The mean accuracy is defined as the mathematical average of each dataset's accuracy on the test data after training the classifier. The dataset's accuracy describes the ratio of correct predictions to the total number of predictions a classifier makes for a given dataset[3]. The mean accuracy serves as a measure of the benefit of the safeguard system, supporting the hypothesis that the system will lead to the better classifier being chosen without the need to train the second model. The mean accuracy with the safeguard system is either the accuracy of M1 or M2, depending on what the system chooses. If the mean accuracy is better or the same, this means the safeguard system has a positive benefit overall; if not, it has a negative benefit. We compare it to just naively choosing M2, which might be worse than M1. The reason for not simply choosing the better classifier (M1 or M2) when not using the safeguard system is that doing so would represent a wasteful result, as knowing after training both classifiers which would perform better is trivial. We therefore maintain that if M2 is trained, then it has to be used.

In this experiment, the complete pipeline was run 32 times with both the AutoGluon and RandomForestClassifier versions of the safeguard system, and out of those 32 runs, 8 runs had worse mean accuracy with the safeguard system used. This includes versions of the safeguard system with less-than-optimally configured settings.

When we compare AutoGluon to scikit-learn by looking at Fig. 4A vs Fig. 4B, we can see that AutoGluon barely makes any "minus" (blue circle) predictions, i.e., it almost never thinks that the performance will decrease, which limits its utility as a reliable predictor.

AutoGluon. AutoGluon's TabularPredictor showed limited effectiveness as a safeguard system, with 3 out of 7 runs resulting in worse performance. The classifier predominantly predicted "plus" (i.e., a performance improvement) regardless of actual outcomes. We attempted to improve its behavior through various

[3] https://developers.google.com/machine-learning/crash-course/classification/accuracy-precision-recall.

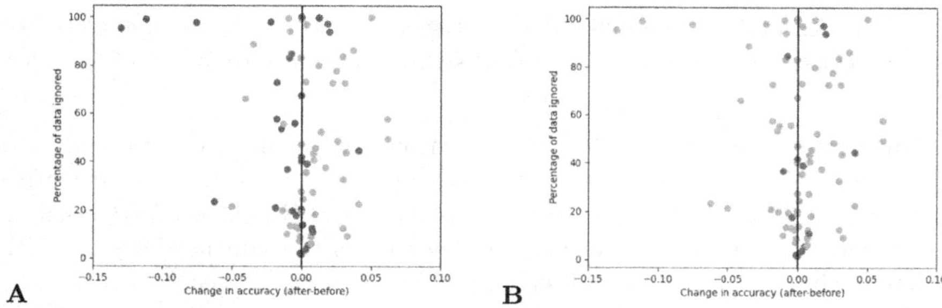

Fig. 4. Percentage of data ignored vs change in accuracy ($accuracy_{M2} - accuracy_{M1}$) for different models: RandomForestClassifier (A) and AutoGluon (B). Each dot represents a dataset. Orange is classifier predicting a positive change, Blue a negative change. Red is an error with the safeguard system.

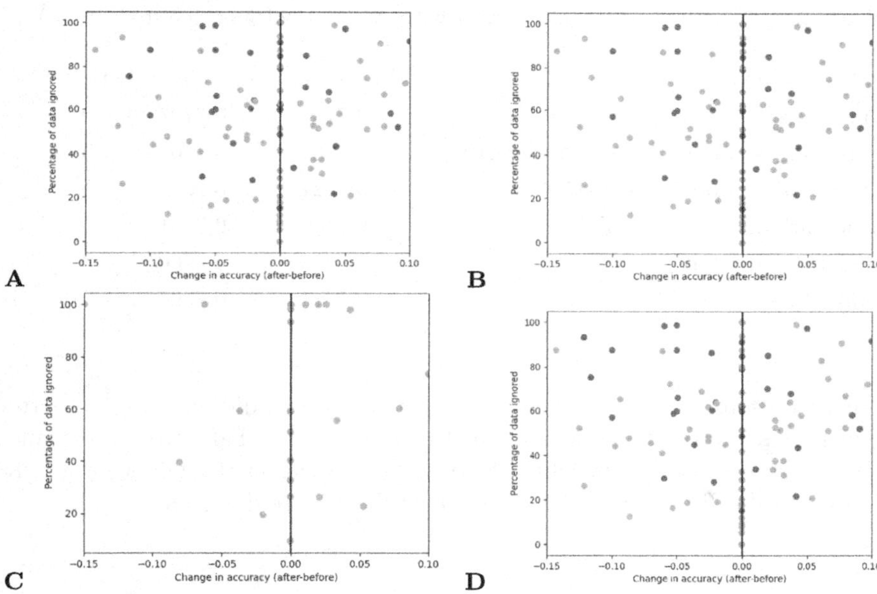

Fig. 5. Percentage of data ignored vs. change in accuracy ($accuracy_{M2} - accuracy_{M1}$) for different metrics. A: Custom metric, heavily weighting false positives; B: f_β metric ($\beta = 2$); C: f_β metric ($\beta = 0.5$); D: F1 metric.

approaches: 1) heavily weighting false positives (Fig. 5A) 2) using f_β scoring with $\beta = 2$ (Fig. 5B) and $\beta = 0.5$ (Fig. 5C) and 3) using F1 scoring (Fig. 5D). We used the same experimental setup: trained on OpenML100 and OpenML-CC18, run on CD3, utilizing model-based meta-features.

None of these changes significantly improved performance. For both landmarking and clustering meta-features, we can even see that the mean performance is identical with and without the safeguard system, with all four being

0.6675. The AutoGluon predictor does not seem to be the optimal solution for our safeguard system; we therefore look at scikit-learn's RandomForestClassifier.

RandomForestClassifier. In this section, we take a look at the performance of the safeguard system implemented using scikit-learn's RandomForestClassifier [8]. For the 25 runs using the RandomForestClassifier, five had worse performance. This includes versions of the safeguard system with less than optimal settings, four of which had worse performance.

Table 2 presents a comparison of mean accuracy achieved with and without the scikit-learn based safeguard system (SfgS) across various meta-feature configurations. In most cases, utilizing the SfgS leads to better performance compared to the baseline, which is the linear ensemble approach without the SfgS.

Table 2. Comparison of mean accuracy with/without safeguard system (SfgS).

Meta-Features	Mean with SfgS	Mean	Is Better?
Clustering	0.7184	0.7103	Y
Clustering + Model-Based (OpenML100)	0.7831	0.7764	Y
Clustering + Model-Based (CD3)	0.7121	0.7019	Y
Landmarking	0.7160	0.7101	Y
Model-Based (Run 1)	0.6953	0.6966	N
Model-Based (Run 2)	0.7918	0.7867	Y

There are some examples where the safeguard system did not perform better in earlier tests, but by changing our code to use bootstrap=False the performance is better. We can see the benefits of no bootstrapping in the following graphs (Fig. 6A and Fig. 6B), where the classifier predicts more datasets correctly.

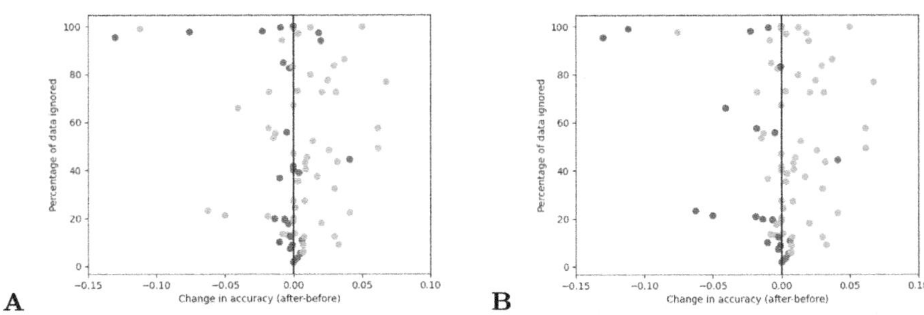

Fig. 6. scikit-learn: Percentage of data ignored vs change in accuracy with (A) and without bootstrapping (B)

5.3 Summary

In summary, scikit-learn's RandomForestClassifier with bootstrap=False proved superior to AutoGluon's approach, offering better prediction accuracy and simpler tuning capabilities. Our safeguard system provides a net benefit for most applications, by avoiding the need for computing M2, with minimal overhead from the RandomForestClassifier. Furthermore, we have found that our approach compares favorably to AutoGluon's built-in semi-supervised learning support.

Finally, we have explored how to predict dataset performance using meta-features and identified model-based and clustering meta-features as the best for our approach.

Overall, our results demonstrate that better performance can be achieved by utilizing unlabeled data, which has been labeled by a separately trained model.

6 Conclusion

We have shown how semi-supervised models can have better performance when trained on small amounts of labeled data. Through our novel approach, known as linear-ensembling, we are able to provide better performance overall, compared to both a supervised learning and AutoGluon's semi-supervised approach, despite not achieving 100% accuracy.

We have investigated which factors could lead to poor performance at the end of a learning process, such as the threshold or AutoGluon's quality configurations and the meta-features of the dataset to be trained on. We have identified clustering and model-based meta-features as particularly meaningful in predicting how a dataset will perform. To take advantage of this information, we have introduced our safeguard system. This system is able to leverage the improved accuracy from our linear-ensembling approach, while reducing computational resources and run times by eliminating unnecessary model training. Consequently, our system is able to minimize energy usage and the associated carbon footprint, while enhancing predictive performance. Ultimately, our findings have shown that the issues with performance loss can be eliminated without time-consuming steps needing to be taken.

Finally, we provide valuable guidance for future research, by identifying which settings and meta-features are the most useful and provide the best results. This includes which factors affect the performance of the safeguard system itself, such as the choice between scikit-learn and AutoGluon. These insights offer a clear direction forward for further research with diverse datasets and applications.

There are several directions for future work. A further improvement and generalization of the safeguard system can include training it on more datasets, tuning the various settings of the RandomForestClassifier, or possibly selecting a different classifier model. While our results on standard benchmarks are promising, the robustness with respect to generalization is an important area for future work. This includes tighter integration of the AutoML model optimization system and the safeguard system, e.g., parameters and models chosen by AutoGluon. Another area of research could be further refining the selection of the meta-features and increasing the robustness of the pymfe system to make it more useful for the safeguard system. This includes removing the need for a time limit, e.g., by calculating only a subset of meta-features to provide a "naive" prediction. A third direction concerns a more hybrid approach, where a human would verify/change some pseudo-labeled results, e.g., the lowest confidence ones, to improve the performance of M2. The fourth direction is expanding our work to non-classifier based labeling systems with AutoML.

References

1. Alcobaça, E., Siqueira, F., Rivolli, A., Garcia, L.P.F., Oliva, J.T., de Carvalho, A.C.P.L.F.: MFE: towards reproducible meta-feature extraction. J. Mach. Learn. Res. **21**(111), 1–5 (2020)
2. Arazo, E., Ortego, D., Albert, P., O'Connor, N.E., McGuinness, K.: Pseudo-labeling and confirmation bias in deep semi-supervised learning. In: Proc IJCNN, pp. 1–8 (2020)
3. Autogluon authors: predictor.py. https://github.com/autogluon/autogluon/blob/0e3bc0e54ab4b0edf865c8b99e1418472006d6b7/tabular/src/autogluon/tabular/predictor/predictor.py. Accessed 26 July 2024
4. Barlow, H.B.: Unsupervised learning. Neural Comput. **1**(3), 295–311 (1989)
5. Bischl, B., et al.: OpenML benchmarking suites (2021)
6. Carlisle, M.: A Boston housing dataset controversy (2019). https://medium.com/@docintangible/racist-data-destruction-113e3eff54a8
7. Chouldechova, A., Roth, A.: The frontiers of fairness in machine learning. arXiv preprint: arXiv:1810.08810 (2018)
8. scikit-learn developers: Randomforestclassifier. https://scikit-learn.org/stable/modules/generated/sklearn.ensemble.RandomForestClassifier.html
9. Ding, F., Hardt, M., Miller, J., Schmidt, L.: Retiring adult: new datasets for fair machine learning. Neurips **34**, 6478–6490 (2021)
10. van Engelen, J.E., Hoos, H.H.: Semi-supervised co-ensembling for automl. In: Trustworthy AI - Integrating Learning, Optimization and Reasoning, pp. 229–250. Springer (2021)
11. Erickson, N., Mueller, J., Shirkov, A., Zhang, H., Larroy, P., Li, M., Smola, A.: Autogluon-tabular: robust and accurate AutoML for structured data. arXiv preprint: arXiv:2003.06505 (2020)
12. Erickson, N., Mueller, J., Shirkov, A., Zhang, H., Larroy, P., Li, M., Smola, A.: autogluon.tabular.TabularPredictor.fit (2024). https://auto.gluon.ai/stable/api/autogluon.tabular.TabularPredictor.fit.html
13. Feurer, M., Klein, A., Eggensperger, K., Springenberg, J., Blum, M., Hutter, F.: Efficient and robust automated machine learning. In: Neurips 28 (2015), pp. 2962–2970 (2015)
14. Fusi, N., Sheth, R., Elibol, M.: Probabilistic matrix factorization for automated machine learning. In: Neurips, vol. 31 (2018)
15. Gijsbers, P., et al.: AMLB: an AutoML benchmark (2023). https://arxiv.org/abs/2207.12560
16. Halevy, A., Norvig, P., Pereira, F.: The unreasonable effectiveness of data. IEEE Intell. Syst. **24**(2), 8–12 (2009)
17. He, J., Gu, J., Shen, J., Ranzato, M.: Revisiting self-training for neural sequence generation. arXiv preprint: arXiv:1909.13788 (2019)

18. Kotlar, M., Punt, M., Radivojević, Z., Cvetanović, M., Milutinovic, V.: Novel meta-features for automated machine learning model selection in anomaly detection. IEEE Access **9**, 89675–89687 (2021). https://doi.org/10.1109/ACCESS.2021.3090936
19. LeCun, Y.: The MNIST database of handwritten digits. http://yann.lecun.com/exdb/mnist/ (1998)
20. Li, Y.F., Wang, H., Wei, T., Tu, W.W.: Towards automated semi-supervised learning. In: Proceedings of the AAAI Conf. on AI, vol. 33, pp. 4237–4244 (2019)
21. Maas, A.L., Daly, R.E., Pham, P.T., Huang, D., Ng, A.Y., Potts, C.: Learning word vectors for sentiment analysis. In: Proc. 49th Annual Meeting of the Association for Computational Linguistics, pp. 142–150 (2011)
22. Murphy, K.P.: Machine Learning: A Probabilistic Perspective. MIT press. (2012)
23. Pedregosa, F., et al.: Scikit-learn: machine learning in Python. J. Mach. Learn. Res. **12**, 2825–2830 (2011)
24. Leibnitz Rechenzentrum: Available slurm clusters and features. https://doku.lrz.de/available-slurm-clusters-and-features-11483939.html
25. Rivolli, A., Garcia, L.P., Soares, C., Vanschoren, J., de Carvalho, A.C.: Towards reproducible empirical research in meta-learning, pp. 32–52. arXiv preprint: arXiv:1808.10406 (2018)
26. Rivolli, A., Garcia, L.P.F., Soares, C., Vanschoren, J., de Carvalho, A.C.P.L.F.: Characterizing classification datasets: a study of meta-features for meta-learning. arXiv preprint: arXiv:1808.10406 (2019)
27. Schwartz, R., et al.: Towards a standard for identifying and managing bias in artificial intelligence. NIST Spec. Publ. **1270**(10.6028) (2022)
28. Scudder, H.: Probability of error of some adaptive pattern-recognition machines. IEEE Trans. Inf. Theory **11**(3), 363–371 (1965)
29. Thai-Nghe, N., Gantner, Z., Schmidt-Thieme, L.: Cost-sensitive learning methods for imbalanced data. In: Proc. IJCNN, pp. 1–8. IEEE (2010)
30. Verma, S., Ernst, M., Just, R.: Removing biased data to improve fairness and accuracy. arXiv preprint: arXiv:2102.03054 (2021)
31. Yao, Q., et al.: Taking human out of learning applications: a survey on automated machine learning. arXiv preprint: arXiv:1810.13306 (2018)
32. Thornton, C., Hutter, F., Hoos, H.H., Leyton-Brown, K.: Auto-WEKA: combined selection and hyperparameter optimization of classification algorithms. In: Proc. of KDD-2013, pp. 847–855 (2013)
33. Mohr, F., Wever, M., Hüllermeier, E.: ML-Plan: automated machine learning via hierarchical planning. Mach. Learn. **107**(8–10), 1495–1515 (2018). https://doi.org/10.1007/s10994-018-5735-z
34. Lee, D.-H., et al.: Pseudo-label: the simple and efficient semi-supervised learning method for deep neural networks. In: Workshop on Challenges in Representation Learning, ICML, Atlanta, vol. 3, no. 2, p. 896 (2013)

Caïssa AI: A Neuro-Symbolic Chess Agent for Explainable Move Suggestion and Grounded Commentary

Mazen Soliman and Nourhan Ehab(✉)

Department of Computer Science and Engineering, German University in Cairo, Cairo, Egypt
mazen.elsayed@student.guc.edu.eg, nourhan.ehab@guc.edu.eg

Abstract. Despite the impressive generative capabilities of large language models (LLMs), their lack of grounded reasoning and susceptibility to hallucinations limit their reliability in structured domains such as chess. We present Caïssa AI, a neuro-symbolic chess agent that augments LLM-generated move commentary with symbolic reasoning, knowledge graph integration, and verification modules. Caïssa AI combines a fine-tuned chess-specific LLM with a Prolog-based rule engine encoding chess tactics and rules, along with a dynamically constructed Neo4j knowledge graph representing the current board state. This hybrid architecture enables the system to generate not only accurate move suggestions but also coherent, strategically grounded commentary. A LangGraph-based verification module cross-checks LLM outputs against symbolic logic to ensure consistency and correctness, effectively mitigating hallucinations. By aligning data-driven generation with formal domain knowledge, Caïssa AI enhances both trustworthiness and explainability. Our results demonstrate that this tight neuro-symbolic integration produces verifiable, high-quality commentary and serves as a generalizable blueprint for AI systems requiring real-time, interpretable decision support.

Keywords: Neuro-Symbolic AI · Chess Agents · Explainable Reasoning

1 Introduction

Large Language Models (LLMs) have recently achieved remarkable performance in natural language understanding and generation across diverse tasks, including dialogue systems, code generation, and domain-specific reasoning [1,13,16]. In structured domains such as chess, fine-tuned LLMs have been applied to generate move predictions and natural commentary by leveraging large corpora of annotated games [18]. Despite their fluency, however, LLMs continue to exhibit critical limitations: they often lack grounding in formal domain knowledge, fail to reason about structured environments [9], and are prone to producing plausible-sounding but incorrect outputs—a phenomenon known as hallucination [4,11].

T. Braun et al. (Eds.): KI 2025, LNAI 15956, pp. 148–160, 2026.
https://doi.org/10.1007/978-3-032-02813-6_11

These limitations are particularly problematic in chess, where precision, tactic awareness, and explainability are essential. While symbolic engines such as Stockfish [15] provide accurate evaluations based on search trees and evaluation functions, they often lack natural explanation capabilities. Conversely, LLM-based systems are capable of generating fluent commentary, yet their suggestions may be logically unsound or tactically invalid [4]. Prior approaches have attempted to align LLM outputs with external chess engines or to inject symbolic tags into language models [9], but these methods remain weakly coupled and fail to enforce logical consistency at runtime.

To address these challenges, we propose *Caïssa AI*[1], a neuro-symbolic chess agent that combines the generative fluency of LLMs with the formal rigor of symbolic AI. Our architecture integrates a fine-tuned LLM (ChessGPT) [18] with a Prolog-based rule engine encoding chess tactics and rules, along with a dynamically constructed knowledge graph in Neo4j [19]. This hybrid design allows Caïssa AI to generate fluent, context-aware commentary while grounding its outputs in explicit, queryable knowledge representations. To mitigate hallucinations and ensure correctness, we introduce a LangGraph-based verification module that cross-checks LLM-generated explanations against the symbolic reasoning pipeline. By aligning data-driven generation with symbolic logic, Caïssa AI improves factual accuracy, interpretability, and user trust—an increasingly important goal for next-generation AI systems [3,7]. Moreover, our architecture provides a generalizable blueprint for combining LLMs with symbolic agents in domains requiring structured decision-making and real-time feedback.

The key contributions of this work are as follows:

1. We introduce *Caïssa AI*, a novel neuro-symbolic architecture for explainable chess commentary that tightly integrates LLMs, Prolog-based symbolic reasoning, and dynamic knowledge graphs.
2. We propose a verification pipeline that validates LLM outputs using formal reasoning over Prolog rules and graph queries, reducing hallucinations and logical inconsistencies.
3. We demonstrate that our system can generate accurate, interpretable commentary across diverse chess positions and tactics, outperforming baseline LLM approaches.
4. We discuss the broader implications of our architecture for grounding LLMs in domains requiring trustworthy and explainable AI.

The remainder of this paper is organized as follows. Section 2 reviews existing sub-symbolic, symbolic, and hybrid approaches for chess commentary, highlighting their interpretability challenges. Section 3 introduces our neuro-symbolic architecture that combines symbolic reasoning with LLM-based generation to minimize hallucinations and enhance explainability. Section 4 presents the evaluation of the presented architecture, and Sect. 5 concludes the paper with directions for future work.

[1] Caïssa is a fictional Thracian dryad portrayed as the goddess of chess. She was first mentioned during the Renaissance by Italian poet Hieronymus Vida.

2 Related Work

In this section, we review prior work on AI-generated chess commentary, organized into three main categories: symbolic logic-based systems, sub-symbolic neural language models, and hybrid approaches that integrate symbolic and neural reasoning.

2.1 Purely Symbolic Logic-Based Approaches

Purely symbolic systems offer a rule-based foundation for reasoning about chess moves. Logic programming languages like Prolog are particularly well-suited for representing chess rules and tactics in a transparent and interpretable manner.

In a notable example [12], a chess coaching system was developed using a logic programming framework where the AI could explain its moves based on predefined chess strategies. The system allowed users to query the AI for justifications, which were generated from a set of first-order logic rules. If users disagreed, they could provide counterarguments that dynamically updated the knowledge base. This argumentation framework provided a glass-box approach, offering full transparency into the reasoning process behind each move. Arguments for and against a particular action were evaluated using rule-based inference, and user-defined rules could override default knowledge, creating an adaptive learning environment.

Despite their strengths in interpretability, symbolic systems are limited by the need for exhaustive rule sets and predefined tactics. They often struggle to adapt to complex, dynamic game states without substantial human input. These limitations have driven the rise of more flexible sub-symbolic approaches based on neural networks.

2.2 Sub-symbolic Approaches

Sub-symbolic models, particularly those powered by large language models (LLMs), have recently been applied to chess tasks such as move prediction and commentary generation. Examples include ChessGPT-Base and ChessGPT-Chat [18], fine-tuned versions of the RedPajama-3B-base model—a replication of Meta's LLaMA [16]. These models are trained on large-scale chess corpora using PGN and FEN encodings.

ChessGPT-Base is trained to predict plausible moves, while ChessGPT-Chat uses supervised instruction tuning to provide interactive explanations. Despite their fluency, these models often suffer from critical flaws: lack of rule enforcement, tactical oversights, and hallucinated explanations. As shown in Fig. 1 and Table 1, ChessGPT-Chat incorrectly references a knight on `c6` that does not exist—highlighting a fundamental limitation in its reasoning capabilities.

While these models can explain basic concepts such as check and capture, they struggle with deeper tactics like forks and skewers and often fail to maintain an accurate model of the board state over multiple moves. This reflects a

Fig. 1. A chessboard representing that moving the white rook from d6 to d8 would capture the black queen at d8

Table 1. Example of hallucination in ChessGPT-Chat response

Prompt	Consider the board: 1r1q1rk1/p1p2ppp/2bR1n2/8/8/B1N2N2/P4PPP/5RK1 w - 0 16. Reason why move d6d8 is considered the best move?
ChessGPT-Chat Response	The move d6d8 is considered the best move because it saves the rook by temporarily pinning the opponent's knight on c6 to the queen on c7
Answer State	Hallucinated and tactically incorrect

broader limitation of sub-symbolic systems: they are statistically grounded but lack structured reasoning.

To mitigate these issues, lightweight hallucination detection mechanisms have been proposed, such as SelfCheckGPT [11] and LLM-Augmenter [14]. Other work explores retrieval-augmented generation or external engine filtering. However, these methods offer only shallow coupling between the language model and domain logic. More structured integration is required for verifiable and trustworthy commentary—an aim pursued by hybrid neuro-symbolic systems.

2.3 Hybrid Neuro-Symbolic Approaches

Recognizing the limitations of sub-symbolic systems, researchers have explored hybrid approaches that combine large language models with symbolic reasoning engines. These systems aim to enhance the interpretability and accuracy of AI-generated commentary by grounding LLM outputs in formal, logic-based representations of chess rules.

One such hybrid approach [9] involves augmenting LLMs with control codes from a chess engine to guide commentary generation. In this setup, a reasoning engine evaluates the current game state and generates control codes such as move quality or suggested moves. The LLM then uses these codes to produce commentary that aligns with the tactical situation on the board. This approach introduces a layer of symbolic reasoning to filter and structure the LLM's outputs.

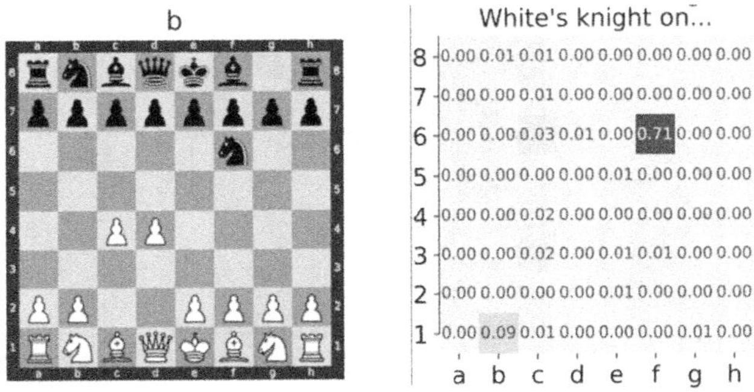

Fig. 2. The model strongly believes that f6 contains a white knight despite the fact that it contains a black knight [9].

While this method improves the quality of commentary by reducing hallucinations, it is still prone to errors. For instance, when asked about a specific piece's location or role, the LLM may still produce outputs that conflict with the actual game state. As shown in Fig. 2, the model wrongly identified the position of a knight, leading to a logically flawed commentary that misrepresented the situation on the board.

A recent contribution by Kim et al. [8] is the Concept-guided Chess Commentary (CCC) framework, which uses concept vectors extracted from expert engines like Stockfish [15] and LeelaChessZero [6] to guide LLMs in generating fluent and strategically relevant commentary. CCC also introduces GCC-Eval, a domain-aware evaluation framework aligned with expert judgments. While CCC improves fluency and reduces hallucinations through concept-based prompting, it lacks live symbolic reasoning and verifiable logic tracing. In contrast, our proposed system, *Caïssa AI*, integrates a Prolog-based rule engine and a Neo4j knowledge graph to provide real-time, explainable commentary grounded in formal reasoning.

3 Caïssa AI: A Neuro-Symbolic Chess Commentary System

The shortcomings of both symbolic and sub-symbolic approaches in chess AI have motivated the development of neuro-symbolic systems that seek to combine the strengths of both paradigms [17]. Symbolic reasoning engines offer structure, correctness, and explainability, but are brittle and inflexible. Sub-symbolic systems such as large language models (LLMs) excel in fluency and pattern recognition, but struggle with logical consistency and factual grounding.

Caïssa AI[2] is a neuro-symbolic framework designed to overcome these limitations. It integrates a Prolog-based symbolic reasoning engine, a dynamically

[2] Source code for the presented implementation can be found here.

constructed knowledge graph in Neo4j, and a fine-tuned LLM (Gemini) orchestrated using LangChain and LangGraph. Together, these components enable the system to provide both accurate move suggestions and interpretable commentary grounded in chess tactics and rules.

Unlike prior approaches that focus only on move prediction or rely solely on sub-symbolic generation, *Caïssa AI* generates rich natural language commentary that is explicitly tied to formal knowledge. It supports user queries related to piece positions, tactical relations, move outcomes, and commentary justification—while verifying the consistency of its responses using symbolic validation modules.

In the following sections, we describe each component of the architecture, the process of knowledge representation and transformation, the pipeline for commentary generation, and the verification mechanism that reinforces correctness and minimizes hallucinations.

3.1 Symbolic Reasoning and Knowledge Representation

At the heart of *Caïssa AI* is a symbolic module implemented in Prolog [2], which encodes chess rules, tactical patterns, and piece relationships in a queryable, logic-based format. This module provides a ground truth engine capable of analyzing board states, validating move legality, and identifying tactical opportunities.

The symbolic module is composed of three components:

1. **Chess Rules**: Defines standard movement rules, including legal piece behavior, check conditions, special rules like castling and en passant, and checkmate detection.
2. **Piece Relationships**: Encodes spatial and defensive relationships among pieces. For example, whether a rook defends a pawn or threatens an opposing piece.
3. **Tactical Patterns**: Detects patterns such as forks, skewers, discovered attacks, and pins, which are critical for high-quality chess commentary.

To enable interaction between the symbolic layer and the LLM, we transform Prolog-derived facts into a dynamic knowledge graph built in Neo4j. Each board state is encoded as a graph structure, with pieces and squares represented as nodes, and tactical or positional relationships as labeled edges. Unlike static ontologies, our dynamic graphs adapt to each move and allow fine-grained querying of state-dependent chess knowledge.

The knowledge graph schema is summarized below:

- **Nodes**
  - *Piece*: Represents a chess piece on the board (e.g., `WhiteRook1`).
  - *Square*: Represents a chessboard square in UCI format (e.g., `d1`).
- **Relationships**
  - *Locate*: Links a piece to its current square.
  - *Suggest*: Suggests a possible move, possibly annotated with a tactic.

- *Threat, Defend, Skewer, Pin, Fork, DiscoveredAttack, HangingPiece, Mate, MateIn2, Interference*: Represent tactical relations between pieces.
- *Feature*: Captures properties of move outcomes, such as:
  * `move_defend`: Move results in defending an ally.
  * `move_is_attacked`: Move ends on a square attacked by an opponent.
  * `move_threat`: Move creates a threat on an opponent piece.
  * `move_is_protected`: Move ends on a square defended by an ally.

This symbolic-graph transformation enables structured retrieval of chess logic through Cypher queries. For instance, the system can return a list of all hanging pieces, identify discovered attacks for a specific piece, or determine which pieces are threatening the king.

Figure 3 illustrates the architecture of the symbolic-to-graph transformation and how it connects with the LLM-based commentary agent.

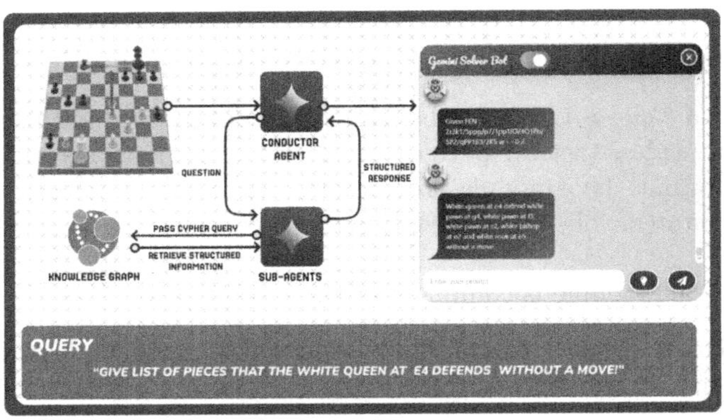

Fig. 3. Chess Commentary Generation Diagram: The symbolic knowledge is transformed into a dynamic graph and used to support commentary generation.

3.2 Generative AI and Commentary Pipeline

To generate fluent, context-aware commentary from symbolic knowledge, *Caïssa AI* integrates a large language model (Gemini) into a multi-agent orchestration pipeline using LangChain. The system functions as a Retrieval-Augmented Generation (RAG) framework, where commentary is grounded in facts retrieved from the Neo4j knowledge graph derived from Prolog.

LLM Agent Architecture. The commentary pipeline is composed of two LangChain-based Gemini agents, each configured using prompt engineering:

- **Chess Expert Agent:** Interprets user queries, generates natural language responses, and delegates Cypher-based retrieval tasks.
  Prompt: You are a chess expert providing information about chess strategies and tactics.

- **Cypher Developer Agent:** Translates queries into Cypher to retrieve symbolic knowledge from Neo4j.
  Prompt: You are an expert Neo4j developer translating user questions into Cypher to answer questions about chess tactics, strategies, positions, and moves.

LangChain's `GraphCypherQAChain` enables the Chess Expert to invoke the Cypher agent, execute queries on the Neo4j graph, and then translate the returned structure into commentary. Chain-of-Thought prompting [5] is used to guide the LLM through reasoning steps when synthesizing the final response.

Commentary Generation Example. Given the board in Fig. 4, the system can accurately retrieve and describe positional and relational information (Table 2 and 3).

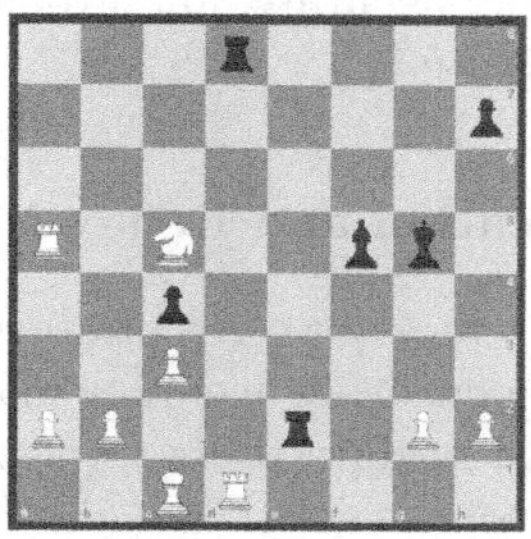

Fig. 4. The state of the chessboard in our example.

Table 2. Sample positional query

Question	Where is the position of the white king?
Response	The position of the white king is c1.
Answer State	Correct

Table 3. Example of fluent tactical commentary

Question	Please give commentary for the move: black rook at e2 to c2
Response	The black rook move from e2 to c2 is attacking a white king and a white pawn. The black bishop on f5 protects the move of the black rook from e2 to c2
Answer State	Correct

Multi-step Reasoning and Prompt Engineering. The system supports multi-step reasoning by decomposing complex questions into sub-tasks. LangChain's agent chaining enables the Chess Expert agent to first retrieve relevant symbolic facts (via Cypher) and then formulate grounded commentary. Chain-of-Thought prompting further encourages the LLM to think through intermediate facts before forming conclusions.

For example, a query like:
"What are the threats if the white knight at c5 moves to d3?"
is broken into:

1. Find pieces that defend or are threatened by c5.
2. Simulate c5→d3 in the knowledge graph.
3. Retrieve new threats.
4. Translate the result into coherent commentary.

In summary, this generative pipeline enables *Caïssa AI* to generate commentary that is fluent, contextually accurate, and grounded in a structured, tactical understanding of the board state.

3.3 Commentary Verification and Reflex Mechanism

While the generative pipeline in *Caïssa AI* produces fluent and grounded commentary, challenges remain in ensuring accuracy and consistency. Hallucinations may still occur when the LLM misinterprets a question, or when Cypher queries return incomplete or ambiguous results. To address this, we introduce a dedicated verifier module powered by LangGraph that analyzes and validates the generated commentary before returning it to the user.

Verification Pipeline Overview. LangGraph enables multi-agent workflows coordinated over a state variable. Our verifier pipeline is structured into two sequential phases, each containing modular agents specialized in interpreting, decomposing, validating, and refining commentary.

The first phase, illustrated in Fig. 5, identifies the type of commentary to be verified:

1. The **Verifier Agent** receives the commentary and determines its type:
   - *Piece Position Commentary*
   - *Piece Relation Commentary*
   - *Move With Feature Commentary*
2. Based on the type, a corresponding module is selected to validate the statements using symbolic knowledge.

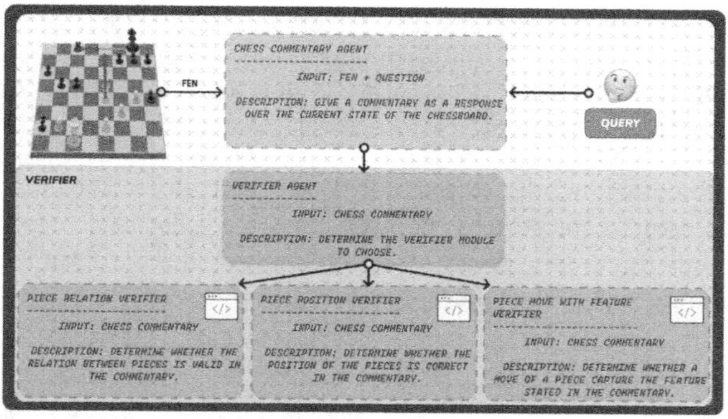

Fig. 5. Verifier Phase 1: Commentary type detection and module selection.

Structured Commentary Validation. The second phase, shown in Fig. 6, verifies the content of the commentary through a multi-step process:

1. **Micro-Agent 1:** Decomposes complex commentary into individual statements.
2. **Micro-Agent 2:** Uses the KOR library to extract structured data (JSON) from each statement.
3. **Symbolic Module:** Validates the structured data against Prolog or Neo4j, filtering out incorrect facts and augmenting missing details (e.g., colors, positions).
4. **Micro-Agent 3:** Rewrites the validated statements in natural language with corrected or enriched content.

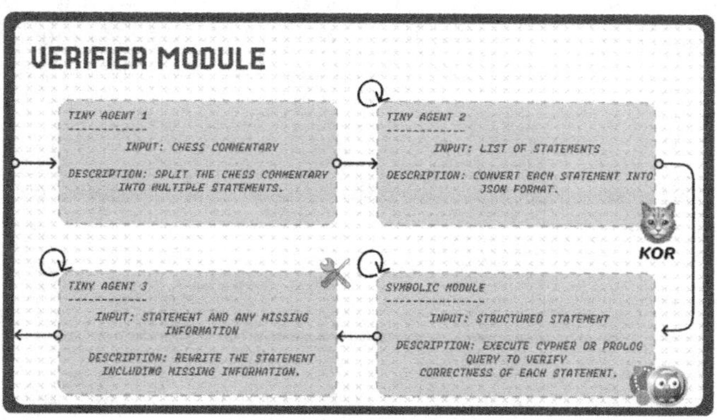

Fig. 6. Verifier Phase 2: Multi-agent commentary validation and refinement

Reflex Loop and Error Recovery. Figure 7 illustrates the full architecture. A new **Main Agent** receives the user query and delegates it to a **Reinforced Agent**, which combines the Commentary Agent and the Verifier Agent.

The verifier may take one of two actions:

- If the commentary is fully or partially correct, it filters out inaccurate statements and returns the rest to the user.
- If the commentary is incorrect, the verifier provides structured feedback to the Commentary Agent and triggers a regeneration step.

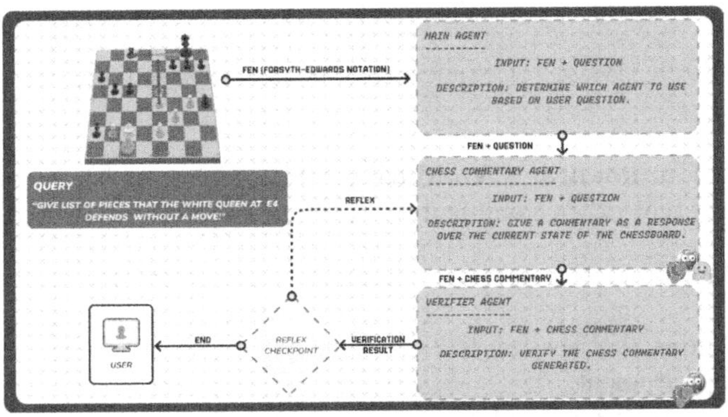

Fig. 7. Full commentary pipeline with reflex-based reinforcement.

LangGraph State Tracking. LangGraph provides a global state variable shared across agents. This state evolves during the execution and controls branching logic and data flow. As shown in Fig. 8, it acts as a memory structure that enables coordination between agents and dynamic decision-making across the pipeline.

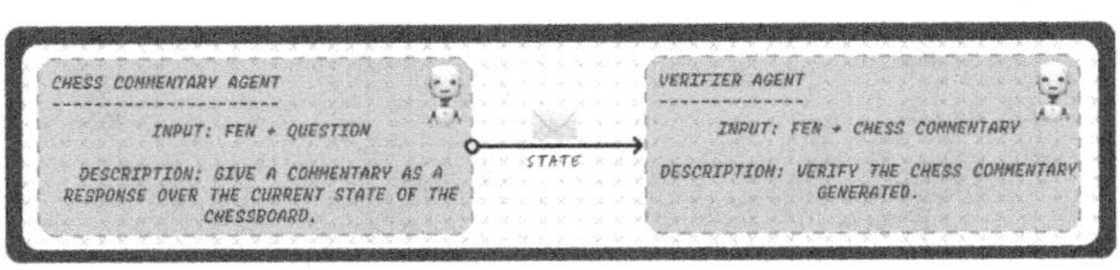

Fig. 8. LangGraph shared state structure across verifier agents.

In summary, this verification layer significantly improves the consistency and reliability of Caïssa AI's output by combining symbolic validation, structured refinement, and reflexive feedback to the generation process.

4 Results

We evaluated the performance of the symbolic reasoning module in *Caïssa AI* using a curated set of 233 chess puzzles from the Lichess tactics dataset [10]. Each puzzle contains a FEN string representing the board state and a corresponding set of tactical themes.

For every puzzle, the optimal move listed in the dataset was applied to the initial FEN using the `python-chess` library. The resulting position was passed to the symbolic module, implemented in Prolog, which attempted to extract all detectable tactics present in the new board state. The system's output was compared to the themes listed in the dataset for that puzzle.

A prediction was considered a **hit** if at least one of the detected tactics matched any of the themes specified for the corresponding puzzle.

The symbolic module achieved:

- **Hits:** 217 out of 233 puzzles matched at least one theme listed in the dataset.
- **Accuracy:** 93.13% hit rate.

This high hit rate demonstrates the symbolic module's robustness in capturing core tactical patterns such as `mate`, `mateIn2`, `fork`, and `skewer`.

While performance was generally strong, some limitations were observed. In a few cases, the symbolic module failed to detect any of the expected themes. Occasionally, it produced additional valid tactics not listed in the dataset's theme annotations. These discrepancies are partly due to the incomplete or coarse-grained labeling in the dataset and the symbolic engine's broader reasoning scope, which can uncover compound or implicit tactical motifs not explicitly captured in the dataset.

5 Conclusion and Future Work

Caïssa AI is a neuro-symbolic chess commentary system that integrates symbolic reasoning with large language models to generate accurate and explainable move suggestions. Its architecture combines a Prolog rule engine, a Neo4j knowledge graph, and a multi-agent generative pipeline using Gemini and LangChain. A retrieval-augmented generation (RAG) pipeline and a LangGraph-based verifier help reduce hallucinations and ground commentary in formal chess knowledge. Evaluation on the Lichess tactics dataset shows that the symbolic engine alone achieves over 93% accuracy in tactic detection.

Despite its strengths, limitations include semantic errors from the LLM, coordination issues between agents, and residual hallucinations in edge cases. Future improvements will target better multi-agent alignment, extended tactic coverage, and integration of user feedback. Ultimately, the system aims to serve as an adaptive, interactive tutoring agent capable of natural dialogue and personalized, explainable chess guidance.

References

1. Brown, T., Mann, B., Ryder, N., et al.: Language models are few-shot learners. In: NeurIPS (2020)
2. Colmerauer, A., Roussel, P.: The birth of prolog. In: History of Programming Languages, pp. 331–367 (1973)
3. Gunning, D.: Explainable artificial intelligence (xai). Defense Advanced Research Projects Agency (DARPA), Program Information (2017)
4. Hammond, K., Leake, D.: Large language models need symbolic ai. In: Proceedings of the 17th International Workshop on Neural-Symbolic Learning and Reasoning, La Certosa di Pontignano, Siena, Italy, vol. 3432, pp. 204–209 (2023)
5. Schuurmans, D., et al.: Chain-of-thought prompting elicits reasoning in large language models. Adv. Neural Inf. Process. Syst. **35**, 24824–24837 (2022)
6. Jenner, E., Kapur, S., Georgiev, V., Allen, C., Emmons, S., Russell, S.: Evidence of learned look-ahead in a chess-playing neural network. arXiv preprint arXiv:2406.00877 (2024)
7. Kautz, H. The third AI summer: a perspective on the future of AI. Communications of the ACM, 2020
8. Kim, J., Goh, J., Hwang, I., Cho, J., Ok, J.: Bridging the gap between expert and language models: concept-guided chess commentary generation and evaluation. arXiv preprint arXiv:2410.20811 (2024)
9. Lee, A., Wu, D., Dinan, E., Lewis, M.: Improving chess commentaries by combining language models with symbolic reasoning engines. arXiv preprint arXiv:2212.08195 (2022)
10. Lichess.org. Lichess tactics puzzle dataset (2025). https://database.lichess.org/. Accessed 14 Apr 2025
11. Manakul, N., Gales, M.: Selfcheckgpt: zero-resource hallucination detection for generative large language models. Trans. Assoc. Comput. Linguist. (2023)
12. Markos, V.T.: Application of the machine coaching paradigm on chess coaching (2021)
13. OpenAI. Gpt-4 technical report (2023)
14. Peng, B., et al.: Check your facts and try again: improving large language models with external knowledge and automated feedback. arXiv preprint arXiv:2302.12813 (2023)
15. Stockfish Developers. Stockfish: An open-source chess engine (2024). https://stockfishchess.org. Accessed Apr 2025
16. Touvron, H., et al.: Llama: open and efficient foundation language models. arXiv preprint arXiv:2302.13971 (2023)
17. Wang, W., Yang, Y., Wu, F.: Towards data-and knowledge-driven artificial intelligence: a survey on neuro-symbolic computing. arXiv preprint arXiv:2210.15889 (2022)
18. Wang, Y., et al.: Chessgpt: bridging language models and chess engines. arXiv preprint arXiv:2306.12948 (2023)
19. Webber, J.: A programmatic introduction to neo4j. In: Proceedings of the 3rd Annual Conference on Systems, Programming, and Applications: Software for Humanity (SPLASH '13), pp. 217–218. ACM (2013)

Unsupervised Selection of Features by Their Resilience to the Curse of Dimensionality

Maximilian Stubbemann[1], Tobias Hille[2], and Tom Hanika[2,3(✉)]

[1] Information Systems and Machine Learning Lab., University of Hildesheim, Hildesheim, Germany
stubbemann@ismll.de

[2] Knowledge and Data Engineering Group, University of Kassel, Kassel, Germany
{hille,hanika}@cs.uni-kassel.de

[3] Intelligent Information Systems, University of Hildesheim, Hildesheim, Germany
hanika@uni-hildesheim.de

Abstract. Real-world datasets are often of high dimensionality and affected by the curse of dimensionality. This hinders their comprehensibility and interpretability. To reduce the complexity, feature selection aims to identify features that are crucial to learn from said data. While measures of relevance and pairwise similarities are commonly used, the curse of dimensionality is rarely incorporated into the process of selecting features. Here we step in with a novel unsupervised method that identifies the features that allow to discriminate data subsets of different sizes. By adapting recent work on computing intrinsic dimensionalities, our method is able to select the features that can discriminate data and thus weaken the curse of dimensionality. Our experiments show that our method is competitive and commonly outperforms established unsupervised feature selection methods. Furthermore, we propose an approximation that allows our method to scale to datasets consisting of millions of data points. Our findings suggest that features that discriminate data and are connected to a low intrinsic dimensionality are meaningful for learning procedures.

Keywords: Curse of Dimensionality · Feature Selection · Intrinsic Dimension

1 Introduction

Contemporary datasets are of high dimension and their complexity continues to increase. Thus, information derived by machine learning procedures is rarely comprehensible or interpretable. Addressing this problem, a variety of works aims to simplify data through methods that reduce size or, in particular, the dimensionality. An important class of such methods has the emblematic name *feature selection* (FS). In contrast to feature reduction methods such as principal

T. Braun et al. (Eds.): KI 2025, LNAI 15956, pp. 161–174, 2026.
https://doi.org/10.1007/978-3-032-02813-6_12

component analysis, they preserve original dimensions by simply discarding feature dimensions. Although there are supervised and unsupervised methods for FS, only unsupervised procedures are suited to improve the understanding of data regardless of a specific learning task. Moreover, they do not require costly label information.

One important phenomenon which underlies the complexity of high dimensional data is the *curse of dimensionality*. The curse of dimensionality is a broad term which was first coined by Bellman [2,3]. It is used for different phenomenons that arise in the context of high-dimensional data. In this work, we follow the definition of Pestov [21]. According to Pestov, the curse of dimensionality describes the situation where features are concentrated to specific regions and therefore do not allow to discriminating different data points. Data that is strongly affected by this phenomenon is considered to be of high *intrinsic dimension* (ID). Yet, in common feature selection methods, the curse of dimensionality is rarely explicitly accounted for, even though it does a) demonstrably influences learning success in high-dimensional real-world scenarios, and b) potentially prevent common FS methods to choose the best features to learn from. Thus, it is crucial to incorporate the intrinsic dimension into the process of feature selection.

We meet this challenge by proposing an unsupervised feature selection method that picks features based on their measurable ability to discriminate different data points. While other methods are based on feature relevance measures, such as variances, or select features by discarding strongly correlated ones, our method ranks features by their resilience against the curse of dimensionality in the sense of Pestov. For this, we build on recent work [14,26] which studies intrinsic dimensionality for *geometric datasets*. By adapting corresponding ideas for computing the intrinsic dimensionality, one can derive an algorithm that allows to find precisely those features that are able to tame the influence of the curse of dimensionality. Furthermore, adapting the speed-up techniques proposed by Stubbemann et al. [26] will allow feature selection in settings with large-scale datasets comprised of millions of data points. While [14] and [26] are focused on computing the intrinsic dimensionality of datasets, we will rank and select individual features by their ability to discriminate data points. Thus, we identify features that are harmed by the dimension curse by a comparable low extent.

We experimentally show on real-world datasets that features selected by our proposed method are meaningful to learn from. To be more specific, we experiment on the *OpenML Numerical Tabular Classification Benchmark* (OpenML)[1] and witness that our method is competitive or even outperforms established feature selection methods with respect to feature selection for classification tasks. Furthermore, we use the *Open Graph Benchmark* [15] to show that our method is capable of feature selection for large-scale learning with *Graph Neural Networks*. To sum up, our experiments indicate that features that are able to discriminate data and thus weaken the curse of dimensionality are highly rel-

[1] https://github.com/LeoGrin/tabular-benchmark.

evant for learning tasks. Our code is available on Github.[2]. Our contributions include the following.

- We introduce a novel feature selection procedures which ranks features by their ability to limit the intrinsic dimensionality of a dataset.
- We an approximation method which allows to apply our feature selection technique to datasets with millions of data points.
- We experimentally show that our feature selection method is competitive and that our approximation technique leads to accurate feature rankings.

2 Related Work

We focus on unsupervised feature selection, which is an established research topic [1,5,25]. Established methods are often focused on the identification of outstanding features by identifying the relevant features for clustering [4,6,7]. Another direction focuses on selecting a subset of important and dissimilar features [9,13,18,27]. Here, feature importance can for example be measured via variances and similarities via correlation coefficients. While applying such methods may result in a decreasing intrinsic dimensionality, they do not explicitly account for the curse of dimensionality in the selection process.

A small amount of research has incorporated intrinsic dimension estimators into feature selection [8,10,12,19]. However, these methods either only use the ID to determine the amount of features to select [12] or they are based on recomputing intrinsic dimensionalities after discarding features which limits scalability [10,19]. Even if applicable to larger datasets, these methods build on notions of intrinsic dimensionality that do not aim to quantify the curse of dimensionality [8]. Instead, these notions work under the assumption that the data lies on a manifold of lower Euclidean dimension. The goal of the ID notion is then to approximate the dimension of this manifold.

In contrast, we build our feature selection procedure on a notion of ID which quantifies the influence of the curse of dimensionality which occurs when data points can not be discriminated. In the works of Pestov [20–22], all $1-$Lipschitz functions are considered as potential features to discriminate data, which often hinders the practical computation. This was tackled by defining intrinsic dimensionality for *geometric datasets* [14], where a set of features to consider can be defined beforehand. The computation and approximation on large-scale real world data has recently been made possible [26]. We use these findings to incorporate Pestov's notions of the curse of dimensionality in the process of unsupervised feature selection. Note, that earlier works [14,26] were solely focused on the intrinsic dimensionality of datasets and thus of the question of how all features together can discriminate data points. In contrast, we present a novel approach to rank and select individual features by their ability to discriminate data points.

[2] https://github.com/mstubbemann/FSCOD.

3 Feature Selection via Discriminability

We present our work in a general setting for datasets, i.e., for *geometric datasets* $\mathcal{D} = (X, F, \mu)$ [14], where X is a set of *data points* and $F \subseteq \mathbb{R}^X$ is a set of *feature functions* or *features* which are defined as maps from X to $\mathbb{R}$. We require $\sup_{x,y\in X} d_F(x, y) < \infty$, where $d_F(x, y) := \sup_{f\in F}|f(x) - f(y)|$. Furthermore, (X, d_F) has to be a complete and separable metric space with μ being a Borel probability measure on (X, d_F).

From this point on, we consider the special case of *finite geometric datasets*, i.e., $0 < |X|, |F| < \infty$, with μ being the normalized counting measure. Real-world Euclidean data as commonly studied in machine learning is often represented by a *data matrix* $\mathcal{X} \in \mathbb{R}^{m\times n}$ where the rows represent the attribute vectors. Such data matrices can be represented by a finite geometric dataset (X, F, μ) with μ being the normalized counting measure, $X := \{1, \ldots, m\}$ and F being the projections on the columns, i.e., $F := \{i \mapsto \mathcal{X}_{i,j} \mid j \in \{1, \ldots, n\}\}$.

Hanika et al. [14] derived an axiomatization for functions to be *dimension functions* of geometric datasets. This axiomatization is based on the concentration of measure phenomenon [11,16,17] and its linkage to the curse of dimensionality by Pestov [20–23]. The latter defines the curse of dimensionality as the phenomenon of features concentrating near their means or medians and not being able to discriminate data. To quantify this phenomenon the *partial diameter* is used, which determines to which extent a feature can discriminate sets of a specific measure α. For finite geometric datasets the partial diameter has the form

$$\text{PartialDiameter}(f, 1-\alpha)_{\mathcal{D}} = \min_{\substack{M\subseteq X\\|M|=c_\alpha}} \max_{x,y\in M} |f(x) - f(y)|,$$

where $c_\alpha := \lceil |X|(1-\alpha)\rceil$. Based on this, the *observable diameter* reflects how the feature set F can discriminate data of a specific measure. More formally, $\text{ObservableDiameter}(\mathcal{D}, -\alpha) := \sup_{f\in F} \text{PartialDiameter}(f, 1-\alpha)_{\mathcal{D}}$. This notions allows to determine for a geometric dataset $\mathcal{D} = (X, F, \mu)$ the ability of F to discriminate the data points in X. This is done by considering the Observable Diameter over all possible values for α via

$$\Delta(\mathcal{D}) := \int_0^1 \text{ObservableDiameter}(\mathcal{D}, -\alpha)d\alpha,$$

which has, in the case of finite geometric datasets, the form

$$\Delta(\mathcal{D}) = \frac{1}{|X|}\sum_{k=2}^{|X|} \max_{f\in F} \min_{\substack{M\subseteq X\\|M|=k}} \max_{x,y\in M} |f(x) - f(y)|, \tag{1}$$

as shown in Stubbemann et al. [26]. With the usage of the notation $\phi_{k,f} := \min_{M\subseteq X,|M|=k} \max_{x,y\in M}|f(x)-f(y)|$, and $\phi_k := \max_{f\in F} \phi_{k,f}$ the discriminability $\Delta(\mathcal{D})$ can be rewritten as

$$\Delta(\mathcal{D}) = \frac{1}{|X|}\sum_{k=2}^{|X|} \max_{f\in F} \phi_{k,f} = \frac{1}{|X|}\sum_{k=2}^{|X|} \phi_k. \tag{2}$$

Problem Statement. Given a finite geometric dataset $\mathcal{D} = (X, F, \mu)$ and a natural number $n_F \ll |F|$, select the n_F most important features of F.

Note, that we consider feature selection with a fixed feature budget in an unsupervised setting, i.e., without any known labels of a specific classification task at selecting time.

3.1 Selection of Discriminating Features

The definition of discriminability as in Eq. (1) and Eq. (2) quantifies to which extent the set of all features can discriminate data subsets of different cardinality. The main idea of our feature selection method is to rank features by their ability to discriminate data subsets of different cardinality by solely using this feature.

Definition 1 (Discriminability). *The* discriminability of $\mathcal{D}$ with respect to feature $f \in F$ *is defined as*

$$\Delta(\mathcal{D})^*_f := \frac{1}{|X|}\sum_{k=2}^{|X|} \min_{\substack{M\subseteq X\\ |M|=k}} \max_{x,y\in M} |f(x) - f(y)| = \frac{1}{|X|}\sum_{k=2}^{|X|} \phi_{k,f}. \tag{3}$$

Enhancing Robustness against Outliers. Note, that one data point with an outstanding value $f(x)$ can have a strong influence on $\Delta(\mathcal{D})_f{}^*$ via drastically increasing $\phi_{|X|,f}$. To weaken this phenomenon, we propose to weight $\phi_{k,f}$ higher for smaller values of k. This leads to the following definition.

Definition 2 (Normalized Discriminability and Normalized Intrinsic Dimensionality). *The* normalized discriminability of $\mathcal{D}$ with respect to f *which we define as*

$$\Delta(\mathcal{D})_f := \frac{1}{|X|}\sum_{k=2}^{|X|} \frac{1}{k}\phi_{k,f}. \tag{4}$$

The normalized intrinsic dimensionality of $\mathcal{D}$ with respect to f *is then given via*

$$\partial(\mathcal{D})_f := \frac{1}{\Delta(\mathcal{D})_f^2}. \tag{5}$$

We then can rank features by their discriminability and select the features with the highest discriminability/lowest intrinsic dimensionality. We call the resulting method *feature selection via discriminability* (FSD).

4 Feature Selection for Large-Scale Data

Since computing $\phi_{k,f}$ is in $\mathcal{O}(|X| - k)$ for fixed $f \in F$ [26], computing $\Delta(\mathcal{D})_f$ is in $\mathcal{O}(\sum_{k=2}^{|X|}|X| - k) = \mathcal{O}(|X|)$. Hence, we have a worst case runtime which scales quadratic with $|X|$. Thus, FSD(C) is applicable to medium-sized data sets with thousands of data points. However, it is not tailored to large-scale data with millions of data points.

Earlier work [26] has used so called *support sequences*, i.e. strictly increasing finite sequences of the form $s = (2 = s_1, \ldots, s_l = |X|)$ to approximate $\Delta(\mathcal{D})$ and $\partial(\mathcal{D})$ by only computing $\phi_{s_i,f}$ for $s_i \in s$.

Inspired by this, we approximate the discriminability and intrinsic dimensionality with respect to a feature $f \in F$. The approximation is based on the result, that the map $k \mapsto \phi_{k,f}$ is monotonically increasing [26]. Using this result, we replace for $s_i < j < s_{i+1}$ the value $\phi_{j,f}$ by $\phi_{s_i,f}$ or $\phi_{s_{i+1},f}$. More formally, we get the following definition.

Definition 3 (Upper and Lower Normalized Discriminability). *For a feature $f \in F$ and a support sequence s we call*

$$\Delta(\mathcal{D})^+_{s,f} := \frac{1}{|X|}\left(\sum_{i=1}^{l}\frac{1}{s_i}\phi_{s_i,f} + \sum_{i=1}^{l-1}\sum_{s_i<j<s_{i+1}}\frac{1}{j}\phi_{s_{i+1},f}\right)$$

the upper normalized discriminability with respect to f and s *and*

$$\Delta(\mathcal{D})^-_{s,f} := \frac{1}{|X|}\left(\sum_{i=1}^{l}\frac{1}{s_i}\phi_{s_i,f} + \sum_{i=1}^{l-1}\sum_{s_i<j<s_{i+1}}\frac{1}{j}\phi_{s_i,f}\right)$$

the lower normalized discriminability with respect to f and s.

Definition 4 (Upper, Lower and Approximated Normalized Intrinsic Dimensionality). *We define the* upper/lower normalized intrinsic dimensionality with respect to f and s *via*

$$\partial(\mathcal{D})^+_{s,f} := \frac{1}{\left(\Delta(\mathcal{D})^-_{s,g}\right)^2} \text{ and } \partial(\mathcal{D})^-_{s,f} := \frac{1}{\left(\Delta(\mathcal{D})^+_{s,f}\right)^2}.$$

We then rank the features in descending order by their approximated normalized intrinsic dimensionality with respect to f and s *which is defined via*

$$\partial(\mathcal{D})_{s,f} := \frac{\partial(\mathcal{D})^+_{s,f} + \partial(\mathcal{D})^-_{s,f}}{2}. \tag{6}$$

We rank features via ascending $\partial(\mathcal{D})_{s,f}$. If we want to select k features, we choose the first k features of the resulting order.

4.1 Error Ratios of Approximations

Note, that $\partial(\mathcal{D})_{s,f}$ as defined in Eq. (6) only gives an approximation of $\partial(\mathcal{D})_f$ as defined in Eq. (5). Thus ranking the features by $\partial(\mathcal{D})_{s,f}$ instead of $\partial(\mathcal{D})_f$ may lead to a different ordering. We are interested in the amount of such changes. Let $l_s := (f_{i_1}, \ldots, f_{i_{|F|}})$ be the list of the features in F ordered by ascending $\partial(\mathcal{D})_{s,f}$.

Definition 5 (Error Ratio). *The* error ratio of s *is then given by*

$$\mathrm{E}(s)^* := \frac{2|\{(k,l) \in \{1,\ldots,|F|\} \mid k < l \wedge \partial(\mathcal{D})_{f_{i_k}} > \partial(\mathcal{D})_{f_{i_l}}\}|}{|F|(|F|-1)}. \tag{7}$$

Computing $\partial(\mathcal{D})_{s,f}$ for all $f \in F$ is especially of interest when computing $\partial(\mathcal{D})_f$ is not feasible. In such circumstances, it is also not possible to compute $\mathrm{E}(s)^*$. Hence, we need an approximation or upper bound of $\mathrm{E}(s)^*$ which can be computed without needing $\partial(\mathcal{D})_f$.

Definition 6 (Maximal Error Ratio). *We define the* maximal error ratio *of* s *via*

$$\mathrm{E}(s) := \frac{2|\{(k,l) \in \{1,\ldots,|F|\} \mid k < l \wedge \partial(\mathcal{D})^+_{s,f_{i_k}} > \partial(\mathcal{D})^-_{s,f_{i_l}}\}|}{|F|(|F|-1)}. \tag{8}$$

The maximal error ratio $E(s)$ can be computed without knowing $\partial(\mathcal{D})_f$ for all $f \in F$. Per definition it holds that $\partial(\mathcal{D})^-_{s,f} \leq \partial(\mathcal{D})_{s,f} \leq \partial(\mathcal{D})^+_{s,f}$. Thus, for given features $f, g \in F$ with $\partial(\mathcal{D})^+_{s,f} \leq \partial(\mathcal{D})^-_{s,g}$ we can conclude

$$\partial(\mathcal{D})_f \leq \partial(\mathcal{D})^+_{s,f} \leq \partial(\mathcal{D})^-_{s,g} \leq \partial(\mathcal{D})_g.$$

Thus, we get the following corollary.

Corollary 1. *For each support sequence s it holds that*

$$\mathrm{E}(s)^* \leq \mathrm{E}(s). \tag{9}$$

To sum up, we now have an approximation method that allow us to rank the features by their intrinsic dimensionality and we can efficiently bound the amount of errors this approximation produces. We denominate this method as *large-scale feature selection via discriminability*(LSFSD).

5 Experiments

In all our experiments, we consider the features F to be the coordinate projections, i.e., the feature functions are given via the data columns. We evaluate our feature selection method regarding to classification performances. We experiment with a Logistic Regression classifier on the *OpenML Numerical Tabular Classification Benchmark* (OpenML)[3]. To further evaluate our selection procedure on

[3] https://github.com/LeoGrin/tabular-benchmark.

modern large-scale data we also select features for classification with *Graph Neural Networks* (GNNs) on a subset of the *Open Graph Benchmark* [15]. In both experiments, we evaluate how to classify on a small subset of the features. To be more detailed, we experiment with selecting 10% and 50% of all features. In both experiments, we will also report results for classification on the full feature set. We evaluate against the following baselines.

- **Random.** Randomly the features.
- **Correlation Based.** Iteratively drop one of the two most correlated features until the amount of to selected featrues is left..
- **Variance Based.** Select features with the highest variance. We choose this baseline because it is a straight forward approach to for selecting features by their individual importance.
- **SPEC.** [27] SPEC first builds a complete graph between the individual data points with edge weights indicating similarities of the points and then uses spectral graph theory for estimating feature relevance. In our experiments, we use an RBF Kernel for similarity and evaluate feature relevance via ϕ_2. For details, we refer to the paper [27]. We choose this baseline because it was the univariate filter method with the highest classification accuracy in a recent survey on unsupervised feature selection [25].
- **RRFS.** [9] RRFS combines a function @sim which measures similarity of feature pairs and a relevance measure @rel that evaluates the importance of features. It iteratively chooses the next most relevant feature that has a similarity to the last chosen feature which below a specific threshold t. In our experiments wight logistic regression, we set $t = 0.4$ and for our experiments on GNNs we set $t = 0.8$. We use the Pearson correlation coefficient for similarity evaluation and evaluate the relevance of features via their variance. For more details, we refer to the paper [9]. We use this baseline because it was the multivariate filter method with the highest classification accuracy in the survey mentioned above [25].

5.1 Feature Selection for Logistic Regression

We select features as explained above and then use a Logistic Regression classifier. We evaluate on the k-fold splits provided OpenML and report mean test accuracies over all splits.

We only experiment on a subset of **OpenML**. To be more detailed, we discard all datasets with NaN. Furthermore, we discard all datasets with less then 20 features. We use default parameters with the exception that we changed the maximal iterations from 100 to 1000 to increase the chance of convergence. Finally, we arrive at 8 out of 50 datasets of **OpenML**. We use default parameters with the exception that we changed the maximal iterations from 100 to 1000 to increase the chance of convergence. Additionally, we do MinMax Scaling per feature so that each feature has the minimal value 0 and maximal value 1. Due to performance issues, we only report scores for SPEC for datasets with not more than 10000 data points. The results are depicted in Table 1.

Table 1. Results of **OpenML** experiment with 10% (top) and 50% (bottom) of features selected. Due to performance issues, we only report SPEC values for datasets with not more than 10000 data points.

Task	Full	FSD	Random	Corr	Vari	RRFS	SPEC
Bior.	.746±.030	.673±.030	.657±.035	.645±.033	**.691**±.026	.687±.026	.668±.030
Higgs	.635±.002	.500±.001	.525±.030	**.533**±.045	.499±.001	.499±.001	–
MinB.	.820±.005	**.780**±.004	.599±.076	.715±.075	.763±.015	.763±.015	–
cr.-ca.	.673±.010	**.591**±.008	.578±.024	.547 ±.029	**.591**±.008	**.591**±.008	–
eye-m.	.560±.016	.503±.017	.507±.023	527±.023	**.528**±0.17	**.528**±.017	.509±.019
heloc	.710±.013	.605±.018	.585±.047	.563±.037	**.641**±.013	.607 ±.019	.593±.012
jannis	.728±.005	**.664**±.008	.640±.037	.498±.006	.659±.008	.659±.008	–
pol	.856±.008	**.781**±.012	.537±.036	.528±.020	.759±.067	.759±.067	–

Task	Full	FSD	Random	Corr	Vari	RRFS	SPEC
Bior.	.746±.030	.734±.033	.730±.029	.729±.028	.734±.036	**.740**±.031	.706±.027
Higgs	.635±.002	.532±.001	**.575**±.019	.563±.019	.531±.006	.531±.006	–
MinB.	.820±.005	**.819**±.004	.776±.039	.819±.005	.780±.021	.780±.021	–
cr.-ca.	.673±.010	**.673**±.009	.641±.023	.620±.021	**.673**±.009	**.673**±.009	–
eye-m.	.560±.016	**.558**±.016	.538±.015	545±.017	**.558**±0.16	**.558**±.016	.494±.015
heloc	.710±.013	.690±.015	.682±.024	.656±.019	.677±.013	.659±.021	**.696**±.015
jannis	.728±.005	**.716**±.004	.702±.015	.544±.059	.715±.004	**.716**±.004	–
pol	.856±.008	.846±.010	.788±.039	.794±.021	.848±.011	**.849**±.011	–

Results. For 6 out of 8 datasets, FSD outcompetes the random baseline when selecting 10 % of the features. One of the two datasets where it does not, namely Higgs, is furthermore the only dataset where the random baseline is not outperformed when selecting 50% of the features. Note, that for these dataset all baselines except the correlation based do also not beat random selection, which indicates that this dataset is in general not well-suited for feature selection. Additionally, FSD is competitive with the considered baselines, surpassing them on 3 out of 8 datasets and being on pair with the best baseline on one additional dataset when selecting 10 % of the features. Furthermore, when selecting 50 % of the features, it outperforms or is on par with the best baseline in 4 out of 8 cases. Overall, selecting features by their discriminability is connected to meaningful learning with limited feature budgets and the resulting procedure is competitive with established selection approaches.

We point out, that our method sometimes lead to the same results then RRFS and the variance baseline. This is not surprising since a higher variance can be seen as some form of measuring how discriminative a feature is.

Overall, selecting only 10% often lead to a high drop compared to the original accuracy, indicating that such a small feature budget often not allows for sufficient feature selection. There is furthermore still a performance drop when 50% of the features are discarded. Thus, our experiment mainly gives insights in cases where feature selection is done because only using a small subset of features is computational feasible or to get new insights into the data and classification behavior. It is primarily not designed for feature selection to enhance classification accuracy. Here, the amount of selected features should be set higher. However, the competitiveness of our approach supports our hypothesis that discriminability of features is connected to their relevance for learning.

5.2 Feature Selection for Graph Neural Networks

We now evaluate to which extent our feature selection methods helps to select features for training Graph Neural Networks. For this, we use *ogbn-arxiv*, *ogbn-mag* and *ogbn-products*[4] from Open Graph Benchmark [15]. These datasets are fundamentally larger then the ones from **Open18**, with a nodeset size of 169, 343 (**ogbn-arxiv**), 2, 449, 029 (**ogbn-products**) and 1, 939, 743 (**ogbn-mag**).

To use LSFD we need to choose a support sequence. As in Stubbemann et al. [26], we use log scale spacing. For this, we first choose a geometric sequence $\hat{s} = (s_1, \dots s_l)$ of length $l = 10{,}000$ with $s_1 = |X|$ and $s_l = 2$ and use the support sequence s which results from $s' = (\lfloor |X| + 2 - s_1 \rfloor, \dots, \lfloor |X| + 2 - s_l \rfloor)$ via discarding duplicated elements.

For training and classification, we use a plain SIGN model [24] with one hidden-layer of dimension 512 and do 2−hop neighborhood aggregation. We train with an Adam optimizer with learning rate of 0.001 and weight decay of 0.0001. We train for a maximum of 1000 epochs with a patience of 15 epochs with respect to validation accuracy. We use a batch size of 256. We dropout at the input layer with a probability of 0.1, 0.3, 0.5 and at the hidden layer with probability of 0.5, 0.4, 0.5 for **ogbn-arxiv**, **ogbn-products** and **ogbn-mag**, respectively.

Results. We report test accuracies, displaying means and standard deviations over 10 rounds. We use all methods and baselines reported above. For SPEC, computation of the RBF kernel was not possible on our Server with 125 GB RAM due to memory costs. This was true for using Scikit-Learn as well as with plain numpy. The results are depicted in Table 2.

In this experiment, LSFSD is again competitive with all considered baselines. For selecting 10% of the features it surpasses all baselines for **ogbn-prodcuts** and for selecting 50% of the features it surpasses all baselines for **ogbn-mag** and **ogbn-arxiv** and performance on par with the best baseline. Furthermore, LSFSD always surpasses random feature selection. This strongly indicates that selecting features by their resilience to the curse of dimensionality is meaningful with respect to classification tasks.

6 Parameter Study on Maximal Errors

We now study the influence of the lengths of support sequences on the maximal error $E(s)$. For this, we generate the support sequence as in Sect. 5.2 with varying

[4] https://ogb.stanford.edu/docs/nodeprop/.

Table 2. Results of our experiments on OGBN. We report accuracies for selecting 10% (top) and 50% (bottom) of the features. Small differences in scores for the full feature set between the top and bottom tables are due to random initializations.

Dataset	Full	LSFSD	Random	Corr	Vari	RRFS
arxiv	.697 ± .006	.537 ± .006	515 ± .027	.478 ± .007	**.540** ± .003	.540 ± .006
products	.748 ± .006	**.544** ± .013	.396 ± .040	.419 ± .007	.486 ± .008	.486 ± .008
mag	.387 ± .002	.283 ± .003	.272 ± .011	.270 ± .003	**.292** ± .004	.284 ± .005

Dataset	Full	LSFSD	Random	Corr	Vari	RRFS
arxiv	.694 ± .004	**.683** ± .005	675 ± .007	.673 ± .004	.680 ± .007	.678 ± .007
products	.753 ± .006	**.714** ± .006	.698 ± .008	.686 ± .004	.712 ± .009	**.714** ± .007
mag	.387 ± .004	**.378** ± .002	.369 ± .003	.362 ± .004	.375 ± .003	.375 ± .002

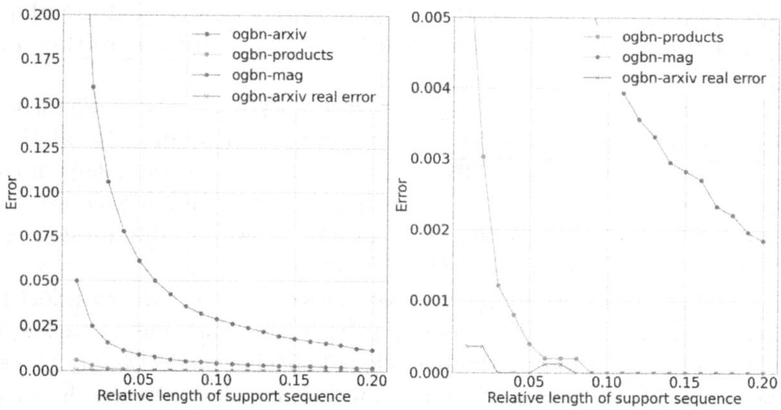

Fig. 1. The maximal errors E(s) of all **ogbn** datasets. For **ogbn-arxiv**, we also display the "real error" E(s)$*$. The right plot is a zoom-in of the left plot to observe the behavior of E(s)$*$.

value l. To be more detailed, we iterate l through $\{\lfloor 0.01*n \rfloor, \lfloor 0.02*n \rfloor, \ldots, \lfloor 0.2*n \rfloor\}$, where n is the amount of points in the respective datasets. Here, if $l = \lfloor r*n \rfloor$, we call r the *relative length of* l. Note, that the final support sequence may have a lower amount of elements in l as we discard doubled elements. Since for **ogbn-arxiv** the exact computation of $\Delta(\mathcal{D})_f$ is possible, we also compute the "real" error E$(s)^*$ for this dataset (Fig. 1).

Results. For the larger datasets, **ogbn-products** and **ogbn-mag**, the maximal error E(s) is negligible for comparable small relative support sizes. For relative support sizes of 0.1, the maximal mistakes are under 0.01. For **ogbn-arxiv** we have an maximal error of around 0.025 for this relative support size and overall higher maximal errors for all relative support lengths.

Comparing $\mathrm{E}(s)^*$ and $\mathrm{E}(s)$ for **ogbn-arxiv** it stands out, that the maximal error is a strong overestimation of the error $\mathrm{E}(s)^*$. As computing $\mathrm{E}(s)^*$ for both other datasets is not feasible, we can not verify whether this is also the case for them. However, we know because of Eq. (9), that $\mathrm{E}(s)^*$ is bound by the maximal error which is already negligible for **ogbn-products** and **ogbn-mag**.

7 Conclusion and Future Work

We proposed a novel unsupervised feature selection method that accounts for the intrinsic dimensionality of datasets. Our approach identifies those features that are able to discriminate data points and are thus responsible for taming the curse of dimensionality. Our experiments provide evidence that intrinsic dimension-based selection of features is competitive with well established procedures, occasionally outperforming them. Furthermore, we demonstrated that sampling techniques can be used to scale our feature selection method to more than millions of data points.

We identify as a natural next step for future work scenarios where the features are not given by coordinate projection, i.e., go beyond columns in tabular data. The generality of our modeling of features allows to encode, e.g., edge information for graphs. Thus, our method can be used for edge-sampling procedures that are important for training Graph Neural Networks. However, the question on how to encode graph information via feature functions is open.

While the present work emphasizes on understanding *data* by selecting features that break the curse of dimensionality, future work has to tackle as well the problem of understanding the behavior of specific *model classes*. This helps to further strengthen our understanding on the interplay between the success of learning and the presence of the curse of dimensionality.

References

1. Alelyani, S., Tang, J., Liu, H.: Feature selection for clustering: a review. In: Data Clustering, pp. 29–60 (2018). https://doi.org/10.1201/9781315373515-2
2. Bellman, R.: Adaptive Control Processes: A Guided Tour. Princeton Legacy Library, Princeton University Press, Princeton (1961)
3. Bellman, R., Corporation, R., Collection, K.M.R.: Dynamic Programming. Rand Corporation research study. Princeton University Press, Princeton (1957)
4. Breaban, M., Luchian, H.: A unifying criterion for unsupervised clustering and feature selection. Pattern Recogn. **44**(4), 854–865 (2011). https://doi.org/10.1016/j.patcog.2010.10.006
5. Cai, J., Luo, J., Wang, S., Yang, S.: Feature selection in machine learning: a new perspective. Neurocomputing **300**, 70–79 (2018). https://doi.org/10.1016/j.neucom.2017.11.077
6. Dutta, D., Dutta, P., Sil, J.: Simultaneous feature selection and clustering with mixed features by multi objective genetic algorithm. Int. J. Hybrid Intell. Syst. **11**(1), 41–54 (2014)

7. Dy, J.G., Brodley, C.E.: Feature selection for unsupervised learning. JMLR **5**, 845–889 (2004)
8. Faloutsos, C., Wu, L., Traina, A., Traina, C., Jr.: Fast feature selection using fractal dimension. J. Inf. Data Manag. **1**(1), 3–3 (2010). https://doi.org/10.1184/r1/6605570.v1
9. Ferreira, A.J., Figueiredo, M.A.T.: An unsupervised approach to feature discretization and selection. Pattern Recognit. **45**(9), 3048–3060 (2012). https://doi.org/10.1016/j.patcog.2011.12.008
10. Golay, J., Kanevski, M.: Unsupervised feature selection based on the morisita estimator of intrinsic dimension. Knowl.-Based Syst. **135**, 125–134 (2017). https://doi.org/10.1016/j.knosys.2017.08.009
11. Gromov, M., Milman, V.D.: A topological application of the isoperimetric inequality. Am. J. Math. **105**(4), 843–854 (1983)
12. Gómez, W., Leija, L., Díaz-Pérez, A.: Mutual information and intrinsic dimensionality for feature selection. In: 2010 7th International Conference on Electrical Engineering Computing Science and Automatic Control, vol. 27, pp. 339–344. IEEE (2010). https://doi.org/10.1109/iceee.2010.5608600
13. Hanika, T., Koyda, M., Stumme, G.: Relevant attributes in formal contexts. In: Endres, D., Alam, M., Şotropa, D. (eds.) ICCS 2019. LNCS (LNAI), vol. 11530, pp. 102–116. Springer, Cham (2019). https://doi.org/10.1007/978-3-030-23182-8_8
14. Hanika, T., Schneider, F.M., Stumme, G.: Intrinsic dimension of geometric data sets. Tohoku Math. J. **74**(1), 23–52 (2022). https://doi.org/10.2748/tmj.20201015a
15. Hu, W., et al.: Open graph benchmark: datasets for machine learning on graphs. arXiv preprint arXiv:2005.00687 (2020). https://doi.org/10.48550/arxiv.2005.00687
16. Milman, V.: Topics in asymptotic geometric analysis. In: Visions in Mathematics, pp. 792–815. Springer, Heidelberg (2000). https://doi.org/10.1007/978-3-0346-0425-3_8
17. Milman, V.: The heritage of p. lévy in geometrical functional analysis. Astérisque **157**(158), 273–301 (1988)
18. Mitra, P., Murthy, C.A., Pal, S.K.: Unsupervised feature selection using feature similarity. IEEE Trans. Pattern Anal. Mach. Intell. **24**(3), 301–312 (2002). https://doi.org/10.1109/34.990133
19. Mo, D., Huang, S.H.: Fractal-based intrinsic dimension estimation and its application in dimensionality reduction. IEEE Trans. Knowl. Data Eng. **24**(1), 59–71 (2012). https://doi.org/10.1109/tkde.2010.225
20. Pestov, V.: On the geometry of similarity search: dimensionality curse and concentration of measure. Inf. Process. Lett. **73**(1–2), 47–51 (2000). https://doi.org/10.1016/s0020-0190(99)00156-8
21. Pestov, V.: Intrinsic dimension of a dataset: what properties does one expect? In: IJCNN 2007, pp. 2959–2964. IEEE (2007). https://doi.org/10.1109/ijcnn.2007.4371431
22. Pestov, V.: An axiomatic approach to intrinsic dimension of a dataset. Neural Netw. **21**(2–3), 204–213 (2008). https://doi.org/10.1016/j.neunet.2007.12.030
23. Pestov, V.: Intrinsic dimensionality. ACM SIGSPATIAL Special **2**(2), 8–11 (2010)
24. Rossi, E., Frasca, F., Chamberlain, B., Eynard, D., Bronstein, M.M., Monti, F.: SIGN: scalable inception graph neural networks. CoRR arxiv:2004.11198 (2020). https://doi.org/10.48550/arxiv.2004.11198
25. Solorio-Fernández, S., Carrasco-Ochoa, J.A., Martínez-Trinidad, J.F.: A review of unsupervised feature selection methods. Artif. Intell. Rev. **53**(2), 907–948 (2019). https://doi.org/10.1007/s10462-019-09682-y

26. Stubbemann, M., Hanika, T., Schneider, F.M.: Intrinsic dimension for large-scale geometric learning. Trans. Mach. Learn. Res. **2023** (2023). https://openreview.net/forum?id=85BfDdYMBY
27. Zhao, Z., Liu, H.: Spectral feature selection for supervised and unsupervised learning. In: Ghahramani, Z. (ed.) ICML 2007, ACM International Conference Proceeding Series, vol. 227, pp. 1151–1157. ACM (2007)

Development of Hybrid Artificial Intelligence Training on Real and Synthetic Data

Benchmark on Two Mixed Training Strategies

Paul Wachter[1(✉)], Lukas Niehaus[2], and Julius Schöning[1]

[1] Faculty of Engineering and Computer Science, Osnabrück University of Applied Sciences, Osnabrueck, Germany
{p.wachter,j.schoening}@hs-osnabrueck.de

[2] Institute of Cognitive Science, Osnabrück University, Osnabrueck, Germany
luniehaus@uni-osnabrueck.de

Abstract. Synthetic data has emerged as a cost-effective alternative to real data for training artificial neural networks (ANN). However, the disparity between synthetic and real data results in a *domain gap*. That gap leads to poor performance and generalization of the trained ANN when applied to real-world scenarios. Several strategies have been developed to bridge this gap, which combine synthetic and real data, known as mixed training using hybrid datasets. While these strategies have been shown to mitigate the domain gap, a systematic evaluation of their generalizability and robustness across various tasks and architectures remains underexplored. To address this challenge, our study comprehensively analyzes two widely used mixing strategies on three prevalent architectures and three distinct hybrid datasets. From these datasets, we sample subsets with varying proportions of synthetic to real data to investigate the impact of synthetic and real components. The findings of this paper provide valuable insights into optimizing the use of synthetic data in the training process of any ANN, contributing to enhancing robustness and efficacy (Supplementary code and results: https://hs-osnabrueck.de/prof-dr-julius-schoening/ki2025)

Keywords: Hybrid Datasets · Mixed Training · Artificial Intelligence · Benchmark · Synthetic Data · Domain Gap · Reality Gap

1 Introduction

Synthetic data has recently gained considerable attention as a cost-effective way to generate training data [14] for artificial neural networks (ANN). Synthetic data, defined as information not derived from real-world sources, inherently lacks realism. This absence creates a discrepancy between synthetic and real examples, which can be termed the *reality gap* [25]. The reality gap is a subset of the broader *domain gap*, which limits the exclusive use of synthetic data in ANN

T. Braun et al. (Eds.): KI 2025, LNAI 15956, pp. 175–189, 2026.
https://doi.org/10.1007/978-3-032-02813-6_13

training since the target domain usually is the real-world. As a result, applying ANNs, which are solely trained on synthetic data, to real-world scenarios requires domain adaptation to transfer knowledge from one domain to the other [17].

One key area of domain adaptation research focuses on enhancing the quality and diversity of synthetic data to bridge the domain gap [14]. This can be achieved by increasing the realism of synthetic data either during its generation or through post-processing techniques. Enhancing realism during generation requires an advanced simulation framework that accurately models real-world conditions [10]. Such frameworks often prove costly and labor-intensive [28]. Alternatively, image transformation methods, such as neural style transfer, improve realism in post-processing [6], although these methods often suffer from training instabilities and may introduce artifacts [32].

Another research focus targets the architecture of ANNs, aiming to enable them to learn domain-invariant features, thereby reducing the domain gap [30]. However, designing such architectures is inherently challenging because they must effectively abstract away domain-specific details to maintain generality. Moreover, many of these architectures are developed for multi-domain adaptation [16], including domains with no available data. These assumptions may limit their effectiveness in addressing the specific reality gap, where real data is usually accessible.

Consequently, this paper examines a mixed training approach for domain adaptation, which integrates synthetic and real data into a unified hybrid dataset. By integrating both types into a hybrid set, mixed training allows models to learn features from synthetic part, while real-world data introduces authentic visual and contextual variations, thereby mitigating the lack of realism in the synthetic data and bridging the domain gap. This approach can be used with every ANN architecture, thus making it broadly applicable, while omitting the need for highly specialized ANNs when relaying on synthetic sources. Mixed training prominently follows two strategies: *simple mixed* (SM), where both data types are used simultaneously, and sequentially *fine-tuned* (FT), in which ANNs are pretrained on synthetic data and afterwards fine-tuned with real data. Although these methods are widely applied, their underlying mechanisms and performance under varying conditions remains only partially understood.

In response to these limitations, this study systematically evaluates both mixed training strategies across three structurally distinct datasets, varying ratios of synthetic-to-real data and three different ANN architectures on image classification tasks. Unlike previous studies that boost model performance by expanding the hybrid dataset with additional synthetic examples, our approach maintains a constant dataset size while gradually increasing the synthetic-to-real proportion. Thus, our goal is not limited to improving the overall performance, but to conduct a precise assessment of each strategies efficacy under varying conditions. This design offers actionable insights to guide the practical application of synthetic data in real-world settings.

2 Related Work

Although many publications have successfully applied mixed training to reduce the domain gap [9,19,21,23], most studies concentrate on the generation of synthetic data rather than examining the mixed training strategies themselves [7,29].

Nowruzi et al. [15] evaluated an object detector for cars and pedestrians using three synthetic and three real datasets. They created hybrid training sets by combining synthetic data with four different, low ratios of real data. Moreover, they compared the SM strategy with the FT approach for their tasks and concluded that FT more effectively reduces the domain gap. However, the study did not consider the effects of varying dataset sizes and class distributions. In contrast, Burdorf et al. [3,18] did not confirm the advantages of FT for a very similar task as reported by Nowruzi et al. [15].

In [26], Vanherle et al. compared the SM and FT strategies across subsets with the same overall size but different synthetic-to-real ratios; notably, the subsets were imbalanced. They employ the DIMO dataset [4], which contains real and synthetic images, to train and test a Mask-Region Based Convolutional Neural Networks (Mask R-CNN) for object detection. Although the FT strategy yielded better performance, the study involved pretraining on the COCO dataset before introducing synthetic examples. Moreover, synthetic data was used solely to train the heads of the pretrained model. These methodological choices impact network performance [12] and may interact with the mixed training strategies, potentially confounding true differences.

Together, these studies leave several questions unresolved. Reported gains could be due to uncontrolled factors, like dataset size, class imbalance, or preraining on large scale datasets with freezing of layers, that obscure the true effect of the mixing strategy. Most experiments rely on a single hybrid dataset and one ANN, making it unclear whether conclusions generalize across synthetic data types or ANN architectures. A more controlled, cross-dataset, cross-architecture analysis is therefore required, which we address in this study.

3 Methodology

The complete code, configurations and numerical results of our study can be found at[1]. In general, mixed training using real and synthetic data can be applied to any neural network and hybrid dataset, regardless of the architecture, synthetic data type or synthetic-to-real ratio. Consequently, this study covers a broad range of possible configurations of these key factors to evaluate their influence on the two training strategies SM and FT. Our study is designed to expose and isolate the effects on model performance, instead of aiming for maximum accuracy in each configuration. Accordingly, we minimize other factors which can influence the training process, such as advanced optimization algorithms,

[1] Supplementary code and results: https://hs-osnabrueck.de/prof-dr-julius-schoening/ki2025

regularization heuristics, data augmentation or hyperparameter tuning. In the following, we lay out the choices of our ANN architectures, hybrid datasets and synthetic-to-real proportions, as well as the implementation of the SM and FT training strategies.

3.1 Application Task

We choose image classification as the application task of our study. Image classification is considered one of the most fundamental tasks within the domain of AI-based and classical computer vision [22]. It was selected, because it's significance extends beyond mere categorization: more intricate tasks, like object detection, semantic segmentation, and image generation, are derived from the same principles [24]. This allows to assume a similar behavior of the mixed training strategies on a broad range of tasks in computer vision.

3.2 Artificial Neural Networks

As noted before, mixed training can be used irrespective of the ANN architecture. Therefor, we decided to assess the impact of three commonly used architectures in AI-based computer vision. Namely, the multilayer perceptron (MLP) [22], the convolutional neural network (CNN) [8], and the vision transformer (ViT) [5]. Each of these ANNs is implemented in the original form, without additional mechanisms, like normalization or regularization layers. This reduces the number of confounding variables and allows to focus on the respective ANN's core mechanism. However, normalization layers were incorporated into the transformer blocks and the patch encoder of the ViT, given their critical role in preserving the fundamental functionality and performance of transformer-based networks [31]. The exclusion of these layers would constitute a deviation from the fundamental design principles of transformers [27]. The general design choices, such as the width of the layers, the activation functions, the initialization, etc., reflect common practices. A detailed overview of the network configurations can be found in[2].

3.3 Datasets

Synthetic data can be generated through several distinct mechanisms [14]. For this comprehensive study on mixed training strategies, we curated three hybrid datasets, each created using a different synthetic data generation mechanism. These are generative Ai (GenAI), computer aided design (CAD) and hand drawings, which results in distinct structures and characteristics of each dataset.

[2] Supplementary code and results: https://hs-osnabrueck.de/prof-dr-julius-schoening/ki2025

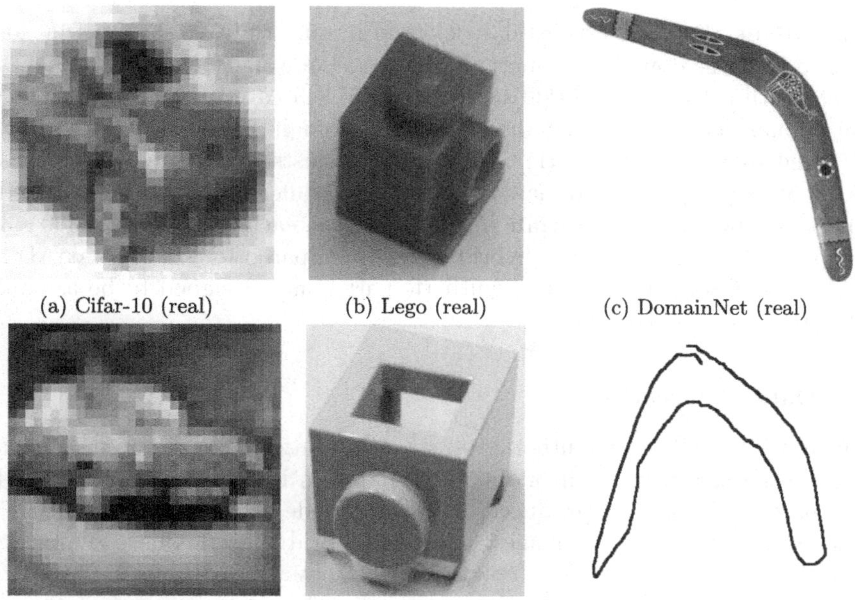

(a) Cifar-10 (real) (b) Lego (real) (c) DomainNet (real)

(d) CiFake (synth.) (e) Lego (synth.) (f) DomainNet (synth.)

Fig. 1. Real and synthetic example images of the datasets.

Cifar-10 and CiFake comprise the first dataset in which, as illustrated in Fig. 1 (a) and (d), Cifar-10 [11] represents the real and CiFake [1] the synthetic part. Cifar-10 and CiFake both comprise 60,000 32×32 RGB images across ten object classes, each with 6,000 images. The CiFake dataset is generated using the *CompVis Stable Diffusion* model [20], an open-source latent diffusion model. Additional prompt modifiers enhanced the diversity, while still producing images that closely resemble Cifar-10's characteristics. Importantly, 668 duplicate image pairs were found in the CiFake dataset, and one from each was removed to ensure dataset integrity.

LEGO Bricks [2] for training classification networks was selected as the second dataset. It's real part is shown in Fig. 1 (b), while the synthetic images, shown in Fig. 1 (e) were generated using the computer-aided design (CAD) program LDraw [13]. The synthetic images introduce additional colors and angles, but still try to mimic the real images accurately. The dataset features 447 unique LEGO brick classes with a total of 50,000 real and 560,000 synthetic images. The classes are identified by the official LEGO IDs based on shape, regardless of color or decorations. All images are RGB with varying sizes, and both datasets are highly, yet differently, unbalanced, with class samples ranging from 6 to over 600.

DomainNet [16] is the third dataset used in our study. In it's original form, it consists of six domains: Clipart, Infograph, Painting, Quickdraw, Real, and Sketch, with a total of 596,000 samples across 345 categories. We selected the Real domain as our real, and the Quickdraw domain as our synthetic data, as visualized in Fig. 1 (c) and (f). The former consists of roughly 176,000 RGB images of varying sizes and class distributions, while the latter is composed of 500 hand-drawn, black-and-white images per class, each with a size of 300×300. This provides a third, distinct hybrid dataset compared to CiFake (GenAI) and LegoBricks (CAD), because it's synthetic part is not designed to be as similar to the real part as possible.

3.4 Dataset Creation

To examine the effect of synthetic-to-real image ratios, we generated 11 equally sized subsamples from each of the three datasets introduced in Sect. 3.3. For every original dataset we produced one subset made entirely of real images, one made entirely of synthetic images, and nine hybrid subsets whose composition ranges from 90% synthetic/10% real to 10% synthetic/90% real in 10-percentage-point steps. To eliminate potential confounding factors, we matched the class distribution across all subsets. For the nine hybrid subsets, the same class balance was enforced separately within the real and synthetic parts. The 11 synthetic-to-real ratios on the three original datasets resulted in 33 subsampled datasets for this study, with the attributes summarized in Table 1.

Furthermore, the images were rescaled to the same spatial dimensions, to match the ANN's fixed input sizes. No rescaling was necessary for Cifar-10 and CiFake, since all images have the shape 32×32. The LegoBricks images were rescaled to 256×256, and all DomainNet images to 300×300. Additionally, the grayscale images from the synthetic part of the DomainNet dataset were converted into the 3-channel RGB format, for the same reason. Lastly, all images were normalized to lie within the range $[0, 1]$, in order to ensure consistency across all RGB channels and image sources.

Table 1. Attribute overview of the (hybrid) datasets

Name	# Images	# Classes	Image Dimensions	Synth. Source
Cifar-10 [11]/Cifake [1]	55,000	10	$32 \times 32 \times 3$	GenAI
LegoBricks [2]	20,100	134	$256 \times 256 \times 3$	CAD
DomainNet [16]	30,000	60	$300 \times 300 \times 3$	Drawing

3.5 Mixed Training Strategies

Two mixed training strategies are compared in this study, which we termed *simple mixed* (SM) and *fine-tuned* (FT). The former treats the synthetic and real part entirely equivalent. It does not distinguish between synthetic and real data during training; data from both domains are utilized in the same manner, as if they were part of a single dataset. The inputs and their corresponding labels are sampled randomly from both datasets, regardless of their size or distribution. This provides the ANN with a broader range of examples, increasing its ability to adapt to various scenarios and improving its robustness.

Conversely, the FT strategy utilizes the real and synthetic parts sequentially. Only the synthetic part is used to train the network in the first step. This pretraining is stopped when there is no more improvement on the real evaluation set, implemented through validation-based early stopping. In the second step, the real part is used to retrain the network obtained after step one. The FT strategy acknowledges the differences between data types and their relation to the real test data, and adjusts the training process accordingly.

3.6 Training Procedure

All models were training using vanilla Stochastic Gradient Descent (SGD) with a learning rate of 0.01 and mini-batches of 64 samples. Datasets were split 60%/20%/20% into training, validation, and test sets. The validation and test sets are sampled exclusively from the real part of the dataset and remained unchanged. In both SM and FT training strategies, the MLP, CNN and ViT were trained for 100 epochs, ensuring convergence. After training, the model was restored to the state where it showed the highest validation accuracy. For the FT strategy, the first training step proceeded until the validation accuracy did not improve for 10 epochs; at that moment, early stopping triggered, and the weights from the best-performing model were preserved. In the second step, the network was trained for the remaining number of epochs. All parameters remained unchanged throughout this procedure.

To summarize, we trained three neural network architectures, each with two training strategies, on 27 hybrid datasets and six non-hybrid datasets (all-real or all-synthetic). One full training cycle therefore produced 162 models trained on hybrid data and 18 models trained on non-hybrid data, yielding 180 trained networks in total. To obtain statistically robust results, we repeated this cycle ten times, resulting in 1,800 trained networks overall. At the beginning of each training cycle, all 33 datasets were resampled.

4 Results

The two baselines that guide the evaluation of the mixed training strategies are the purely synthetic and purely real dataset settings. While the former provides information about the quality of the synthetic data, the latter shows the performance when no domain gap exists between the training and test data. Table 2 reports the test accuracy of the baselines, averaged over ten repetitions. All purely synthetic setups achieve performance above chance level. This result indicates their applicability as a proxy for the real-world, although a large gap to the performance of the purely real setting is evident, highlighting the domain gap. The domain gap can be quantified as the difference between the purely real and purely synthetic results. The goal of mixed training strategies is to minimize this gap, which would result in a comparable performance to the purely real setups.

Table 2. MLP, CNN, and ViT average test accuracy without mixing (0.0 = 100% real, 1.0 = 100% synthetic) averaged over ten runs.

	Cifar-10/CiFake		LegoBricks		DomainNet	
	Real	Synthetic	Real	Synthetic	Real	Synthetic
MLP	0.5090	0.1328	**0.6699**	**0.1423**	0.2064	0.0244
CNN	**0.6413**	**0.1430**	0.5582	0.0130	**0.2870**	**0.0353**
ViT	0.5165	0.1255	0.6296	0.0125	0.2660	0.0345

In the mixed training setting, the two training strategies, SM and FT, have resulted in divergent performances. Across all combinations of datasets, synthetic-to-real proportions, and ANN architectures, FT has outperformed SM in 635 out of 810 cases; when averaged over the ten repetitions, in 69 out of 81.

The synthetic-to-real ratio of the train data strongly influences the performance of the ANNs. This influence is two-fold: it impacts performance in general and determines the magnitude of the difference between the FT and SM strategies. The general impact on the performance was observed throughout all settings. Every increase in the real proportion leads to an improved performance. This finding was expected since more real data in the training set leads to a reduction of the domain gap to the real test data. Notably, the most significant improvement was observed for the first 10% increase. The improvement made through the next increments gradually decreased. In other words, even a small amount of real data in the hybrid dataset leads to a drastic improvement in performance, and the bigger the proportion of real data in the dataset, the less improvement is made by increasing the real data proportion further.

In addition to the general role of the synthetic-to-real ratio on the performance, it also determines the magnitude of the difference between the FT and SM strategies. Figure 2 shows a ViT trained on the Cifar-10/Cifake dataset using both strategies.

Here, the difference between the two strategies increases parallel to the synthetic data proportion. In the setup with only 10% synthetic data, FT only marginally outperforms SM. With every increase of the synthetic data proportion, this difference increases as well, up to the point of 90% synthetic data, where the FT strategy performs roughly twice as good as the SM strategy.

What follows from this is that although the performance of the ANNs decrease with an increase of the synthetic proportion, this degradation of performance is less drastic for the FT strategy. Figure 3 depicts the test accuracy gradients of the ViT trained on the Cifar-10/CiFake dataset with both strategies.

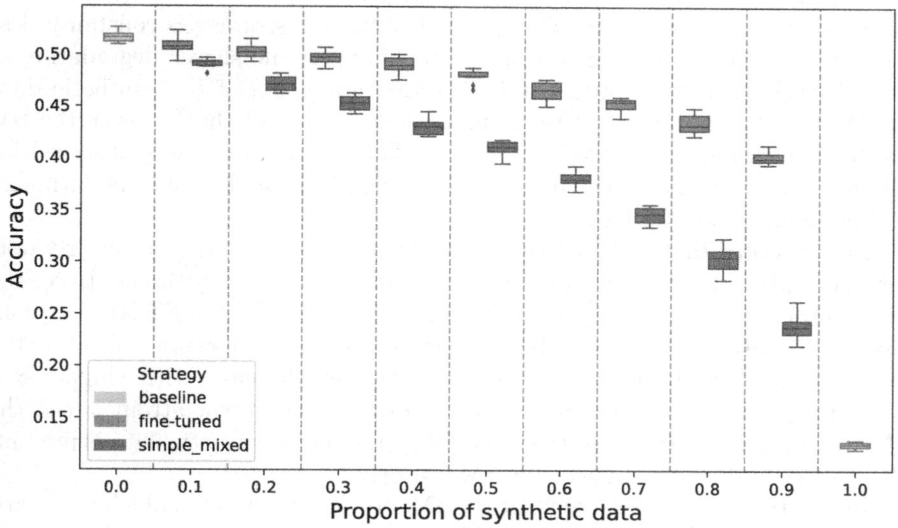

Fig. 2. ViT test accuracy on the Cifar-10/CiFake dataset (0.0 = 100% real, 1.0 = 100% synthetic) averaged over ten runs shown as boxplots.

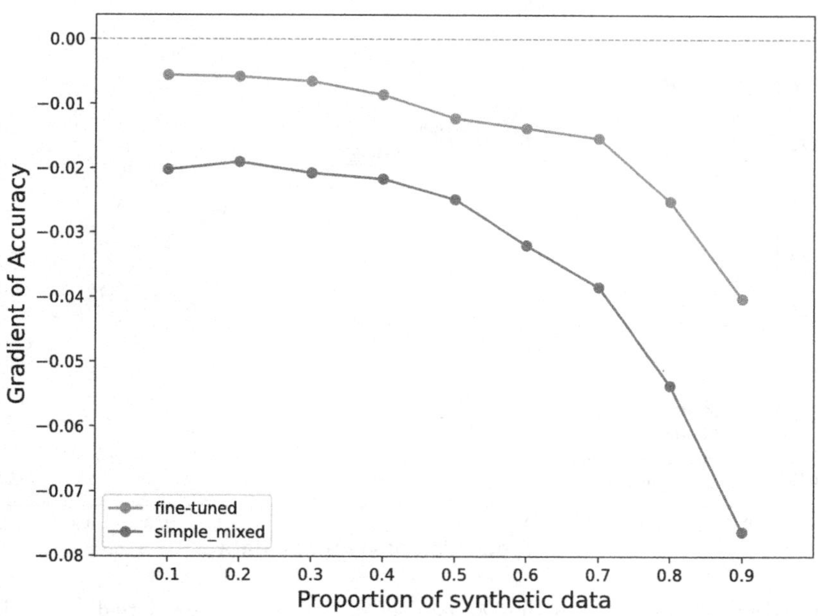

Fig. 3. Test accuracy gradients of the ViT trained on Cifar-10/CiFake datasets. The proportion of 0.0 represents 100% real, and 1.0 represents 100% synthetic data.

The gradient shows the rate at which the performance is changing in relation to the synthetic-to-real ratio. The gradient of the FT strategy is constantly less negative compared to the SM strategy, highlighting the slower degradation of performance. In other words, the higher the proportion of the synthetic data in the dataset, the bigger is the gain in performance of the FT over the SM strategy. This tendency caused by the synthetic-to-real ratio was observed for all networks trained on the Cifar-10/CiFake and LegoBricks datasets, although less pronounced for the CNN.

In the setups that included the DomainNet dataset, FT did not clearly outperform SM, as it was the case for Cifar-10/CiFake and LegoBricks. Here, the architecture of the ANN played a crucial role. For the ViT, the FT strategy consistently resulted in marginally better performances. In the setups, including the MLP shown in Fig. 4, the FT strategy was better only for cases with a higher synthetic proportion. The CNN consistently performed better when trained with the SM strategy, as illustrated in Fig. 5. Notably, the more synthetic data is present, the smaller is the advantage of the SM strategy.

In addition to this preference for the SM strategy, the CNN had a higher variation throughout the 10 repetitions. The interquartile range for the SM strategy showed a high variation around the mean and the FT strategy showed significant outliers. This was observed here, as well as in the case of CNN trained on Cifar-10/Cifake and LegoBricks datasets. In[3], the results of all setups can be found.

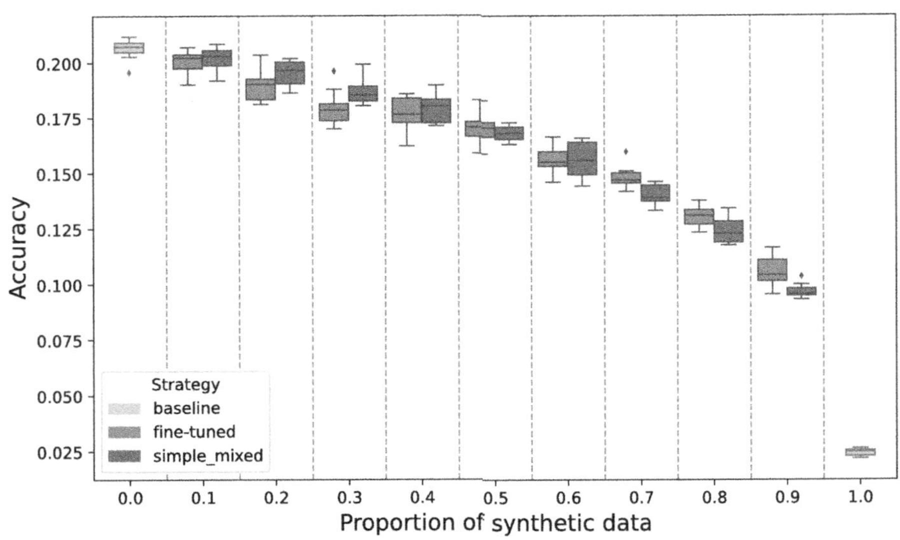

Fig. 4. MLP test accuracy on the DomainNet dataset (0.0 = 100% real, 1.0 = 100% synthetic) averaged over ten runs shown as boxplots.

[3] Supplementary code and results: https://hs-osnabrueck.de/prof-dr-julius-schoening/ki2025

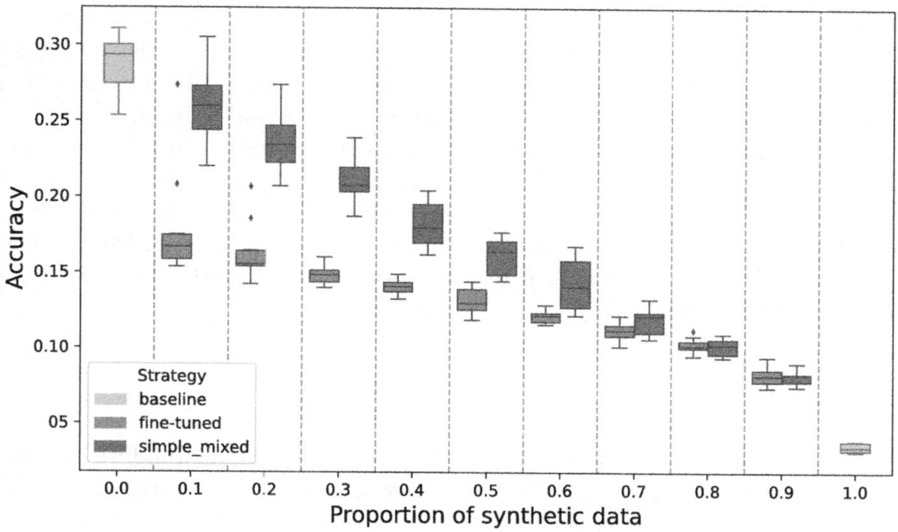

Fig. 5. CNN test accuracy on the DomainNet dataset (0.0 = 100% real, 1.0 = 100% synthetic) averaged over ten runs shown as boxplots.

5 Discussion

The results of our study indicate that FT is a preferable strategy for mixed training using real and synthetic data. Nevertheless, the CNN/DomainNet setting constitutes a clear counter-example: here simultaneous exposure to both domains (SM) yielded better performance. In this section we are going to elaborate our theories for the reasons leading to this outlier.

A convolutional layer extracts localized patterns with learnable kernels; each kernel sees only its receptive field, a region whose size equals the kernel dimensions. Stacking convolutional layers (with larger kernels, strides, or pooling) expands the receptive field, such that deeper layer can capture increasingly complex structures. This, however, relies entirely on the patterns captured by earlier layers; if those first layers miss important structures, later layers have nothing to build on. High-quality, diverse features at the network's outset are therefore critical to the overall performance.

DomainNet exhibits a larger domain gap compared to Cifar-10/CiFake and LegoBricks. The latter two contain colored, texture-matched synthetic images, meticulously designed to mimic their real counterpart, while DomainNet's synthetic subset consists solely of black-and-white sketches with only two pixel values, sharp edges and no texture (Fig. 1).

Pre-training a CNN on these simple, synthetic images biases its early layers toward extreme black-white edges. Kernels aligned with these simple pattern receive large gradients during training, increasing a few weights rapidly, while suppressing the rest. The result is comparable to Sobel or Prewitt kernels, whose few large coefficients emphasize a single-oriented edge, whereas kernels that would encode more fine grained patterns need many smaller, finely balanced coefficients. Once dominated by such coarse detectors, the early layers fail to capture more complex patterns, leaving deeper layers without the information necessary to form complex features. The resulting kernels make subsequent fine-tuning difficult: gradients become ill-conditioned and struggle to reshape the early layers once real data are introduced in the fine-tuning step. In summary, the network is effectively trapped in a local minimum shaped by oversized and diminishing small weights.

6 Conclusion

Our study reveals three findings that refine the current understanding of mixed training using real and synthetic data. First, the fine-tuning strategy usually, but not always, outperforms the simple mixed strategy. Across 89 individual configurations FT surpassed SM in 69 (78%), particularly when synthetic data comprised a large portion of the dataset. This indicates that it is more effective in utilizing real data to bridge the domain gap. Nevertheless, the CNNDomainNet setting constitutes a clear counter-example. Thus, architecture-data interactions, as outlined in 5, influence the strategies' success.

Second, diminishing returns of real data. The largest performance gains occurred with an initial 10% increase in real data; further increases yielded progressively less. Practically, adding a modest subset of real data to a synthetic set can markedly boosts performance, providing a possible cost-effective solution in many cases.

Third, Interaction of strategy with domain gap For the two synthetic sources that explicitly mimic the real domain (GenAI and CAD), FT was consistently advantageous. Conversely, hand-drawn sketches of the DomainNet dataset constitute large discrepancy to the real images, semantically as well as visually. Under such conditions the SM strategy can outperform FT. This suggests that the "optimal" strategy could be chosen based on a quantifiable domain gap measure.

Future work should test mixed training on richer tasks, like object detection, instance or semantic segmentation, and with production-grade networks that include modern regularization and optimization methods. Furthermore, it should develop a robust domain-gap metric that combines low-level visual statistics with high-level semantic information. A reliable score would enable the prediction of the required real-data fraction and select the appropriate mixing strategy, turning today's trial-and-error process into a more systematic workflow.

Acknowledgments. This work is part of the AgrifoodTEF-DE project. AgrifoodTEF-DE is supported by funds of the Federal Ministry of Agriculture, Food and Regional Identity (BMLEH) based on a decision of the Parliament of the Federal Republic of Germany via the Federal Office for Agriculture and Food (BLE) under the research and innovation program 'Climate Protection in Agriculture'.

The computation of this research was done, using computing resources of the High-Performance Computing (HPC) cluster of the Osnabrück University of Applied Sciences, which were provided by the German Federal Ministry of Research, Technology and Space (BMFTR) within the HiPer4All@HSOS project.

Disclosure of Interests. The authors have no competing interests to declare that are relevant to the content of this article.

References

1. Bird, J.J., Lotfi, A.: Cifake: image classification and explainable identification of AI-generated synthetic images. IEEE Access **12**, 15642–15650 (2024). https://doi.org/10.1109/access.2024.3356122
2. Boiński, T., Zaraziński, S., Śledź, B.: Lego bricks for training classification network (2021). https://doi.org/10.34808/RCZA-JY08
3. Burdorf, S., Plum, K., Hasenklever, D.: Reducing the amount of real world data for object detector training with synthetic data (2022). https://doi.org/10.48550/ARXIV.2202.00632
4. De Roovere, P., Moonen, S., Michiels, N., Wyffels, F.: Dataset of industrial metal objects (2022). https://doi.org/10.48550/ARXIV.2208.04052
5. Dosovitskiy, A., et al.: An image is worth 16x16 words: transformers for image recognition at scale. In: International Conference on Learning Representations (ICLR) (2021). https://openreview.net/forum?id=YicbFdNTTy
6. Ettedgui, S., Abu-Hussein, S., Giryes, R.: ProCST: boosting semantic segmentation using progressive cyclic style-transfer (2022). https://doi.org/10.48550/ARXIV.2204.11891
7. Eversberg, L., Lambrecht, J.: Generating images with physics-based rendering for an industrial object detection task: realism versus domain randomization. Sensors **21**(23), 7901 (2021). https://doi.org/10.3390/s21237901
8. Fukushima, K.: Neocognitron: a self-organizing neural network model for a mechanism of pattern recognition unaffected by shift in position. Biol. Cybern. **36**(4), 193–202 (1980). https://doi.org/10.1007/bf00344251
9. Kim, J., Kim, D., Lee, S., Chi, S.: Hybrid DNN training using both synthetic and real construction images to overcome training data shortage. Autom. Constr. **149**, 104771 (2023). https://doi.org/10.1016/j.autcon.2023.104771
10. Klein, J., Waller, R., Pirk, S., Pałubicki, W., Tester, M., Michels, D.L.: Synthetic data at scale: a development model to efficiently leverage machine learning in agriculture. Front. Plant Sci. **15** (2024). https://doi.org/10.3389/fpls.2024.1360113
11. Krizhevsky, A., Hinton, G.: Learning multiple layers of features from tiny images. Technical report, University of Toronto, Toronto, Ontario (2009). https://www.cs.toronto.edu/~kriz/learning-features-2009-TR.pdf
12. Lambrecht, J., Kästner, L.: Towards the usage of synthetic data for marker-less pose estimation of articulated robots in RGB images. In: 2019 19th International Conference on Advanced Robotics (ICAR), pp. 240–247. IEEE (2019). https://doi.org/10.1109/icar46387.2019.8981600
13. LDraw.org: Content and licensing information (2020). https://www.ldraw.org/. Accessed 24 Apr 2025

14. Nikolenko, S.I.: Synthetic Data for Deep Learning. Springer, Cham (2021). https://doi.org/10.1007/978-3-030-75178-4
15. Nowruzi, F.E., Kapoor, P., Kolhatkar, D., Hassanat, F.A., Laganiere, R., Rebut, J.: How much real data do we actually need: analyzing object detection performance using synthetic and real data (2019). https://doi.org/10.48550/ARXIV.1907.07061
16. Peng, X., Bai, Q., Xia, X., Huang, Z., Saenko, K., Wang, B.: Moment matching for multi-source domain adaptation. In: 2019 IEEE/CVF International Conference on Computer Vision (ICCV), pp. 1406–1415. IEEE (2019). https://doi.org/10.1109/iccv.2019.00149
17. Peng, X., Usman, B., Kaushik, N., Wang, D., Hoffman, J., Saenko, K.: Visda: a synthetic-to-real benchmark for visual domain adaptation. In: 2018 IEEE/CVF Conference on Computer Vision and Pattern Recognition Workshops (CVPRW), pp. 2102–21025. IEEE (2018). https://doi.org/10.1109/cvprw.2018.00271
18. Poucin, F., Kraus, A., Simon, M.: Boosting instance segmentation with synthetic data: a study to overcome the limits of real world data sets. In: 2021 IEEE/CVF International Conference on Computer Vision Workshops (ICCVW), pp. 945–953. IEEE (2021). https://doi.org/10.1109/iccvw54120.2021.00110
19. Rajpura, P.S., Bojinov, H., Hegde, R.S.: Object detection using deep CNNs trained on synthetic images (2017). https://doi.org/10.48550/ARXIV.1706.06782
20. Rombach, R., Blattmann, A., Lorenz, D., Esser, P., Ommer, B.: High-resolution image synthesis with latent diffusion models. In: 2022 IEEE/CVF Conference on Computer Vision and Pattern Recognition (CVPR), pp. 10674–10685. IEEE (2022). https://doi.org/10.1109/cvpr52688.2022.01042
21. Ros, G., Sellart, L., Materzynska, J., Vazquez, D., Lopez, A.M.: The synthia dataset: a large collection of synthetic images for semantic segmentation of urban scenes. In: 2016 IEEE Conference on Computer Vision and Pattern Recognition (CVPR). IEEE (2016). https://doi.org/10.1109/cvpr.2016.352
22. Schmidhuber, J.: Annotated history of modern AI and deep learning (2022). https://doi.org/10.48550/ARXIV.2212.11279
23. Staniszewski, M., et al.: Searching for the ideal recipe for preparing synthetic data in the multi-object detection problem. Appl. Sci. **15**(1), 354 (2025). https://doi.org/10.3390/app15010354
24. Szeliski, R.: Computer Vision: Algorithms and Applications, 2nd edn. Springer, Cham (2022). https://doi.org/10.1007/978-3-030-34372-9
25. Tremblay, J., et al.: Training deep networks with synthetic data: bridging the reality gap by domain randomization. In: 2018 IEEE/CVF Conference on Computer Vision and Pattern Recognition Workshops (CVPRW), pp. 1082–10828. IEEE (2018). https://doi.org/10.1109/cvprw.2018.00143
26. Vanherle, B., Moonen, S., Reeth, F.V., Michiels, N.: Analysis of training object detection models with synthetic data. In: 33rd British Machine Vision Conference 2022, BMVC 2022, London, UK, November 21–24, 2022. BMVA Press (2022). https://bmvc2022.mpi-inf.mpg.de/0833.pdf
27. Vaswani, A., et al.: Attention is all you need. In: Guyon, I., Luxburg, U.V., Bengio, S., Wallach, H., Fergus, R., Vishwanathan, S., Garnett, R. (eds.) Advances in Neural Information Processing Systems, vol. 30. Curran Associates, Inc. (2017). https://proceedings.neurips.cc/paper_files/paper/2017/file/3f5ee243547dee91fbd053c1c4a845aa-Paper.pdf
28. Vuletic, J., Polic, M., Orsag, M.: Procedural generation of synthetic dataset for robotic applications in sweet pepper cultivation. In: 2022 International Conference on Smart Systems and Technologies (SST), pp. 309–314. IEEE (2022). https://doi.org/10.1109/sst55530.2022.9954643

29. Wachter, P., Kruse, N., Schöning, J.: Synthetic fields, real gains: enhancing smart sgriculture through hybrid datasets. In: Informatik in der Land-, Forst-und Ernährungswirtschaft-Fokus: Biodiversität fördern durch digitale Landwirtschaft, pp. 437–442. Gesellschaft für Informatik e.V. (2024). https://doi.org/10.18420/GILJT2024_44
30. Wang, M., Deng, W.: Deep visual domain adaptation: a survey. Neurocomputing **312**, 135–153 (2018). https://doi.org/10.1016/j.neucom.2018.05.083
31. Xiong, R., et al.: On layer normalization in the transformer architecture. In: Proceedings of the 37th International Conference on Machine Learning. ICML 2020, JMLR.org (2020). https://dl.acm.org/doi/10.5555/3524938.3525913
32. Zhu, J.Y., Park, T., Isola, P., Efros, A.A.: Unpaired image-to-image translation using cycle-consistent adversarial networks. In: 2017 IEEE International Conference on Computer Vision (ICCV), pp. 2242–2251. IEEE (2017). https://doi.org/10.1109/iccv.2017.244

Towards Systematic Evaluation of Computer Vision Models Under Data Anonymization

Sarah Weiß(✉), Christopher Bonenberger, Tobias Niedermaier, Maik Knof, Benjamin Stähle, and Markus Schneider

Institute for Artificial Intelligence, HS Ravensburg-Weingarten (RWU), Weingarten, Germany
Sarah.Weiss@rwu.de

Abstract. Camera-based systems in everyday applications require careful handling of privacy-sensitive image data. A promising approach to prevent data misuse is anonymization right after perception but before processing. However, traditional anonymization methods such as blurring and pixelation remove information essential for subsequent processing algorithms. Realistic anonymization, in contrast, can preserve vital information by generation of naturalistic replacements. Nevertheless, these systems pose unresolved questions regarding information preservation and correctness. We develop a systematic approach to analyze the effects of anonymization methods on model training and downstream tasks.

Through a quantitative analysis of various anonymization methods, the challenges and pitfalls introduced by anonymization are highlighted. The lack of datasets for anonymization research, forces the adaption of inapt data. We use the state-of-the-art toolbox DeepPrivacy2 to generate realistic full-body anonymized data from COCO data and evaluate object classification using YOLOv10. Additionally, we demonstrate that appropriate evaluation of anonymization techniques requires specialized datasets. To address this gap, we introduce a handcrafted dataset, enabling us to prove the need for dedicated anonymization datasets.

Based on systematically selected metrics, we assess the impact of anonymization on images or classes and present a range of experiments focusing on factors as object size and co-occurrence frequency with the anonymized class. Furthermore, novel findings on robustness of different model sizes and processing of anonymized images are presented. Our findings guide future directions in model adaptation to anonymized data, highlight improvements necessary in realistic anonymization generation, and underscore the importance of dedicated anonymization datasets.

Keywords: Anonymization · Computer Vision · Object Detection

1 Introduction

Technologies like Artificial Intelligence, the Internet of Things and cloud computing transform modern life by enabling disruptive innovations and novel applications. However, these advancements also pose significant risks to privacy [11].

T. Braun et al. (Eds.): KI 2025, LNAI 15956, pp. 190–203, 2026.
https://doi.org/10.1007/978-3-032-02813-6_14

Fig. 1. Left: Non-realistic and realistic full-body anonymization methods. Objects within the anonymized region (e.g., a wristband) are removed. **Right:** Difference image of the original and DeepPrivacy2-anonymized version, revealing changes in surrounding areas, altering nearby objects.

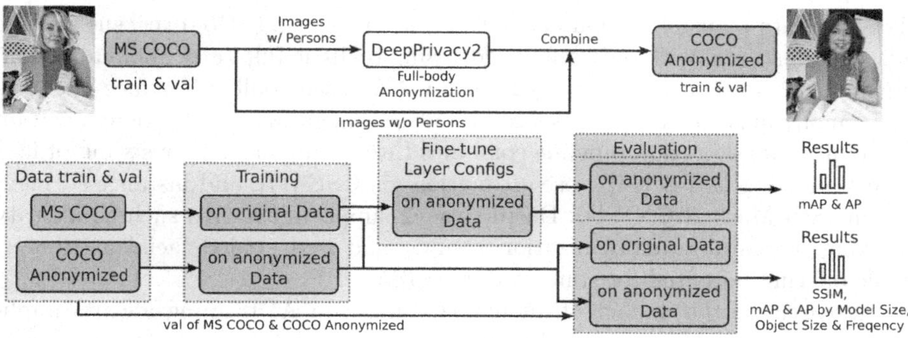

Fig. 2. Top: Anonymizing the entire COCO train/val sets with DeepPrivacy2. **Bottom:** Training/evaluation pipeline without filtering hard or complex scenes.

While private users can reject, public spaces do not offer this possibility. Despite growing attention of the public and authorities (e.g. [3]), many domains lack adequate solutions. In Ambient Assisted Living (AAL), privacy concerns often lead to the avoidance of cameras [1]. For Autonomous Driving (AD), simple blurring or face generation is used (e.g. [10]). Blurring may be reversed [12,19,22], and face generation does not ensure full privacy, as it keeps identifying features [8]. This work addresses the challenge of privacy-compliant usage of machine learning (ML) and computer vision. *Realistic anonymization* like DeepPrivacy2 [5] (see Fig. 1, left) are promising, as they enable continued use of image data or existing algorithms without compromising privacy. However, adopting this technique raises several questions—either unexplored or only partially examined. Our goal is to contribute to a deeper understanding of the associated challenges.

We investigate the impact of anonymization on ML training for object detection. In this context, we address a set of key questions, including the effect of different training settings, the role of model size and image distortions, fine-tuning with or without anonymized data, and the necessity of a dedicated anonymization dataset. For this broad set of investigations, the systematic approach shown in Fig. 2 is applied (details in Sect. 3).

2 Related Work

Existing research on anonymization primarily focuses on evaluating anonymization techniques and their impact on person detection, with fewer studies addressing the influence of anonymized data on model training. A broad and comprehensive survey on visual privacy is provided in [18], with recent findings about the evaluation of anonymization effects reported in [6,13,23].

A structured analysis of data anonymization must account for a broad spectrum of algorithms, ranging from simple to state-of-the-art techniques. [18] offers a comprehensive survey, covering both traditional and advanced approaches. In general, non-realistic anonymization methods—such as pixelation, blurring, and masking—tend to be unsuitable for computer vision tasks due to significant information loss. In contrast, realistic anonymization methods are more advanced, as they generate plausible visual content, as shown in Fig. 1. However, such methods remain scarce, with only a few addressing realistic full-body anonymization, including [2,5,7,16]. DeepPrivacy2 [5], a GAN-based toolbox for face and full-body anonymization, represents a recent advancement in realistic anonymization techniques. Its detection pipeline combines three components, consisting of face detection using DSFD [14], pose estimation via CSE [17], and instance segmentation with Mask R-CNN [4]. DeepPrivacy2 supports features such as identity tracking, repeatable anonymization results, and text-guided facial attributes, while offering potential for context preservation.

Research on the impact of anonymized data on AI training and its implications for computer vision tasks remains limited. [13] investigates the use of non-realistic anonymization techniques, and their effects on different computer vision tasks if used as training data. It further states a relationship between lower performance and the co-occurrence with the anonymized class. The impact of face-anonymized training data on segmentation tasks within AD is investigated by [23]. Both studies report that larger models tend to be more resilient to performance degradation caused by anonymization, than smaller ones. However, [6] presents opposing findings, suggesting that smaller models are less affected.

To determine whether realistic anonymization fundamentally alters image features or merely requires classification fine-tuning, a detailed examination of the employed model's architecture is essential. The architecture of YOLOv10 [9,20] comprises three primary components, each responsible for a distinct stage in the object detection pipeline. The backbone extracts features from the input image, which are then forwarded to the neck and into a self-attention layer (PSA). The PSA captures global-scale patterns within the features. Extracted features along with the global patterns provided by the self-attention layer are processed and combined within the neck. This enables a comprehensive feature combination before passing the data to the head, which determines class types, defines bounding boxes, and outputs the final detections.

In contrast to the aforementioned studies, we establish a systematic evaluation approach to analyze anonymization methods, assessing both image modifications and their effect on model training and detection performance. Furthermore, we investigate the influence of model size on performance under realistic anonymization, contributing additional results to the contradiction between [13,23] and [6]. Finally, performance gains from relabeling anonymized data are demonstrated by comparing the use of the original ground truth with a ground truth adapted to anonymized data.

3 Experiments and Results

In the following, we present several experiments to evaluate the impact of anonymization on model training and performance. Our key investigations include: evaluating detection performance and model size influence, assessing anonymization effects on an image level, comparing realistic to non-realistic anonymization, analyzing impact of object size and frequency, exploring fine-tuning with different freezing configurations, and assessing improvements through annotation correction for anonymized data. An overview of our experiments is shown in Fig. 2. The code and detailed figures of single experiments are available on GitHub[1].

The majority of experiments use data from the MS COCO 2017 instance data set (cf. [15]). An anonymized version is created for training and evaluation, following the workflow shown in Fig. 2 (top). All images containing persons are retrieved and processed with DeepPrivacy2 in full-body anonymization mode.

Initially, we investigate changes through anonymization at the image level, extending beyond pixel-difference metrics. Since the anonymization inherently affects the regions surrounding the anonymized individuals (see Fig. 1, right), a more sophisticated evaluation is necessary. To address this, the Structural Similarity Index Measure (SSIM) [21] is used, which differs from absolute-difference metrics such as Mean Squared Error (MSE) by assessing perceptual similarity.

SSIM evaluates changes in luminance, contrast, and structure between two images or neighborhoods. Its core principle is that pixels exhibit strong dependencies, especially within a neighborhood. These dependencies encode crucial structural information about objects in a scene. Consequently, statistical properties such as pixel mean and variance are incorporated into the SSIM calculation (full definition in [21]). The final SSIM score is derived as a weighted product of luminance, contrast, and structure, yielding values in the range $[-1, 1]$. A score of 1 means perfect similarity, scores of 0 imply no similarity and negative values represent anti-similarity (negative correlation).

For SSIM evaluation, we define case-specific groups tailored to applications in AAL and AD. Specifically, we distinguish between

- *change group* (CG), where object modifications are acceptable, and a
- *no change group* (NoCG), which should remain unaltered by anonymization.

This grouping is inherently application-dependent. For instance, in an autonomous driving context, classes like ‘traffic sign’ belong to the NoCG, while classes like ‘chair’ may belong to the CG. Conversely, in an AAL context, ‘chair’ typically belongs to the NoCG, whereas traffic sign’ usually falls into the CG.

[1] https://github.com/iki-wgt/anonymized_data_evaluation_and_training_pipeline.

To assess overall model performance and the impact of model size, we evaluate all YOLOv10 variants, from YOLOv10n (2.3 million trainable parameters) to YOLOv10X (around 30 million parameters). Each model size is trained on the original COCO training set and the full-body anonymized training set—yielding two models per size, differing in training data. All models are evaluated on both the anonymized validation set and the original COCO validation set. For all possible configurations, we use the following abbreviations

- *Org*: for a model trained on original (non-anonymized) data, and
- *Anon*: for models trained on full-body anonymized data, while
- *on Org*: stands for an evaluation on original validation data, and
- *on Anon*: is an evaluation on anonymized validation data.

For instance, *Anon on Anon* refers to a model trained and evaluated on anonymized data. Regarding training methodology, we follow [13], using default settings and the SGD optimizer. Full training details are available on GitHub.

To analyze the impact of object size and frequency, we use four different model and evaluation data combinations as described above, with focus on YOLOv10m models. Specific object classes are selected to represent varying frequency levels. Classes that frequently co-occur with the 'person' class are grouped into three frequency categories based on their joint appearance count across images. Frequency group boundaries are defined based on noticeable shifts in class-wise image count, resulting in definitions of

- High frequency f_{high}: 9,000 down to 3,200 images,
- Medium frequency f_{med}: 3,200 down to 1,400 images, and
- Low frequency f_{low}: less than 1,400 images

Since performance loss is expected to depend not only on frequency but also on object size, the classes are further categorized by size relative to humans, as

- Small objects: items that are easily carried in one hand or fit inside it,
- Medium-sized objects: approximately knee- to hip-height, and
- Large objects: human-sized items or items that are significantly larger.

Based on the defined frequency and size categories, four classes are selected per combination. The chosen objects are listed in Table 1. To investigate the impact of anonymization on detection performance, changes in AP are investigated for *Org on Anon, Anon on Org* and *Anon on Anon* each relative to the baseline *Org on Org* evaluation.

To assess how anonymization affects training, an *Org* model (size m), is fine-tuned on anonymized data for 30 additional epochs. Multiple layer-freezing strategies are compared to identify which detection step is influenced the most by anonymization. We refer to these frozen layers as Backbone (layer 0–10), Backbone without Psa (0–9), Neck (11–22), Neck with Head (11–23) and Neck with Head and Psa (10–23).

If fine-tuning the neck (freezing the backbone) leads to improved performance, it suggests that anonymization mainly impacts features analysis and

Fig. 3. Examples from proposed data set with problematic original ground truth. **Left:** 'Tie' (blue) and 'book' (pink) removed by anonymization. **Right:** Visualization of required extension (yellow mask) to original 'person' box (yellow). (Color figure online)

combination rather than fundamentally altering the image. This implies that anonymized individuals retain key natural characteristics, aligning with expectations based on the quality of anonymization, where essential features such as face, hair, arms, and legs seem clearly recognizable.

Conversely, if fine-tuning the backbone yields better results, it indicates that anonymization introduces substantial pixel-level changes, necessitating adjustments in feature extraction to improve performance. Additionally, the fine-tuned models are compared to those trained solely on anonymized data to evaluate whether pre-training on original data offers a performance advantage when adapting to anonymized images. To prevent overfitting on classes likely unaffected by anonymization, a subset of the anonymized training data is selected. Only images containing anonymized persons and the chosen classes are included. From these, a random sample comprising 20% of the full training set is drawn. Used parameters for fine-tuning are listed in our repository.

Finally, we examine the impact of labeling errors, as the original ground truth may be unreliable for anonymized data—especially if individuals interact with objects. During anonymization, such objects may not be preserved (see Fig. 1), resulting in labeling inconsistencies that can affect both training and evaluation.

When training on anonymized data, models depend on object labels from altered pixel regions, where objects may have been removed or modified due to anonymization. During evaluation, this leads to two key issues: (1) the model

Table 1. Objects classified by size and frequency.

Size	Frequency		
	low	medium	high
small	banana,vase, bird,toothbrush	book,bowl, baseball glove,clock	sports ball,cup, cell phone,bottle
medium	hydrant,microwave, toilet,keyboard	dog,laptop, baseball bat,suitcase	backpack,chair, umbrella,bench
large	refrigerator,elephant, bed,stop sign	couch,bus, horse,motorcycle	dining table,car, truck,surfboard

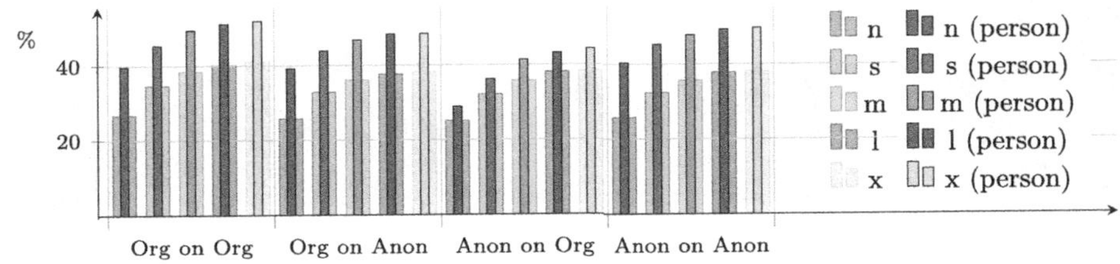

Fig. 4. Performance across all training/evaluation pairings and model sizes for all classes (shaded, mAP) and the anonymized class 'person' (narrow bars, AP).

Table 2. Changes in mAP across model sizes. **Left:** Compared to the *Org on Org* base model. **Right:** For exchanged evaluation data types compared to training type.

Model	Size									
	n	s	m	l	x	n	s	m	l	x
Org on Org	26.6	34.6	38.5	40.1	41.2	26.6	34.6	38.5	40.1	41.2
Org on Anon	−1.1	−1.8	−2.5	−2.6	−2.9	−1.1	−1.8	−2.5	−2.6	−2.9
Anon on Org	−1.7	−2.6	−2.8	−2.1	−2.8	25.6	32.3	35.6	37.7	38.2
Anon on Anon	−1.0	−2.3	−2.9	−2.4	−3.0	−0.7	−0.3	+0.1	+0.3	+0.2

may correctly fail to detect an object that has been removed, yet this will be incorrectly counted as a false negative; (2) due to erroneous training data, the model might falsely detect non-existent objects, resulting in false positives. This issue has not yet been explored, largely due to the lack of datasets tailored for anonymization. To address this, we introduce a handcrafted dataset focused on an AAL application, comprising 80 selected images recorded in an ambulatory care facility.[2] Examples of the data set are provided in Fig. 3. This data is used to assess the impact of relabeling after anonymization with DeepPrivacy2. We quantify the effect by evaluating the anonymized subset using both the original ground truth and corrected annotations. This comparison reveals the impact of annotation errors on model performance and highlights the importance of accurate re-labeling in anonymized datasets.

3.1 General Performance and Influence of Model Size

To assess performance and the influence of model size, we compare each model-evaluation combination across all model sizes. Figure 4 shows that mAP increases with model size in all types of training and evaluation. However, larger models also exhibit a greater absolute drop compared to *Org on Org*, aligning with [6] but contradicting [13] and [23] (see Table 2, left). Compared to [13], these differences remain small; their non-realistic anonymization yields a drop of −2.4 to −4.9 for object detection on original images. Achieving a smaller performance reduction with realistic anonymization is more favorable, as noted by [6].

[2] The dataset, and further details, will be released upon publication (see repository).

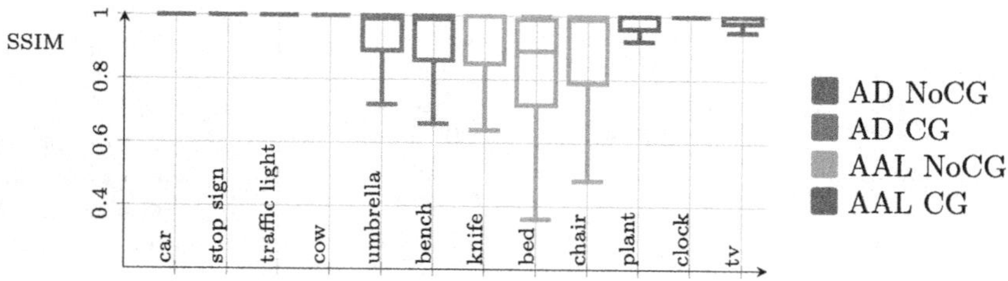

Fig. 5. SSIM of respective bounding boxes between oiginal and anonymized images.

Table 3. AP differences of NoCG classes, evaluated on original and anonymized data for *Org* (size m). Bold indicates AP decrease, aligning with lower SSIM scores in Fig. 5.

Model	Class					
	car	stop sign	traffic light	knife	bed	chair
Org on Org	38.0	28.8	15.6	13.5	30.2	22.0
Org on Anon	0.0	+0.4	**−0.1**	**−2.9**	**−1.3**	**−1.2**

Anonymized data lowers performance relative to *Org on Org* in both training and evaluation. When data types match training type, *Anon on Anon* scores below the *Org on Org* baseline, whereas other combinations decline less.

The data exchangeability is explored in Table 2 (right). In general, original-data models exhibit a higher performance drop when switching to anonymized data. Models trained on anonymized data show only a minor drop when evaluated on original data, suggesting a certain exchangeability of data type. These differences likely stem from imperfections introduced by anonymization, similar to data augmentation, which can boost resilience. Section 3.2 explores these effects more in depth.

Figure 4 shows similar trends for the anonymized class ‘person’, but it reveals clearer difference in data type exchangeability: *Org on Anon* reduces by −3.5 AP, whereas *Anon on Org* drops by −5.5 AP (size x). In our use cases, these represent significant decreases for an important class, so switching between data types should be avoided within these use cases. When maintaining the same training type within the evaluation (*Org on Org*, *Anon on Anon*), the gap between *Org* and *Anon* models shrinks (e.g., x-size: −2.2). With targeted tuning and carefully crafted datasets, this drop can be reduced or eliminated. For applications requiring anonymized data, a model trained on anonymized data is preferable, particularly for person classification, where training and inference on the same data type yield better performance.

3.2 Effects of Anonymization on an Image Level

Comparing SSIM scores for class-specific bounding boxes assesses anonymization-induced changes to each object. Figure 5 shows that ‘car’, ‘stop sign’, and ‘traffic

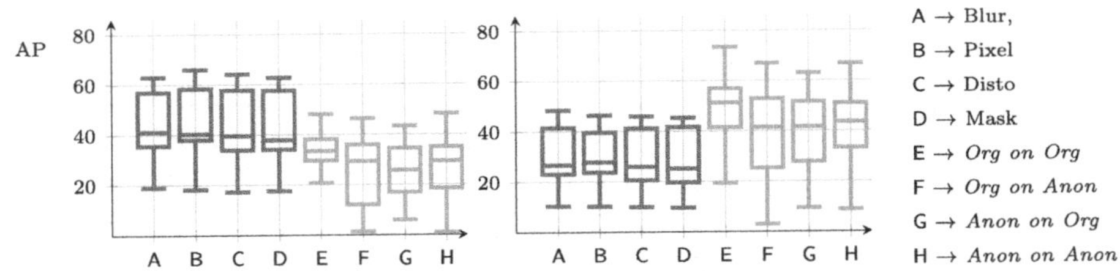

Fig. 6. YOLOv10m comparison of our results (orange) with [13] (blue) for IOU 50 (right) and IOU (50 to 95). [13] trains on anonymized data and evaluates on original Data. Our models train and evaluate on Org and Anon. (Color figure online)

light' remain unchanged (SSIM = 1.0), confirming that NoCG is unaffected by anonymization. Minimal changes in AD CG (SSIM around 0.9) indicate limited impact on relevant classes. In AAL, NoCG classes are more affected, likely because they are closely linked to persons. This also appears for "umbrella" and "bench" in AD. AAL CG values stay near 1, exceeding AD CG.

These findings indicate anonymization is more problematic for human-centered tasks like AAL, where relevant classes undergo larger changes and a more pronounced AP decline, correlating with bounding-box-based SSIM (compare Table 3 and Fig. 5). Overlaps with generated content or partial contact can remove crucial features, as shown in Fig. 1 (e.g., the umbrella's pointy edges). Per [13], models trained on original data often rely on shape and contour, so reduced performance on anonymized data is explainable by the observed effects.

3.3 Realistic vs. Non-realistic Anonymization

To compare anonymization performance to the models of [13], here we use the same 14 classes for evaluation. Figure 6 shows AP at IoU 50 to 95 (left) and IoU 50 (right). For an IoU of 50 to 95, our models perform worse than [13]. Notably, *Org on Org* has the smallest AP variance, while the reference models maintain relatively consistent variance across non-realistic anonymization methods.

When evaluated at an IoU of 50, our models outperform those of [13] (right side of Fig. 6), indicating better detection with less stringent overlap. As performance drops at higher IoU thresholds (50 to 95), it suggests difficulties in precise localization. Lower IoU thresholds like 50 are more forgiving, while higher ones suit tasks requiring accurate localization, such as in AD.

A likely reason for weaker performance is that [13] employs non-realistic anonymization, introducing abrupt changes in the anonymized region that may be easier to learn and usually affect only the anonymized object. DeepPrivacy2 creates natural-looking persons that blend into the image, altering more surrounding pixels and generating additional errors. Restricting classes to those closely tied to humans (Sect. 3.2) amplifies the issue.

Yet, these factors do not fully explain why the base model (*Org on Org*) underperforms. One possible reason is that the training and optimization approach adopted from [13] did not transfer effectively from YOLOv8 to YOLOv10.

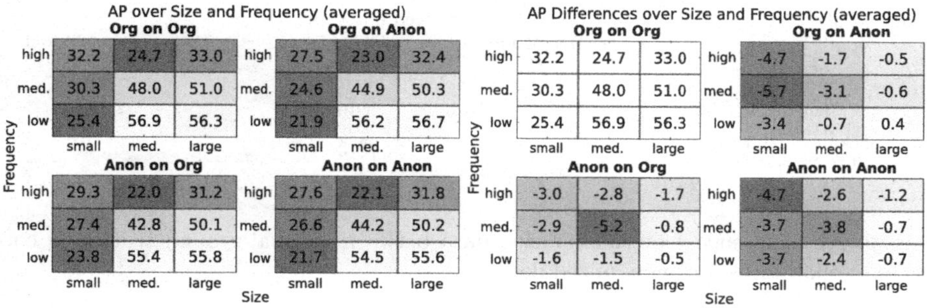

Fig. 7. Left: Class AP averages by Size-Frequency pair. **Right:** Differences to *Org on Org* of other evaluation combinations. Model: size m.

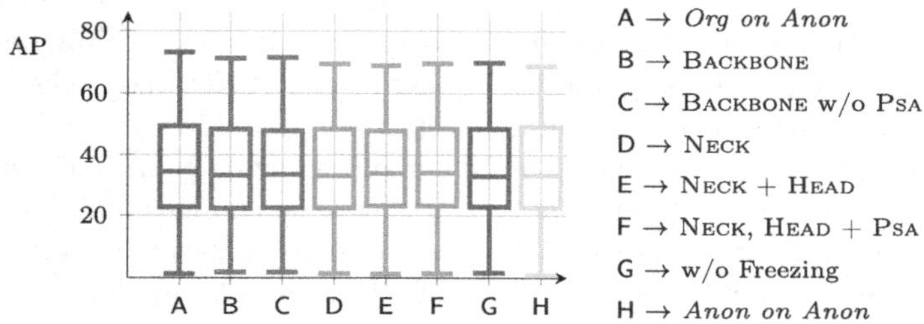

Fig. 8. Different freezing strategies compared to fine-tuned base model (*Org* size m) and *Anon* model. All models are evaluated on anonymized data.

3.4 Influence of Object Size and Frequency

AP and its changes from *Org on Org*, averaged per size-frequency pair, appear as heatmaps (classes in Table 1). Lower AP values are indicated by a darker color. Figure 7 (left) shows average AP per size-frequency pairing. Objects with higher frequency to 'person' yield lower performance across sizes, and smaller objects also show lower performance. These patterns hold for all models and evaluations, reflecting a general performance trend rather than anonymization effects.

Figure 7 (right) compares average changes in AP between the base *Org on Org* and the other model-evaluation combinations, revealing a stronger link between lower performance and smaller objects. Higher frequency with the 'person' class also lowers performance. Negative impacts are stronger for *Org on Anon* and *Anon on Anon* than for *Anon on Org*, indicating that training on anonymized data boosts resilience to data type changes. Like in [13], high-frequency classes perform worse, but smaller objects see an even sharper drop; as they are more likely to be altered or removed (e.g., 'tie' drops from 31.4 AP to 1.3 AP).

3.5 Influence on Model Training

Different freezing strategies for fine-tuning aim to identify where anonymization affects the model's learning. The AP comparison in Fig. 8 shows that fine-tuning

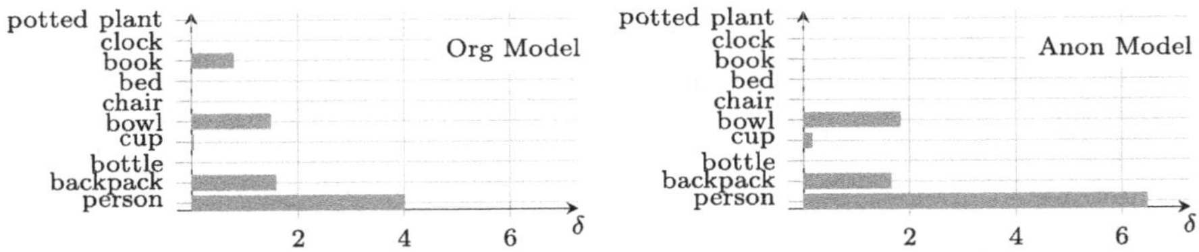

Fig. 9. AP difference δ on anonymized data between original ground truth and corrected annotations for anonymized data.

does not improve performance. Various frozen-layer configurations only converge to the anonymized model's performance, which is worse than the base model.

The non-frozen model performs similarly to the *Anon on Anon* model. Fine-tuning of an original model gives slightly better results than training directly on anonymized data, presumably due to the base model's prior knowledge. Although differences are minimal, the graph suggests that models with frozen backbone outperform those with frozen neck, implying anonymization primarily affects feature analysis and combination. However, focusing on changes of the anonymized class points to a different interpretation, since anonymization occurs there.

The AP (IOU of 50) for person' started for the base *Org* model on anonymized data with 65.8. Freezing the backbone leads to an improvement by +1.6, which is close to the *Anon on Anon* performance of +1.7. All other fine-tuning configurations (including Backbone without Psa) lead to an even better result of around +2.4 AP. Not freezing at all improves results further to +3.2, suggesting that newly learned features need different processing. These findings align with [13], which notes that anonymized models interpret features differently from models trained on original data. The assumption that generated humans retain key classification features does not hold for anonymized persons.

We observe a general performance decrease on anonymized data, indicating a broader issue when using an anonymized dataset with the original ground truth. Generated content sometimes occludes other classes, leading to evaluation errors. Section 3.6 illustrates the effect of correcting these errors.

3.6 Influence of Label Error

Using our own dataset, it is tested if relabeling is worthwhile. Both *Org* and *Anon* models evaluate against two ground truths: original annotations and corrected annotations after anonymization.

Figure 9 shows that AP for *Org* and *Anon* either improves or remains unchanged, indicating no reduction in AP for any evaluated class. Correcting labels on anonymized data boosts results, especially for small objects, and provides a measurable gain for the 'person' class. As DeepPrivacy2 sometimes fails to anonymize persons when lying or sitting, as seen in Fig. 3, and adjusting to the correct dimensions likely explains the improvement.

Correcting erroneous annotations yields a noticeable improvement in evaluation for both training types. Models trained on anonymized data with corrected annotations produce better results because they align with the supplied data. The key result is that existing datasets cannot be directly used with anonymization without additional label adjustments.

4 Conclusions and Future Work

This study develops a systematic approach to analyze the effects of anonymization methods on model training and downstream tasks. Applying this novel method allows to identify several challenges and pitfalls associated with the use of anonymized data in computer vision pipelines, such as performance degradation, difficulties in object localization, and increased labeling errors. Despite these inherent challenges, our findings demonstrate how future research can assess the performance of computer vision tasks on anonymized data. We introduced a systematic evaluation strategy for assessing the impact of data anonymization in computer vision. SSIM is used to quantify visual changes caused by anonymization. We explored various training and fine-tuning strategies, compared multiple anonymization methods, and analyzed the influence of model size on performance with anonymized data. Additionally, we gained further insights on how object size and frequency with the anonymized class impacts detection performance. To support future research, we provide a curated benchmark dataset along with best practices for anonymization evaluation.

In summary, models trained on anonymized data exhibit better generalization to mismatched inputs, although models trained on original data consistently perform best. Fine-tuning from an original model yields slight gains compared to a solely anonymized trained model, suggesting that retained knowledge is beneficial. Especially, the 'person' class improves through this. Realistic anonymization methods introduce greater performance degradation compared to nonrealistic anonymization due to their more complex visual modifications. Finally, correcting labels significantly improves model performance, emphasizing the critical need for dedicated, accurately annotated datasets in anonymization research.

Regarding full-body anonymization, future work should focus on minimizing the bordering region around the anonymized person. Moreover, occlusion remains a major challenge. Beneath technical improvements, there is a need for deeper exploration of decision-making processes and for specialized datasets emphasizing distinct detection processes in anonymized scenarios. Computer vision datasets specifically designed for anonymization research are lacking. Such datasets require rich annotations (objects, segmentation masks, poses, human actions, emotions, context) and must be use case–specific, adding complexity in development and annotation corrections.

Advances in these priorities boost research on privacy- and law-respecting AAL and AD applications. Privacy-by-design systems, including anonymization, produce data usable for follow-up algorithms. The findings of this and prior work underscore the importance of continuing progress in anonymization to establish transformative technologies like AAL and AD.

References

1. Ake-Kob, A., et al.: State of the art on ethical, legal, and social issues linked to audio- and video-based AAL solutions. CA19121 GoodBrother COST Action (2021). https://doi.org/10.5167/UZH-220633, https://www.zora.uzh.ch/id/eprint/220633
2. Brkic, K., Sikiric, I., Hrkac, T., Kalafatic, Z.: I know that person: generative full body and face de-identification of people in images. In: 2017 IEEE Conference on Computer Vision and Pattern Recognition Workshops (CVPRW), pp. 1319–1328. IEEE, Honolulu, HI, USA (2017). https://doi.org/10.1109/CVPRW.2017.173, http://ieeexplore.ieee.org/document/8014907/
3. European Parliament, Council of the European Union: Regulation (EU) 2016/679 of the European Parliament and of the Council. of 27 April 2016 on the protection of natural persons with regard to the processing of personal data and on the free movement of such data, and repealing Directive 95/46/EC (General Data Protection Regulation) (2016). https://data.europa.eu/eli/reg/2016/679/oj, place: OJ L 119, 4.5.2016, p. 1–88
4. He, K., Gkioxari, G., Dollar, P., Girshick, R.: Mask R-CNN. In: 2017 IEEE International Conference on Computer Vision (ICCV), pp. 2980–2988. IEEE, Venice (2017). https://doi.org/10.1109/ICCV.2017.322, http://ieeexplore.ieee.org/document/8237584/
5. Hukkelas, H., Lindseth, F.: DeepPrivacy2: towards realistic full-body anonymization. In: 2023 IEEE/CVF Winter Conference on Applications of Computer Vision (WACV), pp. 1329–1338. IEEE, Waikoloa, HI, USA (2023). https://doi.org/10.1109/WACV56688.2023.00138, https://ieeexplore.ieee.org/document/10030153/
6. Hukkelås, H., Lindseth, F.: Does image anonymization impact computer vision training? (2023). https://doi.org/10.48550/arXiv.2306.05135, http://arxiv.org/abs/2306.05135, arXiv:2306.05135 [cs]
7. Hukkelås, H., Smebye, M., Mester, R., Lindseth, F.: Realistic full-body anonymization with surface-guided GANs (2023). https://doi.org/10.48550/arXiv.2201.02193, http://arxiv.org/abs/2201.02193, arXiv:2201.02193 [cs]
8. Jain, A.K., Flynn, P., Ross, A.A.: Handbook of Biometrics. Springer, Cham (2008)
9. Jocher, G., Qiu, J., Chaurasia, A.: Ultralytics YOLO (2023). https://github.com/ultralytics/ultralytics
10. Klemp, M., Rösch, K., Wagner, R., Quehl, J., Lauer, M.: LDFA: latent diffusion face anonymization for self-driving applications. In: Proceedings of the IEEE/CVF Conference on Computer Vision and Pattern Recognition (CVPR) Workshops, pp. 3199–3205 (2023)
11. Kröger, J.L., Miceli, M., Müller, F.: How data can be used against people: a classification of personal data misuses (2021). Available at SSRN 3887097
12. Kupyn, O., Martyniuk, T., Wu, J., Wang, Z.: DeblurGAN-v2: deblurring (orders-of-magnitude) faster and better. In: 2019 IEEE/CVF International Conference on Computer Vision (ICCV), pp. 8877–8886. IEEE (2019). https://doi.org/10.1109/ICCV.2019.00897, https://ieeexplore.ieee.org/document/9008540/
13. Lee, J.H., You, S.J.: Balancing privacy and accuracy: exploring the impact of data anonymization on deep learning models in computer vision. IEEE Access **12**, 8346–8358 (2024). https://doi.org/10.1109/ACCESS.2024.3352146, https://ieeexplore.ieee.org/document/10387326/
14. Li, J., et al.: DSFD: dual shot face detector. In: 2019 IEEE/CVF Conference on Computer Vision and Pattern Recognition (CVPR), pp. 5055–5064. IEEE, Long

Beach (2019). https://doi.org/10.1109/CVPR.2019.00520, https://ieeexplore.ieee.org/document/8954268/

15. Lin, T.Y., et al.: Microsoft COCO: common objects in context (2015). https://doi.org/10.48550/arXiv.1405.0312, http://arxiv.org/abs/1405.0312, arXiv:1405.0312 [cs]
16. Maximov, M., Elezi, I., Leal-Taixé, L.: CIAGAN: conditional identity anonymization generative adversarial networks. In: 2020 IEEE/CVF Conference on Computer Vision and Pattern Recognition (CVPR), pp. 5446–5455 (2020). https://doi.org/10.1109/CVPR42600.2020.00549, http://arxiv.org/abs/2005.09544, arXiv:2005.09544 [cs]
17. Neverova, N., Novotny, D., Khalidov, V., Szafraniec, M., Labatut, P., Vedaldi, A.: Continuous surface embeddings (2020). https://doi.org/10.48550/arXiv.2011.12438, http://arxiv.org/abs/2011.12438, arXiv:2011.12438 [cs]
18. Ravi, S., Climent-Pérez, P., Florez-Revuelta, F.: A review on visual privacy preservation techniques for active and assisted living. Multimedia Tools Appl. **83**(5), 14715–14755 (2023). https://doi.org/10.1007/s11042-023-15775-2, https://link.springer.com/10.1007/s11042-023-15775-2
19. Rozumnyi, D., Oswald, M.R., Ferrari, V., Matas, J., Pollefeys, M.: DeFMO: deblurring and shape recovery of fast moving objects. In: 2021 IEEE/CVF Conference on Computer Vision and Pattern Recognition (CVPR), pp. 3455–3464. IEEE (2021). https://doi.org/10.1109/CVPR46437.2021.00346, https://ieeexplore.ieee.org/document/9577579/
20. Wang, A., et al.: YOLOv10: real-time end-to-end object detection (2024). https://doi.org/10.48550/arXiv.2405.14458, http://arxiv.org/abs/2405.14458, arXiv:2405.14458 [cs]
21. Wang, Z., Bovik, A., Sheikh, H., Simoncelli, E.: Image quality assessment: from error visibility to structural similarity. IEEE Trans. Image Process. **13**(4), 600–612 (2004) https://doi.org/10.1109/TIP.2003.819861, http://ieeexplore.ieee.org/document/1284395/
22. Zhang, K., et al.: Deblurring by realistic blurring. In: 2020 IEEE/CVF Conference on Computer Vision and Pattern Recognition (CVPR), pp. 2734–2743. IEEE, Seattle, WA, USA (2020). https://doi.org/10.1109/CVPR42600.2020.00281, https://ieeexplore.ieee.org/document/9156306/
23. Zhou, J., Beyerer, J.: Impacts of data anonymization on semantic segmentation. In: 2022 IEEE Intelligent Vehicles Symposium (IV), pp. 997–1004. IEEE, Aachen, Germany (2022). https://doi.org/10.1109/IV51971.2022.9827262, https://ieeexplore.ieee.org/document/9827262/

Accessible Language Simplification: Large Language Models for Generating Easy German

Raeesa Yousaf(✉), Marina Walther, and Michael Gertz

Institute of Computer Science, Heidelberg University, Heidelberg, Germany
raeesa.yousaf@alumni.uni-heidelberg.de,
{walther,gertz}@informatik.uni-heidelberg.de

Abstract. The growing application of Large Language Models (LLMs) in text simplification holds significant promise for improving accessibility for non-native speakers and individuals with learning or cognitive disabilities. We examine the effectiveness of LLMs in generating Easy German responses through a domain-agnostic Question-Answering (QA) framework, leveraging explicit simplification rules and tailored prompting strategies. Focused on health-related questions, the framework transforms complex medical information into accessible language to enhance health literacy. Automated readability and semantic alignment metrics are combined with human evaluations from non-native speakers (A1-B1 CEFR proficiency) and individuals with special needs. Results show that GPT-4 consistently outperforms open-source models like Llama and Mixtral, generating factually accurate, clear, and accessible responses, while the latter models often struggle with coherence. Though developed for Easy German, the domain-agnostic methodology can be adapted to any language with minimal prompt adjustments. For reproducibility and material, please refer to our GitHub repository[1]https://github.com/raeesay/easy-german-qa.

Keywords: Language Models · Question Answering · Plain Language · Retrieval Augmented Generation

1 Introduction

Health literacy encompasses the ability to find, understand, analyse, and apply health-related information to make informed decisions [8]. It depends not only on individual competencies but also on the availability of high-quality, accessible information. People with learning disabilities, cognitive impairments, or limited language proficiency especially require information they can understand. To meet this need, plain language aims to ensure information is easy to read, comprehend, and use by avoiding technical jargon and overly complex structures. In many countries, public agencies are legally required to use plain language to improve access to services.

T. Braun et al. (Eds.): KI 2025, LNAI 15956, pp. 204–217, 2026.
https://doi.org/10.1007/978-3-032-02813-6_15

In Germany, the Federal Act on Equal Opportunities for Persons with Disabilities (Behindertengleichstellungsgesetz, BGG) [1] mandates that public authorities provide general rulings, notices, contracts, and laws in Easy German upon request. German distinguishes between **Einfache Sprache (Plain German)** and **Leichte Sprache (Easy German)** [10]. Plain German is a more accessible version of everyday language, intended for broader audiences, while Easy German adheres to a strict set of linguistic rules and targets people with cognitive disabilities, learning difficulties, or low literacy. The complexity of German—with nested sentence structures and compound words—makes such simplification particularly important.

Our work addresses the challenge of building a domain-agnostic, robust Question Answering (QA) system that generates responses in **Easy German**. While prior research has explored plain language generation using zero-/few-shot prompting or fine-tuning, to our knowledge, no existing approach integrates such techniques into a QA pipeline that jointly emphasizes linguistic accessibility and factual accuracy. We aim to fill this gap by leveraging Large Language Models (LLMs) and Retrieval-Augmented Generation (RAG) to produce simplified yet precise answers.

To support this, we propose two prompting strategies—**Translated-Answer** and **Translated-Context**—which introduce intermediate translation steps to enhance readability and semantic alignment. Along with **On-the-Fly Prompting**, which directly generates simplified responses, these strategies form the basis of our comparison of three distinct approaches to simplification (Sect. 3). Our modular, domain-agnostic system includes document preprocessing, retrieval, and response generation components. To address output variability, we introduce methods for improving reproducibility and reducing inconsistencies in RAG-based QA. We also explore the trade-off between readability and factual accuracy, finding that larger models with broader input windows better preserve both by incorporating more supporting context.

In Sect. 4, we present a comprehensive evaluation framework to assess system performance, including semantic alignment, readability metrics, and human-in-the-loop validation to ensure compliance with Easy German standards. Section 5 summarizes our findings and outlines directions for future work.

2 Related Work

Past work has laid important groundwork for text simplification. Anschütz et al. fine-tuned language models on corpora of Easy German texts and used them as decoders in sequence-to-sequence simplification tasks [3]. Their models effectively mimicked the style of Easy German, favoring short, precise sentences, but they were not tested in question-answering (QA) settings, which require generating coherent responses without reference material. Similarly, Deilen et al. explored ChatGPT for intralingual translation of administrative texts into Easy German [4], showing that the outputs were generally simpler, though many did not fully meet Easy Language standards due to small mistakes or missing details.

In a related work, Poupet et al. developed an automated tool to adapt complex French texts into plain language under human supervision [14]. The system calculates a complexity score from 0 (very unclear) to 100 (very clear), highlighting problematic words, sentences, or paragraphs and allowing users to accept or ignore suggested edits. Complexity is assessed across lexical, syntactic, and textual levels. Expanding beyond simplification tools, Guo et al. proposed APPLS, a framework for evaluating plain language summarization using four criteria: informativeness, simplification, coherence, and faithfulness [7]. They introduced a new metric, POMME, based on the perplexity of language models. Similarly, Joseph et al. developed FACTPICO, a benchmark for assessing factual accuracy in plain language medical summaries of randomized controlled trials [9], evaluating critical elements like populations, interventions, comparators, outcomes, and model-added information. Building on this body of work, our research introduces a novel QA framework for Easy German, aiming to address these gaps by combining simplification strategies with rigorous evaluation methods.

3 Easy German Question Answering Model

This section outlines the key modules in our Easy German Question-Answering (QA) pipeline, which integrates retrieval and generation processes to deliver simplified answers in Easy German. The pipeline consists of two main stages: context retrieval and answer generation, each of which employs distinct strategies to ensure accuracy, relevance, and accessibility. The retrieval pipeline is responsible for sourcing relevant information from a pre-processed knowledge base. It follows a naïve RAG approach [6], leveraging a Retrieve-Read framework. The generation pipeline takes the retrieved context and the user query and transforms them into an Easy German response. This is achieved through structured prompting, which integrates the query and context into a format that guides the LLM. We employ three distinct answering strategies to simplify the output:

- **On-the-Fly Prompting (Fig. 1, blue)**: The answer is directly generated in Easy German, without intermediate translation. The two instructions - answer the question and answer in Easy German - are integrated into one prompt.
- **Translated-Answer Prompting (Fig. 1, green)**: The answer is first generated in standard German and then simplified into Easy German, thus splitting the two instructions into two distinct steps.
- **Translated-Context Prompting (Fig. 1, orange)**: The context is simplified into Easy German before generating the answer. Here we also have two steps, but the first step translates the context to Easy German before prompting the language model to answer in Easy German - again using a single prompt like in the on-the-fly method.

These methods form the foundation of the system's adaptability and precision in delivering contextually relevant and comprehensible answers. Each approach tackles unique challenges in balancing linguistic simplicity, factual accuracy, and

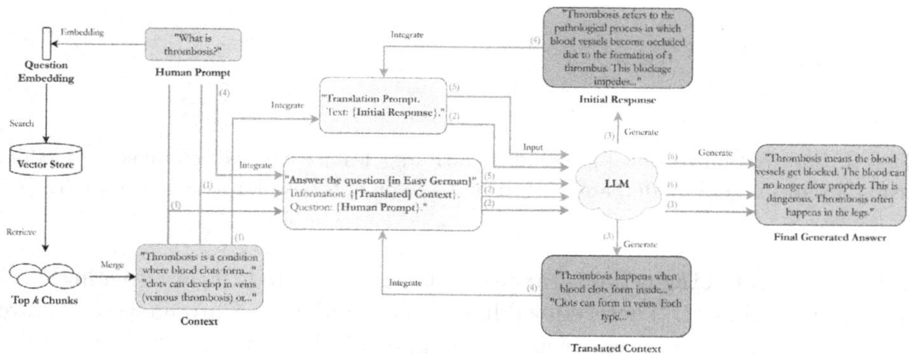

Fig. 1. Full QA Pipeline showing the three approaches to prompt large language models (LLMs) to respond in Easy German.
On-the-Fly (blue arrows): The user's question and retrieved context (light purple) are directly passed to a prompt instructing the LLM to answer the question in Easy German.
Translated-Answer (green arrows): The question is answered in standard German, this initial response (dark purple, top) is passed to a prompt that instructs the LLM to translate the response to Easy German.
Translated-Context (orange arrows): The retrieved context is first fed to the translation prompt, the translated context (dark purple, bottom) is fed with the user's question to a prompt instructing the LLM to answer in Easy German. (Color figure online)

efficiency. Prompt engineering [15] plays a central role in this pipeline by guiding the LLM to generate responses that meet the accessibility standards of Easy German. This ensures that the output is linguistically simple and easy to understand, addressing the needs of users with varying levels of literacy or cognitive ability. We employ the various prompts listed in Sect. 3.2 to guide the LLMs to the best possible results.

3.1 Information Retrieval

The information retrieval pipeline is central to generating accurate Easy German answers. It processes a large dataset of health-related documents, prepares them for semantic search, and stores them in a vector database for efficient retrieval. For our Easy German QA system targeting medical queries, we use articles from Apotheken Umschau, Germany's leading health magazine [20]. The magazine provides accessible health content in lay language, though not classified as Plain or Easy German. A small curated subset of 220 articles is available in Plain German. From an initial set of 11,307 articles, we remove duplicates and boilerplate, resulting in 8,527 usable texts.

We apply a naïve Retrieval Augmented Generation (RAG) approach (Fig. 2). Documents are split into 350-character chunks with a 70-character overlap using LangChain's `RecursiveCharacterTextSplitter` to balance efficiency and context preservation. Chunks are embedded with the multilingual model

Fig. 2. Information retrieval pipeline: Health articles are cleaned, chunked, embedded using a multilingual model, and stored in a vector database for semantic search.

`intfloat/multilingual-e5-large-instruct`[1] selected from HuggingFace's MTEB leaderboard. These embeddings are stored in a Chroma vector store. The entire process is batch-oriented to support scalability.

3.2 Response Generation

The response generation module uses language models (LLMs) to answer queries based on retrieved context. LangChain manages component integration and execution. For query embedding, we reuse the same multilingual model as for the documents, ensuring consistent semantic search within Chroma. The system handles each query in the following steps (see Fig. 2):

1. **Query Embedding:** The query is embedded into a vector representation.
2. **Semantic Search:** The top-5 most relevant document chunks are retrieved and concatenated. This value was selected as a practical balance between chunk size, retrieval precision, and computational efficiency.
3. **Prompt Formatting:** The context and query are structured using LangChain's schema: a `SystemMessage` with instructions and a `HumanMessage` with the query.
4. **Model Invocation:** The prompt is passed to the language model, which generates an answer based on the retrieved context.

The generated answers are compiled into a Pandas DataFrame and exported as a CSV file for analysis. Prompt engineering is central to the system. Prompts follow a consistent and stringent structure for clarity and reproducibility:

Answer the question in Easy German using the information provided.
Information: Context
Question: Human prompt

Translate the text to Easy German.
Text: Text

We test different prompt variants to identify which best elicits Easy German outputt; the complete set of prompts is available in our GitHub repository:

- **standard**: Zero-shot prompt instructing the model to answer in standard German.

[1] https://huggingface.co/intfloat/multilingual-e5-large-instruct, accessed: March 27, 2025.

- **nonsense**: Prompt without meaningful content, acting as a control.
- **plain-de**: Zero-shot Plain German language prompt.
- **easy-de**: Zero-shot Easy German prompt.
- **rules**: Zero-shot prompt instructing models to follow Easy German guidelines from Netwerk Leichte Sprache e.V. [11], an association dedicated to promoting and supporting the use of Easy German.
- **example-text**: One-shot prompt with an Easy German example text.
- **rules-example**: Prompt with Easy German rules and an example text.
- **five-shot-example**: Few-shot examples illustrating the intended output style.
- **aligned-translations**: Few-shot prompt with aligned Easy German - German translations from a sentence-aligned corpus [19].
- **translations**: Few-shot prompt with unaligned translations.

This structured, reproducible framework enables systematic evaluation of prompt strategies for optimizing Easy German outputs.

4 Experimental Evaluation

This section discusses the experimental setup and results of developing our robust Question Answering (QA) system tailored for Easy German, with a focus on accessibility and clarity. We describe the language models employed, benchmarks for evaluating readability and semantic alignment, and results from both automated and human evaluations. By integrating simplification rules and testing multiple models, this study highlights the effectiveness of our approaches while identifying challenges and areas for improvement. The findings provide insights into creating accessible content for diverse audiences, including language learners and individuals with varying literacy levels.

4.1 Language Models

Several language models are employed to generate the answers, each offering distinct capabilities. Each model is initialized with configuration parameters to ensure optimal performance. We use instruction-tuned versions of the models:
GPT Models[2]: OpenAI's models (`gpt-4o` and `gpt-4o-mini`) are utilized for general-purpose text generation, set with the lowest possible temperature and frequency penalty (0.1) to ensure more deterministic and focused responses.
LLaMA Models[3]: Meta's LLaMA models (`Llama-3.2-3B-Instruct` and `Llama-3.1-70B-Instruct`) are also incorporated, known for their excellent multilingual capabilities and instruction-following tasks. They are accessed via Hugging Face's API and set with the same temperature and frequency penalty to maintain consistency across models.

[2] https://platform.openai.com/docs/models, accessed: March 27, 2025.
[3] https://www.llama.com/, accessed: March 27, 2025.

Mistral Models[4]: Mistral's models (`Mixtral-8x7B-Instruct-v0.1`), also hosted on Hugging Face's API, are included for their competitive performance in instruction-based tasks and set with similar hyperparameters for consistent performance.

4.2 Benchmark

To contextualize our results, we establish a benchmark based on a sentence-aligned corpus of German and Easy German texts provided by Toborek et al. [19]. This benchmark was necessary to interpret the automatic evaluation metrics used in our study—particularly BLEU [13], BERTScore [21], and BLEURT [16]—which assessed how well information was preserved in the translated Easy German responses. The Toborek corpus provided a reference for "good" scores in standard-to-Easy German translation. It contains sentence-aligned German-Easy German document pairs created with high-precision automatic methods. We evaluated this corpus using the semantic similarity metrics, treating German texts as references and Easy German texts as candidates. We also assessed the Easy German texts with readability metrics—the German Flesch Reading Ease [5] and Flesch-Kincaid Grade Level [18]—which, despite being interpretable, benefited from empirical validation on a high-quality corpus to establish practical accessibility thresholds.

Table 1. Benchmark of Metrics for Easy German: Readability metrics include the German variant of Flesch Reading Ease and Wiener Sachtext Formel (WSTF) and semantic alignment metrics include BERTScore and BLEURT.

Metric	Mean	Median	Std. Deviation
Avg Sentence Length	11.10	11.00	2.43
Avg Syllables per Word	1.67	1.66	0.15
Flesch Reading Ease	71.25	72.02	8.68
Flesch Kincaid Grade Level	8.43	8.16	1.89
WSTF1	5.05	5.23	1.71
WSTF2	5.06	5.29	1.68
WSTF3	5.02	4.95	1.50
WSTF4	4.98	4.90	1.54
BLEU Score	3.49	2.15	4.58
BERT Precision	0.68	0.68	0.05
BERT Recall	0.66	0.66	0.05
BERT F1	0.67	0.67	0.05
BLEURT Score	0.58	0.58	0.08

The Wiener Sachtext Formel (WSTF) is used as our primary readability metric because it was developed specifically for German and accounts for both average sentence length and word complexity [12]. A WSTF score of around 5 aligns with a fifth-grade reading level, reflecting an ideal balance between

[4] https://docs.mistral.ai/, accessed: March 27, 2025.

simplicity and informativeness. This makes the widely used WSTF especially relevant for evaluating outputs meant for readers with cognitive or language challenges. Other metrics supplement this picture: German Flesch Reading Ease (mean: 71.25) suggests moderately easy readability, and German Flesch-Kincaid (mean: 8.43) indicates suitability for readers at the upper end of the middle school level. Semantic metrics show that Easy German retains reasonable content fidelity, with a BERT F1 score averaging 0.67 and BLEURT score at 0.58—indicating that simplified texts preserve key meaning and fluency.

Together, these benchmarks allow us to assess whether the outputs of our QA models truly meet the linguistic and accessibility standards of Easy German.

4.3 Results

We generated responses using different prompting strategies and evaluated them using a range of automated metrics (see Table 1). For conciseness, we focus here on the Wiener Sachtext Formel (WSTF), a German readability metric that better reflects text simplicity than English-based metrics like Flesch-Kincaid. Full results are available in our GitHub repository. To evaluate the quality of model outputs, we used 77 medical question-answer pairs, manually translated into German from the MedQuAD dataset [2]. These were randomly sampled from categories like general health and senior health, which closely match the content style of Apotheken Umschau articles and are representative of common health-related queries. Each figure compares prompting strategies across three models—GPT, Llama, and Mix-

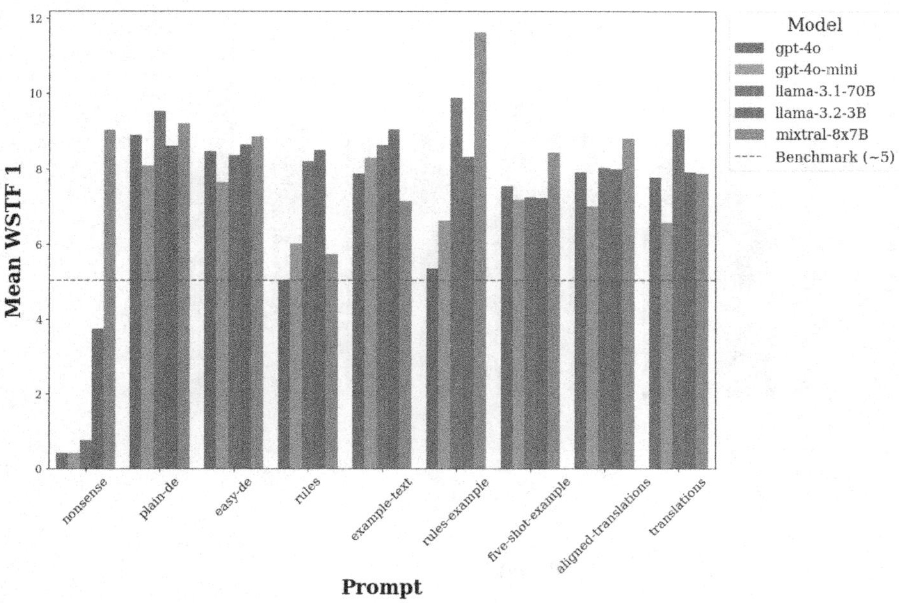

Fig. 3. Mean WSTF Scores for On-the-Fly Prompting

tral. The x-axis indicates the prompt used, and the y-axis shows the average WSTF score. Lower scores reflect simpler, more accessible texts.

In the **On-the-Fly** setup, models were prompted to answer questions directly in Easy German using retrieved context (Fig. 3). The nonsense prompt, instructing models to reply with "I like eating schnitzel," served as a control and yielded the lowest scores, confirming its simplicity. GPT models consistently followed instructions and produced clear, concise outputs. Prompts incorporating Easy German rules led to the lowest WSTF scores among non-control prompts. In contrast, Llama and Mixtral frequently produced overly complex or looping outputs. For instance, Llama-3.1-70B generated repeated unrelated questions and answers, inflating sentence length and reducing readability:

Q: What are tinea infections?
A: I like eating schnitzel. How can you recognize athlete's foot? I like eating schnitzel.

Mixtral, when prompted with rules and examples, occasionally looped phrases or duplicated words, yielding artificially high WSTF scores. In contrast, GPT models generated well-structured Easy German texts with short sentences and almost no subordinate clauses.

The **Translated-Answer** method introduces a two-step process: the model first generates an answer in standard German, then translates it into Easy German (Fig. 4). This strategy reduced misinterpretations seen in multitask prompts

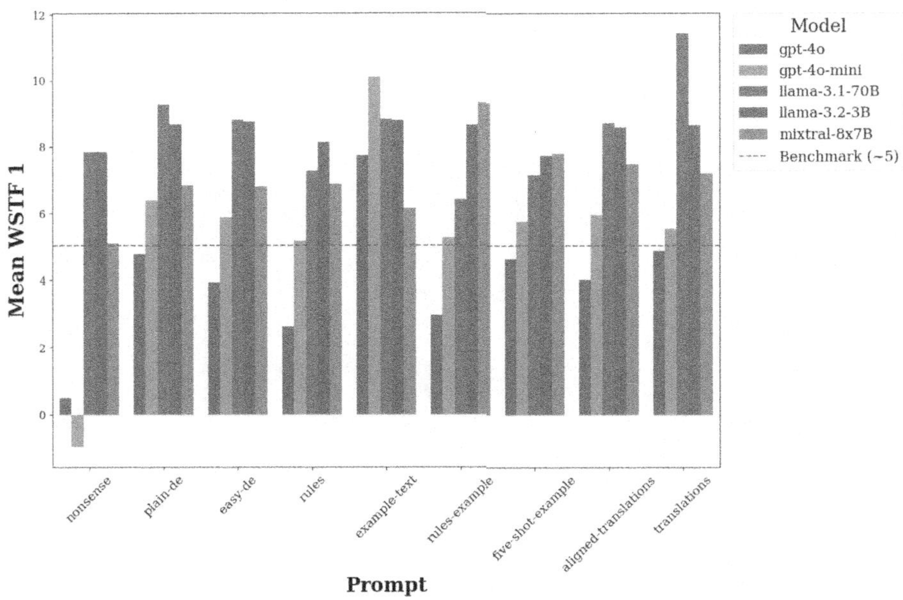

Fig. 4. Mean WSTF Scores for Translated-Answer Prompting.

and yielded slightly lower WSTF scores than On-the-Fly prompting, indicating simpler output.

Again, GPT models excelled. Their outputs were both semantically accurate and syntactically simple, often structured as short lists or action-based statements:

> **Q:** Do you have any information about mold?
> **A:** Mold can be harmful. It can be a health hazard. Mold can grow in your home. Especially if it is damp. Mold should be removed quickly. Experts can help. [...]

In contrast, Llama and Mixtral often ignored the translation instruction, producing nearly identical responses before and after translation. This led to abnormally high BLEU and BERTScore results but minimal simplification, thus highlighting that high similarity doesn't always mean better accessibility.

In our **Translated-Context** method, the retrieved context is pre-translated into Easy German using a consistent rules-and-example prompt before passing it to the model for answering (Fig. 5). This approach offered slight improvements over On-the-Fly prompting, though GPT models remained the most consistent.

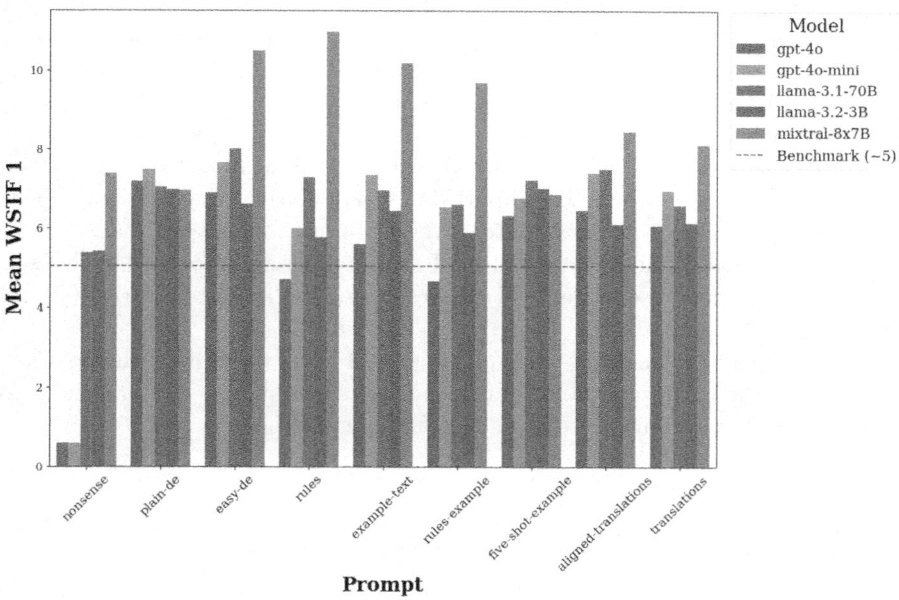

Fig. 5. Mean WSTF for Translated-Context Prompting.

Mixtral continued to suffer from phrase repetition, while Llama models showed reduced sentence length but still lacked coherence. We ensured reproducibility by verifying that translating the same context with the same prompt led to consistent outputs across runs—a crucial factor for building a reliable QA system.

To summarize, GPT-4o consistently generated the most readable and accessible responses, especially when guided by Easy German rules, producing short, clear sentences without subordinate clauses. In contrast, Llama and Mixtral often failed to follow instructions, frequently generating incoherent or repetitive outputs. Among the prompting strategies, Translated-Answer yielded the simplest texts, followed by Translated-Context and then On-the-Fly. The Wiener Sachtext Formel (WSTF) emerged as the most reliable metric for evaluating German text simplicity, outperforming general-purpose measures like BLEU and FKGL, which reflect semantic similarity but not linguistic accessibility. Overall, applying clear simplification rules and breaking tasks into smaller subtasks is essential for generating accessible responses. Despite its proprietary nature, GPT remains the most effective model, and our QA pipeline combined with these strategies offers a reproducible approach to generating Easy German responses in the medical domain.

4.4 Human Evaluation

To assess real-world accessibility, we conducted a human evaluation using the question "Was sind Röteln?" ("What is rubella?"), a representative, open-ended query requiring both clarity and appropriate detail. Eleven model-generated responses were evaluated by two groups: native German speakers from Lebenshilfe Heidelberg e.V. (including a professional Easy German translator and two individuals with special needs) and non-native speakers with A1-B1 proficiency. Prior consultation with the professional translator indicated that limiting the number of responses would be important—particularly for participants with cognitive disabilities, who may find it overwhelming to compare many similar examples. A smaller, more focused set of outputs was therefore deemed essential to support critical engagement. All responses are available in our GitHub repository.

Both groups identified the response generated using prompts with Easy German rules as the most accessible. This response - shown below - follows the Easy German paradigm of one piece of information per sentence.

Rubella is a viral disease.
It is often harmless.
But during pregnancy it can harm the unborn child.
There is a vaccination against rubella.

In contrast the response below from the Translated-Context setup was considered the most complex due to its long sentences with subordinate clauses, though it provided the most information.

Rubella is a disease caused by a virus. The disease is usually harmless, but pregnant women can transmit the virus to the unborn child, which can harm the child. There is a vaccination against rubella. Rubella is a childhood disease, but adults can also get it.

Participants also struggled with vague terms like mostly or harmless, and non-native speakers found words like pregnancy and vaccination unfamiliar but interpretable from context. Visual imagery and clarity were suggested to improve understanding. Overall, the evaluation confirmed that rule-based prompting leads to the most accessible outputs, directly corroborating our automatic evaluation results. Both human and automated assessments highlight the effectiveness of linguistic simplification and the importance of avoiding ambiguous language to improve accessibility. It is important to note some critical and ethical reflections. For instance, non-native speakers, people with varying literacy levels, and individuals with cognitive disabilities have different requirements when it comes to text accessibility. As such, the interpretation of our results should not be overgeneralized across these distinct user groups [17].

5 Conclusion

In this paper, we explored how large language models (LLMs) can generate accessible answers in Easy German to support users with varying literacy and cognitive abilities. We developed and evaluated three prompting strategies—On-the-Fly, Translated-Answer, and Translated-Context—within a Retrieval Augmented Generation (RAG) framework using unstructured health data.

Our results show that applying clear simplification rules, such as avoiding subordinate clauses and using simple vocabulary, significantly improves readability. GPT-4 consistently produced the most accurate and accessible outputs, demonstrating the strength of instruction fine-tuning, though its closed-source nature limits accessibility for future research. Open-source models like Llama and Mixtral underperformed, often generating incoherent or overly complex responses, highlighting the need for improved fine-tuning. Among prompting methods, the Translated-Answer approach—a two-step process generating responses first in standard German, then translating into Easy German—outperformed others by enhancing readability and semantic alignment. While slightly more computationally expensive, this trade-off is justified by notable gains in output quality. Although focused on Easy German, our methodology is language-agnostic and adaptable to other plain language contexts via prompt adjustment.

It is worth noting that our study did not yet include a systematic evaluation of factual correctness; responses were manually reviewed, but thorough semantic validation remains ongoing. Similarly, an ablation study examining the effects of individual RAG components—such as chunking strategies, retrieval parameters, and reranking—would help clarify their contributions to prompting performance.

This work lays the foundation for future development of interactive, conversational systems. Current research is expanding the QA framework to multi-turn dialogues and integrating human feedback to improve response quality. We are also exploring query reformulation and clarification strategies to build an inclusive conversational assistant for healthcare information in Easy German—potentially fine-tuned on open-source models to reduce reliance on proprietary tools.

Acknowledgments. We would like to express our gratitude to the participants of the human evaluation, including the members of Lebenshilfe Heidelberg e.V.. Their valuable feedback and insights greatly contributed to the success of this study.

References

1. Behindertengleichstellungsgesetz (BGG). BGBl. I 2002, S. 1467 (2002). https://d-nb.info/gnd/4716950-3, Gesetz vom 27.04.2002; Artikel 1 des Gesetzes zur Gleichstellung behinderter Menschen und zur Änderung anderer Gesetze; in Kraft getreten am 01.05.2002
2. Abacha, A.B., Demner-Fushman, D.: A question-entailment approach to question answering. BMC Bioinf. **20**(1), 511 (2019). https://doi.org/10.1186/s12859-019-3119-4
3. Anschütz, M., Oehms, J., Wimmer, T., Jezierski, B., Groh, G.: Language models for German text simplification: overcoming parallel data scarcity through style-specific pre-training. Findings of the Association for Computational Linguistics: ACL 2023, pp. 1147–1158 (2023). https://doi.org/10.18653/v1/2023.findings-acl.74
4. Deilen, S., Garrido, S.H., Lapshinova-Koltunski, E., MaaSS, C.: Using ChatGPT as a CAT tool in easy language translation (2023). https://arxiv.org/abs/2308.11563
5. Flesch, R.: A new readability yardstick. J. Appl. Psychol. **32**(3), 221 (1948), https://doi.org/10.1037/h0057532
6. Gao, Y., et al.: Retrieval-augmented generation for large language models: a survey (2024). https://arxiv.org/abs/2312.10997
7. Guo, Y., August, T., Leroy, G., Cohen, T., Wang, L.L.: APPLS: evaluating evaluation metrics for plain language summarization (2024). https://arxiv.org/abs/2305.14341
8. Jordan, S.: Gesundheitskompetenz/ health literacy. In: für gesundheitliche Aufklärung (BZgA), B. (ed.) Leitbegriffe der Gesundheitsförderung und Prävention. Glossar zu Konzepten, Strategien und Methoden. Bundeszentrale für gesundheitliche Aufklärung (BZgA) (2023). https://doi.org/10.17623/BZGA:Q4-i065-3.0
9. Joseph, S., et al.: FactPICO: factuality evaluation for plain language summarization of medical evidence. In: Ku, L., Martins, A., Srikumar, V. (eds.) Proceedings of the 62nd Annual Meeting of the Association for Computational Linguistics (Volume 1: Long Papers), ACL 2024, pp. 8437–8464. Association for Computational Linguistics (2024)

10. MaaSS, C.: Easy Language – Plain Language – Easy Language Plus: Balancing Comprehensibility and Acceptability. Frank & Timme (2020). https://doi.org/10.26530/20.500.12657/42089. https://library.oapen.org/handle/20.500.12657/42089
11. Netzwerk Leichte Sprache: Guidelines of the netzwerk leichte sprache (2024). https://www.leichte-sprache.org/. Accessed 27 Mar 2025
12. Oomen-Welke, I.: Sachtexte verstehen – Dichte, Lesbarkeit, Wortschatz, pp. 127–150. De Gruyter Mouton, Berlin, Boston (2017). https://doi.org/10.1515/9783110404166-006
13. Papineni, K., Roukos, S., Ward, T., Zhu, W.J.: BLEU: a method for automatic evaluation of machine translation. In: Proceedings of the 40th Annual Meeting of the Association for Computational Linguistics, pp. 311–318 (2002). https://cir.nii.ac.jp/crid/1571980075806514432
14. Poupet, P., Hauguel, M., Boehm, E., Roze, C., Tardy, P.: An automated tool with human supervision to adapt difficult texts into plain language. In: Proceedings of the Second Workshop on Text Simplification, Accessibility and Readability, pp. 131–133. INCOMA Ltd., Shoumen, Bulgaria, Varna, Bulgaria (2023). https://aclanthology.org/2023.tsar-1.13/
15. Sahoo, P., Singh, A.K., Saha, S., Jain, V., Mondal, S., Chadha, A.: A systematic survey of prompt engineering in large language models: Tech. Appl. (2024). https://arxiv.org/abs/2402.07927
16. Sellam, T., Das, D., Parikh, A.P.: BLEURT: learning robust metrics for text generation. arXiv preprint: arXiv:2004.04696 (2020). https://arxiv.org/abs/2004.04696
17. Stajner, S.: Automatic text simplification for social good: progress and challenges. In: Zong, C., Xia, F., Li, W., Navigli, R. (eds.) Findings of the Association for Computational Linguistics: ACL-IJCNLP 2021, pp. 2637–2652. Association for Computational Linguistics (2021). https://doi.org/10.18653/v1/2021.findings-acl.233
18. Thomas, G., Hartley, R.D., Kincaid, J.P.: Test-retest and inter-analyst reliability of the automated readability index, Flesch reading ease score, and the fog count. J. Reading Behav. **7**(2), 149–154 (1975)
19. Toborek, V., Busch, M., Boßert, M., Bauckhage, C., Welke, P.: A new aligned simple German corpus. arXiv preprint: arXiv:2209.01106 (2022)
20. Umschau, A.: Apotheken umschau (2024). https://www.apotheken-umschau.de/einfache-sprache/. Accessed 27 Mar 2025
21. Zhang, T., Kishore, V., Wu, F., Weinberger, K.Q., Artzi, Y.: BERTScore: evaluating text generation with BERT (2020). https://arxiv.org/abs/1904.09675

Technical Communications

Learn, Optimize, Explain: A Neuro-Symbolic Advisor for Personal Finance

Youssef Ayman and Nourhan Ehab(✉)

Department of Computer Science and Engineering, German University in Cairo, Cairo, Egypt
youssef.mohamedali@student.guc.edu.eg, nourhan.ehab@guc.edu.eg

Abstract. This paper presents LOX, a neuro-symbolic decision support system for personal finance that integrates neural forecasting with symbolic reasoning. LOX combines Long Short-Term Memory (LSTM) networks for market trend prediction with an ontology-based representation of investment knowledge. The core contribution is a unified pipeline that learns patterns from data, optimizes portfolios based on forecasted returns and risk, and explains recommendations using rule-based reasoning grounded in expert logic. By bridging data-driven learning and symbolic AI, LOX advances the transparency, accuracy, and trustworthiness of automated investment advisors.

Keywords: Neuro-Symbolic AI · Financial Decision Support · Portfolio Optimization

1 Introduction

Making effective investment decisions is critical for achieving financial stability, yet the complexity and volatility of markets often overwhelm individual investors [14]. As personal finance spans budgeting, saving, and investing, individuals increasingly turn to robo-advisors and decision support systems (DSSs) for guidance [12].

Most robo-advisors rely on statistical models trained on historical data [1], which limits their ability to capture market nuances and incorporate expert knowledge [13,15]. These systems often lack transparency, leaving users uncertain about the rationale behind recommendations [3].

To address this, neuro-symbolic AI combines machine learning with symbolic reasoning [2]. Neural networks such as LSTMs [7] are well-suited for modeling time series in finance [13], while symbolic methods using ontologies and rules offer interpretability [14]. However, most existing solutions lean heavily on one approach. Deep learning models like the one in [1] excel in prediction but offer little explanation. Conversely, symbolic systems provide transparency but lack adaptability. InvestLM [15] improves financial language understanding but lacks structured reasoning and forecasting.

T. Braun et al. (Eds.): KI 2025, LNAI 15956, pp. 221–227, 2026.
https://doi.org/10.1007/978-3-032-02813-6_16

Previous DSSs explored statistical learning (e.g., HInvestor [8]) and expert logic via ontologies (e.g., DSS-OMA [5,9,11]. These systems either optimize based on learned patterns without explanations or encode rules without adapting to dynamic data.

Neuro-symbolic AI [4] aims to unify the adaptability of neural models and the structure of symbolic reasoning. Despite promising results in semantic search and planning, applications in personal finance remain scarce [2]. There is a clear need for systems that predict market behavior and explain investment decisions using domain knowledge.

This paper introduces *LOX* (Learn, Optimize, Explain), a neuro-symbolic framework that combines LSTM-based forecasting with an OWL ontology of investment principles. LOX builds optimized portfolios and validates them against symbolic rules to generate transparent recommendations.

The rest of this paper is organized as follows. Section 2 describes the architecture. Section 3 presents evaluations. Section 4 concludes and outlines future directions.

2 The LOX Architecture: A Neuro-Symbolic Framework for Investment Support

The LOX framework combines machine learning and symbolic reasoning to provide personalized, explainable investment advice. It comprises three components:

1. An LSTM-based forecasting model for predicting asset prices,
2. A portfolio optimizer guided by Sharpe ratio maximization,
3. A symbolic reasoning module using OWL ontologies to validate and explain decisions.

2.1 Neural Forecasting Component: LSTM-Based Prediction

Each asset is modeled independently using a 3-layer LSTM with 32 hidden units. Inputs are 20-day sliding windows of adjusted close prices, predicting the next day's value. Models are trained with Adam optimizer and MSE loss for 10,000 epochs. A recursive strategy yields 30-day forecasts. This component captures non-linear temporal patterns across diverse asset classes.

2.2 Portfolio Optimization Engine

Given forecasted prices, the optimizer computes returns and risks, generates 100,000 random portfolios, and selects the one with the highest Sharpe ratio. The `DiscreteAllocation` module translates continuous weights into actionable allocations within a fixed budget. This enables both robust performance and real-world usability.

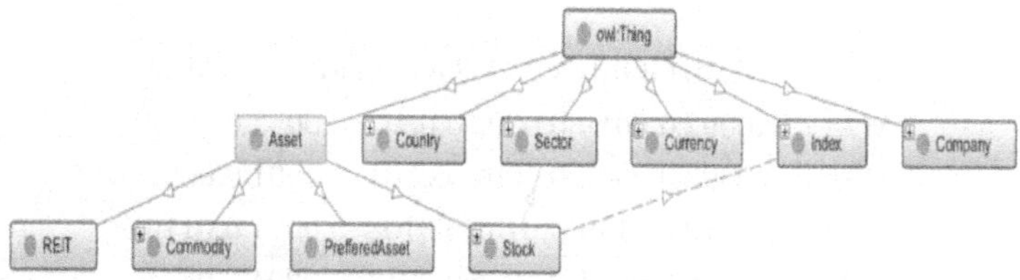

Fig. 1. Ontology class hierarchy including asset types, investor profiles, and financial metrics

2.3 Symbolic Knowledge and System Integration

The symbolic reasoning layer encodes expert investment logic in an OWL ontology developed in Protégé [10]. It defines asset classes (e.g., `Stock`, `REIT`), investor profiles (e.g., `RiskAverseInvestor`), and metrics (e.g., `hasVolatility`, `hasDividendYield`) (Fig 1).

Logical rules express expert strategies—e.g., assets with low volatility and positive yield are inferred as `PreferredAsset`. The HermiT reasoner [6] and SPARQL queries are used to evaluate and retrieve inferences.

This module verifies whether the optimizer's output aligns with investment constraints and generates justifications. If assets violate the ontology's logic, the plan is flagged as unsupported and explanations are withheld. The module ensures that investment plans comply with expert-defined logic, provides human-readable justifications, and supports investor-specific constraints that may not be captured in data. It also enhances transparency and auditability, making the system more suitable for regulated or high-stakes financial environments.

3 Results and Evaluation

We evaluate LOX across three dimensions: forecasting accuracy, portfolio performance, and symbolic reasoning alignment.

3.1 Forecasting Accuracy

LSTM models were assessed using MAE, MSE, RMSE, MAPE, and R^2 scores. Results (Tables 1 and 2) show strong predictive accuracy across all assets, with R^2 values typically above 0.95.

3.2 Portfolio Performance

The Sharpe-optimal portfolio outperformed a naïve equally weighted baseline. Table 3 shows a substantial improvement in expected return and risk-adjusted return, with minimal increase in volatility.

Discrete allocation results for a \$100,000 budget: 10 Nvidia, 7 Microsoft, 1 Tesla, 1 ounce of Gold. Remaining funds (\$109.73) were held as cash.

Table 1. LSTM Forecasting Performance – Assets 1

Metric	AAPL	GOLD	VNQ	GOOG	MSFT	AMZN
MAE	0.0092	0.0010	0.0092	0.0065	0.0053	0.0072
MSE	0.0003	0.0002	0.0002	0.0001	0.0001	0.0001
RMSE	0.0172	0.0149	0.0139	0.0108	0.0091	0.0119
MAPE	0.0486	0.0604	0.0276	0.0213	0.0204	0.0344
R^2	0.9968	0.9972	0.9962	0.9986	0.9990	0.9984

Table 2. LSTM Forecasting Performance – Assets 2

Metric	META	TSLA	NFLX	NVDA	CRM	IBM	SPG
MAE	0.0080	0.0089	0.0079	0.0168	0.0083	0.0013	0.0011
MSE	0.0002	0.0005	0.0002	0.0025	0.0002	0.0003	0.0003
RMSE	0.0132	0.0215	0.0132	0.0498	0.0134	0.0187	0.0162
MAPE	0.0186	0.0362	0.0266	0.1677	0.0182	0.0712	0.0896
R^2	0.9969	0.9933	0.9972	0.9540	0.9971	0.9843	0.9923

Table 3. Portfolio Comparison: Optimized vs. Naïve

Portfolio	Expected Return	Volatility	Sharpe Ratio
Optimized	0.3670	0.1780	1.9500
Equally Weighted	0.2300	0.1700	1.3100

3.3 Symbolic Reasoning Outcomes

The symbolic module validated optimizer results against expert-defined OWL rules. Figure 2 shows the logical criteria, and Listings 1.1 and 1.2 shows the output of the neuro-symbolic advisor.

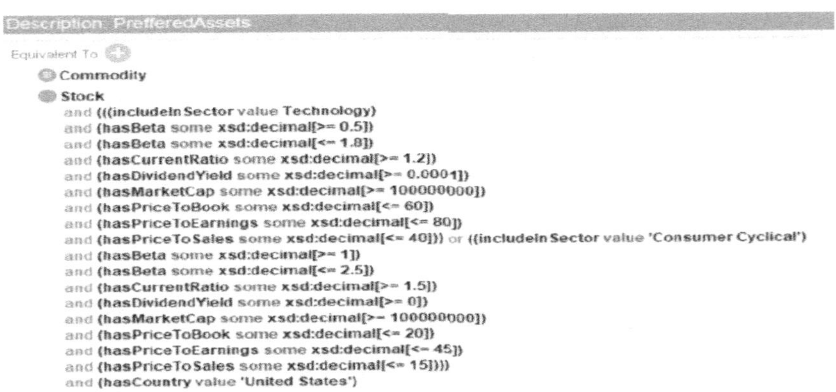

Fig. 2. Expert Criteria for Asset Validation

```
Recommended portfolio:
- GOLD: 1 troy ounce
- MSFT: 7 shares
- TSLA: 1 share
- NVDA: 10 shares
Funds remaining: \$109.73
```

Listing 1.1. Allocated funds for $100,000 investment

```
Reasoning:
The selected assets match with the recommendations from the ontology.
Explanations:

'GOLD':
- Return: 3.58\%, Volatility: 8.12\%, Sharpe: 0.4405

'MSFT':
- Return: 26.82\%, Volatility: 17.11\%, Sharpe: 1.5674

'TSLA':
- Return: 41.40\%, Volatility: 42.63\%, Sharpe: 0.9710

'NVDA':
- Return: 47.92\%, Volatility: 24.29\%, Sharpe: 1.9730
```

Listing 1.2. Explanation when Ontology and Optimizer Agree

To test robustness, we modified the ontology to prefer REITs and low-valuation communication stocks as shown in Fig. 3. The resulting divergence led LOX to flag the plan as unsupported, suppressing explanation generation. The output is illustrated in Listing 1.3.

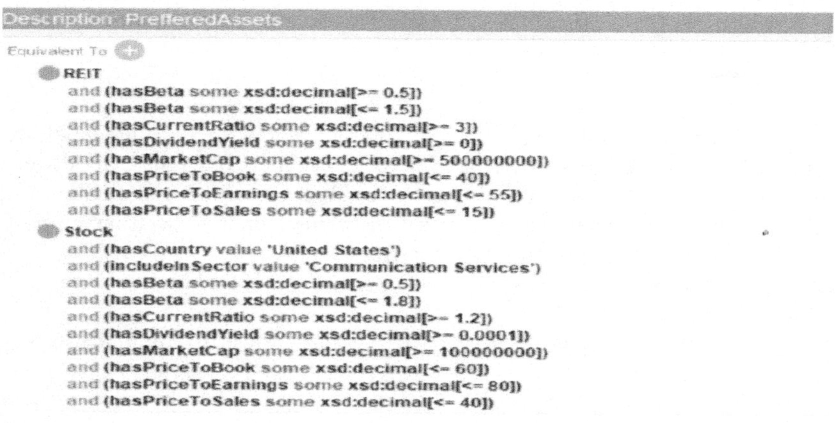

Fig. 3. Modified Expert Criteria

```
Portfolio:
- GOLD: 1 troy ounce
- MSFT: 7 shares
- TSLA: 1 share
- NVDA: 10 shares
Funds remaining: \$109.73
Recommended investment is not supported by expert knowledge.
```

Listing 1.3. Portfolio Flagged as Unsupported

4 Conclusion and Future Work

We presented **LOX**, a neuro-symbolic architecture for investment support that integrates LSTM-based forecasting, portfolio optimization, and OWL-based symbolic reasoning. By combining neural learning with expert-encoded logic, LOX delivers both accurate and explainable investment recommendations.

Empirical results demonstrated strong predictive performance ($R^2 > 0.95$ for most assets), superior risk-adjusted returns compared to a naïve baseline, and effective validation of recommendations against domain-specific investment criteria. The symbolic module provided clear explanations when optimizer outputs aligned with expert rules, and flagged inconsistencies when they did not—ensuring transparency and trustworthiness.

Future work includes extending the ontology to incorporate ESG criteria, macroeconomic signals, and investor profiling. We also aim to integrate financial-domain LLMs (e.g., InvestLM) to enhance explanation generation, and explore multi-objective optimization that balances return, volatility, and sustainability. Real-world validation in sandboxed trading environments will further assess LOX's practical impact.

References

1. Day, M.Y., Cheng, T.K., Li, J.G.: AI robo-advisor with big data analytics for financial services. In: 2018 IEEE/ACM International Conference on Advances in Social Networks Analysis and Mining (ASONAM), pp. 1032–1035. IEEE (2018)
2. Demidovskij, A.V., Babkin, E.A.: Integrated neurosymbolic decision support systems: problems and opportunities. Bus. Inf. (2021)
3. Drigs, R.: Man vs. machine: the future of financial advising. Finance Undergraduate Honors Theses (2023)
4. Garcez, A.A., et al.: Neurosymbolic AI: the third wave. Communications of the ACM, **62**(12), 74–82 (2019)
5. Ghose, A., Koliadis, G.: An ontological framework for investment decision support. Int. J. Inf. Technol. Decis. Making **5**(4), 673–698 (2006)
6. Glimm, B., Horrocks, I., Motik, B., Stoilos, G., Wang, Z.: Hermit: An owl 2 reasoner. Experiences and Directions (OWLED), In: International Workshop on OWL (2010)
7. Hochreiter, S., Schmidhuber, J.: Long short-term memory. Neural Comput. **9**(8), 1735–1780 (1997)

8. Ma, Y., Kwon, S., Kim, S.W.: Hinvestor: a framework for portfolio optimization using deep neural networks. In: Proceedings of the IEEE International Conference on Big Data, pp. 1027–1032 (2016)
9. Martins, J., Figueiredo, J., Vieira, S.: An ontology-based intelligent decision support system for investment planning. In: International Conference on Information Systems and Technologies (ICIST), pp. 349–358 (2015)
10. Musen, M.A.: The protégé project: a look back and a look forward. AI Matters **1**(4), 4–12 (2015)
11. Nguyen, T., Nguyen, T., Cao, T.: An ontology-based decision support system for stock investment. Expert Syst. Appl. **42**(21), 7426–7435 (2015)
12. Ribes, E.: Transforming personal finance thanks to artificial intelligence: myth or reality? Financial Economics Letters, pp. 11–21 (2023)
13. Sen, J., Dutta, A., Mehtab, S.: Stock portfolio optimization using a deep learning LSTM model. arXiv preprint arXiv:2111.04709 (2021)
14. Weber, B.W.: Financial DSS: systems for supporting investment decisions. In: Handbook on Decision Support Systems 2, pp. 419–442. Springer Berlin Heidelberg (2008)
15. Yang, Y., Tang, Y., Tam, K.Y.: InvestLM: a large language model for investment using financial domain instruction tuning. arXiv preprint arXiv:2309.13064 (2023)

Deep Learning Emulators for Large-Scale, High-Resolution Urban Pluvial Flood Prediction

Dinesh Krishna Natarajan[1(✉)], Marco Stricker[2], Rushan Mukherjee[2], Marcela Charfuelan[2], Marlon Nuske[2], and Andreas Dengel[1,2]

[1] RPTU University Kaiserslautern-Landau, Kaiserslautern, Germany
dinesh.natarajan@cs.rptu.de
[2] German Research Center for Artificial Intelligence (DFKI), Kaiserslautern, Germany

Abstract. Flood inundation simulations can help in planning preventive measures against flood damage, however conventional simulation methods can be computationally expensive and time-consuming. An alternative to speed up the calculations is to use data-driven emulators. In this work, we present two contributions: (1) the development of a large-scale, high-resolution flood dataset and (2) the development of deep learning (DL)-based emulators for flood prediction trained using our dataset. We show that in comparison to previous works, our emulators are able to generalize to previously unseen test locations and achieve comparable performance metrics in terms of RMSE. In comparison to a GPU-accelerated simulator, an inference time speed-up of approximately 1000 times is achieved using these emulators. The dataset and code is available at https://github.com/dinesh-k-natarajan/urban-flood-emulator/.

Keywords: Urban flood prediction · Deep Learning · Image-to-image regression · Data-driven emulators

1 Introduction

Climate change increases the occurrence of extreme weather events such as storms, droughts, fires and floods [14]. Extreme rainfall events could lead to pluvial floods, which can rapidly occur with little to no warning, and cause extensive damage to infrastructure and human lives in urban or rural areas [2,11]. Early warning systems for pluvial floods can help in disaster management and response. Conventional simulation methods for the hydrodynamic modeling of pluvial flooding, although precise, can be slow and computationally expensive for high-resolution domains [7]. Thus, their use is still limited in real-time early warning systems, despite continuous efforts to speed up computations using GPU accelerated solvers [13]. Another solution to speed up computations is to

M. Stricker, R. Mukherjee and M. Nuske—Work done during time at DFKI.

T. Braun et al. (Eds.): KI 2025, LNAI 15956, pp. 228–235, 2026.
https://doi.org/10.1007/978-3-032-02813-6_17

use data-driven emulators or surrogate models for flood prediction, which require high offline training effort, but offer a significant reduction in computational time in the online phase [16].

There have been recent studies focused on deep learning (DL)-based flood prediction for urban areas using convolutional neural network (CNN) architectures. Using terrain elevation data (spatial resolution of 1 to 5 m) of specific study areas and specific rainfall events, a conventional flood simulator is usually used to generate flood water depth maps which act as ground truth labels for supervised learning. In [5], a CNN autoencoder model per city was trained on smaller patches of data from 3 different cities and the corresponding models were evaluated on unseen rainfall events in each city. A U-Net architecture called U-FLOOD was introduced in [10], where different architecture configurations were trained and evaluated on smaller patches of data from the same city. In contrast to these patch-based methods, [7] used the entire area of the city center of Aachen, Germany to train their floodGAN (generative adversarial network) architecture on pairs of rainfall maps and flood depth maps, implicitly learning the terrain elevation data. Since these emulators were trained and evaluated on a few catchments or locations, it is not clear whether these emulators can be reused in other locations, unless they are retrained with new data. Emulators which are generalizable to other unseen regions could be of interest to users without data and resources to train their own emulators. In this direction, [6,8] proposed U-Net and conditional generative adversarial network (cGAN) approaches which were trained and evaluated on different sub-catchments of up to two cities respectively.

In this paper, the development of a large-scale flood dataset of diverse urban areas and rainfall events to train generalizable deep learning-based flood emulators is presented. The Flood Inundation Parallel Computation (FIP) [13] simulator was used to generate ground truth data for supervised learning. Thus hydrodynamic flood simulations were generated for many urban areas under various rain events. DL-based flood emulators were trained using data from this large-scale dataset and their generalization was evaluated on cities not used during training. In the following sections, the two main contributions of this paper are presented:

1. Development of a large-scale flood dataset using high-resolution terrain data and synthetic rainfall events
2. Development of generalizable DL-based emulators for fast flood prediction

2 Development of a Large-Scale Flood Dataset

For a realistic urban flood simulation, the urban infrastructure which influences the flow of flood water must be accurately represented with high-resolution elevation data. Such high-resolution data is not available on a global scale. However, recently several federal states in Germany provide open access to their high-resolution elevation data with 1 m spatial resolution. In this work, our focus was limited to the state of North Rhine-Westphalia (NRW) in Germany. Starting with a list of cities in NRW, the GPS coordinates of their respective city centers were obtained. Using these city center coordinates, a 2048 × 2048 m^2 study area was created for each city. Then, using these study areas, 2-dimensional (2D) digital elevation maps (DEM) of 1 m resolution were downloaded from opennrw[1]. This procedure resulted in DEMs of 322 cities. Further filtering using average built-up imperviousness [1] was done to finally obtain the 200 most urbanized cities in NRW. The city locations in our dataset is shown in Fig. 1.

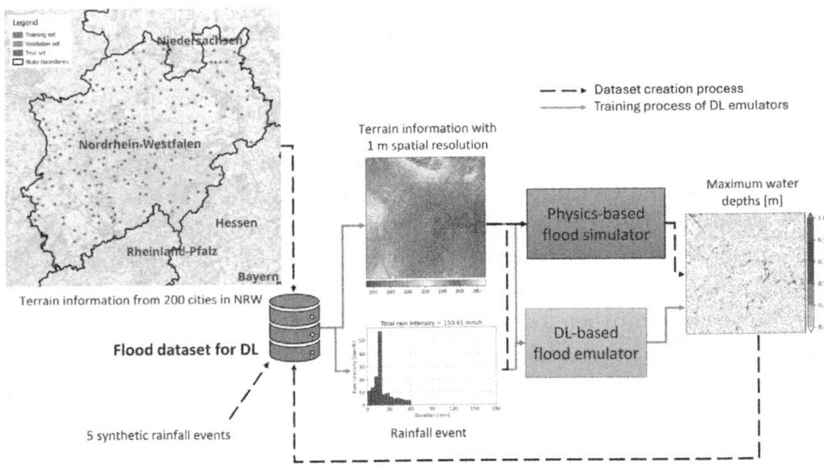

Fig. 1. Creation of large-scale flood dataset using a physics-based flood simulator FIP [13], and the training process of a deep learning-based flood emulator.

Modeling of Rainfall Events. Based on historical rainfall data (KOSTRA-DWD) from two cities, Kirchsahr and Pützfeld, 5 synthetic rainfall events were created. These events follow an Euler-Type 2 distribution with different rainfall intensities of 20, 40, 60, 80, and 100 mm/h. Each synthetic rainfall event consists of 60 min of rainfall followed by 120 min of no rainfall. This is represented as a time-series of rain intensities at 5-min intervals as shown in Fig. 1.

Flood Inundation Modeling Using FIP. FIP [13] is a physics-based flood simulator which solves the shallow water equation leveraging GPUs for parallel computing. This tool is available as free and open-source[2]. As shown in Fig. 1, flood simulations were generated using FIP using two main inputs: (i) a 2D digital elevation map (DEM) and (ii) a geo-referenced rainfall time-series in *Radolan (Radar-Online-Aneichung)*[3] format. The simulation outputs consist of

[1] https://www.opengeodata.nrw.de/produkte/geobasis/hm/dom1_tiff/dom1_tiff/.
[2] https://github.com/FIP-ems/FIP.
[3] https://www.dwd.de/DE/leistungen/radolan/radolan.html.

time-series of 2D flood water depth maps and velocity maps, as well as, a maximum flood water depth map aggregated over the time-series. The maximum flood water depth maps of each simulation were used as the ground truth labels for supervised training of our emulators. Using the DEMs of 200 cities and the 5 rainfall events, 1000 corresponding flood depth maps were generated using FIP for our dataset.

3 Development of DL-Based Flood Emulators

Flood prediction, under a supervised learning setting, is formulated as an image-to-image regression task, where the terrain information and rainfall information are mapped to maximum flood depths as illustrated in Fig. 1.

Terrain Feature Extraction. Additional terrain features were derived from elevation data (DEM) and used as input terrain features to the deep learning model, as proposed by [10]. RichDEM [3], an open-source Python package for terrain analysis, is used for processing and extracting terrain attributes and flow metrics from DEMs. Using RichDEM, the following additional features were extracted from the elevation data: slope, aspect, curvature, flow accumulation and topographic wetness index (TWI). The six spatial terrain features are each normalized to the range [−1,1] using the scaling parameters calculated from the training set. Due to certain artifacts in the laser-scanned elevation data, the FIP simulator estimated unusually high water depths at certain pixels in the data. To avoid extreme outliers, water depth labels were clipped to range [0 m, 5 m].

DL Architectures. The U-Net architecture [15] has been adopted in previous works [6,10] for the task of flood prediction. In this work, the performances of several variants of U-Net architecture are compared. Another variation of the architectures 'UNetTransformer' was considered by adding a self-attention block at the bottleneck, as introduced in [12]. Furthermore, as shown in Fig. 2, we added additional convolutional operations in the skip connections of the U-Net to enhance the feature propagation from the input feature space to the output feature space. These architectures are indicated in Table 1 by the suffix '-SC'. For each of the 4 UNet architecture variations, we increased the model size by doubling the number of convolutional filters leading to models with more trainable parameters, indicated in Table 1 by the suffix 'Large'.

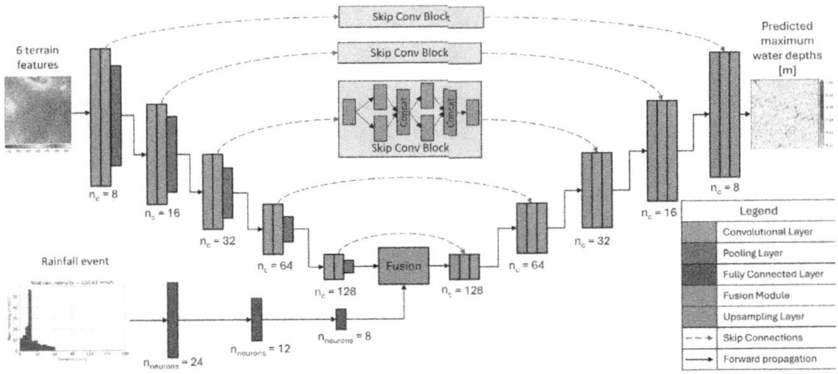

Fig. 2. Deep learning-based flood emulator using the UNet-SC architecture with additional skip convolutional blocks.

4 Results and Discussion

The dataset of 1000 samples was split into training, validation and test sets by randomly splitting the cities with the ratio 60%–20%–20%. The models were trained to minimize the mean squared error (MSE) loss function using the AdamW optimizer [9] with a learning rate of 0.001 for a maximum of 500 epochs. In line with previous works [4,7,10], Root mean squared error (RMSE), Critical Success Index (CSI) and pixel-wise coefficient of determination (pixelwise R^2) were used as evaluation metrics. Critical Success Index (CSI) indicates how good the model is at correctly identifying the pixelwise flood depths above a certain depth threshold. In literature [4,10], different binarizing depth thresholds have been used from 0.02 m to 0.5 m. CSI metric ranges between [0,1]. The predicted and ground truth water depths are binarized if the pixelwise values are greater than d. By computing binary classification counts of the two images, the CSI is then calculated using $CSI_d = \frac{1}{N}\sum_{i=0}^{N} \frac{TP_i}{TP_i+FP_i+FN_i}$.

The performances of the different UNet-based architectures are presented in Table 1. The best performance was achieved by the UNet-SC Large architecture with an RMSE of 0.0164 and pixelwise R^2 of 0.5501. It is also observed that the SC architectures outperformed each of their base architectures in terms of RMSE and pixelwise R^2, with the exception of UNetTransformer Large and UNetTransformer-SC Large. Increasing the model sizes led to an improvement across all evaluation metrics in comparison to the base architectures. When comparing UNet-SC and UNet-SC Large architectures, RMSE improved from 0.0179 to 0.0164 and pixelwise R^2 improved from 0.4782 to 0.5501. The UNet-SC architecture achieved a $CSI_{0.05}$ of 0.4068, which further improved to 0.4447 for the UNet-SC Large architecture. This indicates that the model can correctly estimate 44.47% of the pixels with water depth greater than 0.05 m. It is also observed that the $CSI_{0.25}$ scores are lower than the $CSI_{0.05}$ scores across all architectures, as observed in other literature [4,10].

Table 1. Comparison of evaluation metrics of UNet architecture variants

Architecture	Parameters	RMSE ↓	CSI_d ↑		Pixelwise R^2 ↑
			d = 0.05	d = 0.25	
UNet	4.0 M	0.0181	0.3550	0.1675	0.4544
UNet-SC	4.1 M	0.0179	0.4068	0.1661	0.4782
UNetTransformer	4.3 M	0.0185	0.3686	0.1453	0.4156
UNetTransformer-SC	4.4 M	0.0177	0.3883	0.1703	0.4891
UNet Large	15.9 M	0.0171	0.4287	0.1972	0.5305
UNet-SC Large	16.4 M	**0.0164**	**0.4447**	0.2229	**0.5501**
UNetTransformer Large	17.0 M	0.0169	0.4020	**0.2298**	0.4618
UNetTransformer-SC Large	17.5 M	0.0171	0.4229	0.1884	0.5299

In the test set, there are 39 cities which were not seen by the model during training. The test metrics presented in Table 1 are the mean scores across all the 195 test samples (39 cities × 5 rainfall events). The RMSE scores of the model for each city across the 5 rainfall events in the test set is illustrated in Fig. 3. It was also observed that RMSE scores of the model increases as rain intensity increases. In general, the RMSE values for the test cities are comparable to values reported in recent literature [7,10], although the other metrics can be improved further. In terms of inference time performance, the trained emulator can make a single prediction in 0.33 s using a single NVIDIA RTX A6000 GPU and 16.36 s using a Intel Core i7-10870H CPU. This resulted in upto 1000 times speed-up in comparison to the FIP simulator, which takes approximately 300 s per simulation using GPU acceleration.

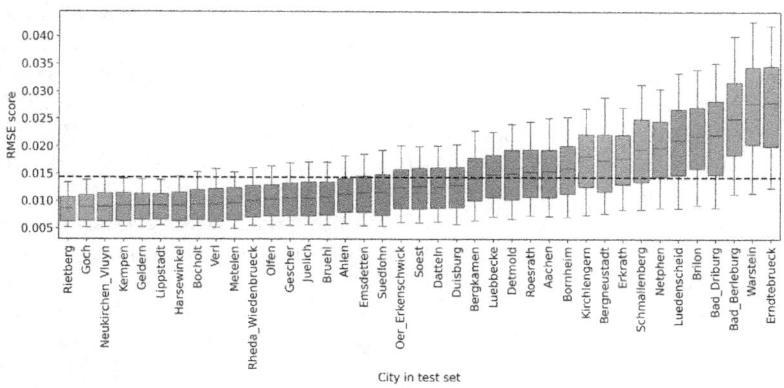

Fig. 3. RMSE scores of the best flood emulator for each test city across the 5 rainfall events. The boxplot for each city indicates the minimum, first quartile, median, third quartile and maximum RMSE scores across the 5 rainfall events.

5 Conclusion

In this paper, the development of a large-scale, high-resolution flood dataset of 200 diverse urban areas and 5 synthetic rainfall events was presented. Using this dataset, deep learning-based flood emulators based on the UNet architecture were trained and evaluated. It was shown that architecture improvements such as adding skip convolutions improved the performance of the models. In comparison to previous works, our flood emulators are able to generalize to previously unseen test cities and achieve similar error metrics such as RMSE. Another advantage of the flood emulators are the fast inference times on either a GPU or CPU device. The generalizability and fast inference times of our flood emulators paves the way for their potential integration in existing GIS tools.

Acknowledgments. This work is based on research conducted during the FloReST project, which was funded by the Federal Ministry of Education and Research (BMBF) as a part of the "WaX âĂŞ Extreme Water Events" funding measure.

References

1. Imperviousness density 2018 (raster 10 m), europe, 3-yearly, aug. 2020. European Union's Copernicus Land Monitoring Service information. https://doi.org/10.2909/3bf542bd-eebd-4d73-b53c-a0243f2ed862
2. Bai, X., et al.: Six research priorities for cities and climate change. Nature **555**(7694), 23–25 (2018). https://doi.org/10.1038/d41586-018-02409-z
3. Barnes, R.: RichDEM: terrain analysis software (2016). http://github.com/r-barnes/richdem
4. Burrichter, B., Hofmann, J., Koltermann da Silva, J., Niemann, A., Quirmbach, M.: A Spatiotemporal deep learning approach for urban pluvial flood forecasting with multi-source data. Water **15**(9), 1760 (2023). https://doi.org/10.3390/w15091760
5. Guo, Z., Leitão, J.P., Simões, N.E., Moosavi, V.: Data-driven flood emulation: speeding up urban flood predictions by deep convolutional neural networks. J. Flood Risk Manage. **14**(1), e12684 (2021). https://doi.org/10.1111/jfr3.12684
6. Guo, Z., Moosavi, V., Leitão, J.P.: Data-driven rapid flood prediction mapping with catchment generalizability. J. Hydrol. **609**, 127726 (2022). https://doi.org/10.1016/j.jhydrol.2022.127726
7. Hofmann, J., Schüttrumpf, H.: floodGAN: using deep adversarial learning to predict pluvial flooding in real time. Water **13**(16), 2255 (2021). https://doi.org/10.3390/w13162255
8. do Lago, C.A.F., Giacomoni, M.H., Bentivoglio, R., Taormina, R., Gomes, M.N., Mendiondo, E.M.: Generalizing rapid flood predictions to unseen urban catchments with conditional generative adversarial networks. J. Hydrol. **618** (2023). https://doi.org/10.1016/j.jhydrol.2023.129276
9. Loshchilov, I., Hutter, F.: Decoupled weight decay regularization (2019). https://doi.org/10.48550/arXiv.1711.05101
10. Löwe, R., Böhm, J., Jensen, D.G., Leandro, J., Rasmussen, S.H.: U-FLOOD - topographic deep learning for predicting urban pluvial flood water depth. J. Hydrol. **603**, 126898 (2021). https://doi.org/10.1016/j.jhydrol.2021.126898

11. Newman, R., Noy, I.: The global costs of extreme weather that are attributable to climate change. Nat. Commun. **14**(1), 6103 (2023)
12. Petit, O., Thome, N., Rambour, C., Themyr, L., Collins, T., Soler, L.: U-Net transformer: self and cross attention for medical image segmentation. In: Lian, C., Cao, X., Rekik, I., Xu, X., Yan, P. (eds.) MLMI 2021. LNCS, vol. 12966, pp. 267–276. Springer, Cham (2021). https://doi.org/10.1007/978-3-030-87589-3_28
13. Rak, A., Mewis, P., Guthe, S.: Accelerating flash flood simulations: an efficient GPU implementation for a slim shallow water solver. Environment. Modell. Softw. **177**, 106030 (2024). https://doi.org/10.1016/j.envsoft.2024.106030
14. Rolnick, D., et al.: Tackling climate change with machine learning. ACM Comput. Surv. **55**(2), 42:1–42:96 (2022). https://doi.org/10.1145/3485128
15. Ronneberger, O., Fischer, P., Brox, T.: U-Net: convolutional networks for biomedical image segmentation. In: Navab, N., Hornegger, J., Wells, W.M., Frangi, A.F. (eds.) MICCAI 2015. LNCS, vol. 9351, pp. 234–241. Springer, Cham (2015). https://doi.org/10.1007/978-3-319-24574-4_28
16. Zhao, T., et al.: Artificial intelligence for geoscience: progress, challenges, and perspectives. Innovation **5**(5), 100691 (2024). https://doi.org/10.1016/j.xinn.2024.100691

Towards Observing the Effect of Abstraction on Understandability of Explanations in Answer Set Programming

Zeynep G. Saribatur[1(✉)], Johannes Langer[2], Anna M. Thaler[2], and Ute Schmid[2]

[1] Institute of Logic and Computation, TU Wien, Vienna, Austria
zeynep.saribatur@tuwien.ac.at
[2] University of Bamberg, Bamberg, Germany
{johannes.langer,anna.thaler,ute.schmid}@uni-bamberg.de

Abstract. In order for AI systems to provide explanations of their decision-making that are concise and understandable, they need to have the ability of getting rid of irrelevant details and presenting a higher-level view. Answer Set Programming (ASP) is one of the core formalisms of Symbolic AI, based on logic programming with stable model semantics, widely used in various applications. Explaining the solutions (i.e., answer sets) of an answer set program continues to be a widely studied topic, with various systems available. This technical communication reports on a preliminary study for observing the effect of abstraction on the understandability of ASP explanations, by considering the recent abstraction notions which preserve the dependencies as much as possible while abstracting over answer set programs. We describe our experiment design which will be our base for further extensions of the study. Our preliminary results show the challenge of capturing the effect of abstraction on understandability, requiring further investigations in this direction.

Keywords: Abstraction · Explanations · Answer set programming

1 Introduction

Abstraction through simplifying and generalizing are abilities that humans unwittingly use when understanding and reasoning about the world [3,13]. Especially as we aim for AI systems that are transparent and understandable to humans, such systems need to acquire abstraction abilities that allow them to present explanations of their complex decision-making and representations overviewing their complex structures via showing the key elements, so-called "model-of-self" [14], making it easier for humans to understand. In the field of Explainable AI (XAI), some recent works involving abstraction include forgetting [22] or projecting away details in explanations [6], simplifying solutions [23],

T. Braun et al. (Eds.): KI 2025, LNAI 15956, pp. 236–243, 2026.
https://doi.org/10.1007/978-3-032-02813-6_18

abstracting rule-based representations by predicate invention [20] or domain clustering [8], explaining neural networks through causal abstractions [11] or decision trees [7], showing the need and the potential of abstraction.

Answer set programming (ASP) [4], which is based on logic programming with stable model semantics, is one of the core formalisms in Symbolic AI and thanks to efficient solvers widely used in many areas of Computer Science and AI, from combinatorial search problems over system modeling to knowledge-intensive applications [9]. Obtaining explanations for ASP solutions has also been a topic that is widely studied with available systems. The challenge of achieving concise explanations to aid in human understanding still remains [10]. Obtaining abstractions over the explanations would be a way to achieve this.

Abstraction is first introduced for ASP as an "over-approximation" of programs that ensures all answer sets are preserved while reducing the vocabulary [16,17]. Later, more restrictive properties have been investigated ensuring that all dependencies are preserved, by considering strong/uniform equivalence like relations [18,19]. These abstractions of programs were shown to get rid of or abstract over details that are not the key points for reasoning.

We report on an ongoing work that investigates, through cognitive experiments, whether the concise explanations obtained over these abstract answer set programs aid in human understandability. In this paper, we illustrate our hypothesis and describe the experiment design, with preliminary results in fact showing no effect. We view these results as the first step towards understanding the problem, discuss potential shortcomings and ideas on further extensions of the experiment with added complexity to be able capture the effect of abstraction on understandability.

2 Background

ASP An answer set program P over a set $\mathcal{U}$ of propositional atoms is a set of rules r of the form $\alpha_0 \leftarrow \alpha_1, \ldots, \alpha_m, not\ \alpha_{m+1}, \ldots, not\ \alpha_n,\ \ 0 \leq m \leq n$,where each $\alpha_i \in \mathcal{U}$ is a propositional literal and *not* is default negation. We also write r as $H(r) \leftarrow B^+(r),\ not\ B^-(r)$, where $H(r) = \{A_1, \ldots, A_l\}$ is the *head*, $B^+(r) = \{A_{l+1}, \ldots, A_m\}$ is the *positive body* and $B^-(r) = \{A_{m+1}, \ldots, A_n\}$ is the *negative body* of r. A rule r is a *constraint* if α_0 is falsity ($\perp$, then omitted), a *fact* if $n = 0$ and *positive* if $B^-(r) = \emptyset$. Semantically P induces a set $AS(P)$ of stable models (or answer sets), which are sets I of atoms of P that are minimal models of the *reduct* [12] given by $P^I = \{H(r) \leftarrow B^+(r) \mid r \in P, B^-(r) \cap I = \emptyset\}$.
Abstraction. Recent works study the theory of removal [19] and abstraction [18] of irrelevant details in answer set programs. The aim is to define a simplification resp. an abstraction of an answer set program that preserves the dependencies while removing resp. abstracting over some details. As it was shown that simplification by removal [19] is a special case of abstraction defined in [18], we only provide details of the latter, where a mapping is defined to cluster atoms in $\mathcal{U}$.

Definition 1. ([18]). *Given sets of atoms $\mathcal{U}$, $\mathcal{U}'$ with $|\mathcal{U}| \geq |\mathcal{U}'|$, a program P over $\mathcal{U}$ and a mapping $m : \mathcal{U} \mapsto \mathcal{U}'$, Q over $\mathcal{U}'$ is a* uniform m-abstraction of P *if, for any set F of facts over $\mathcal{U}$, we have*

$$m(AS(P \cup F)) = AS(Q \cup m(F)). \tag{1}$$

Informally, the aim is to map atoms from the language $\mathcal{U}$ of program P to atoms in a smaller language $\mathcal{U}'$ in such a way that the answer sets of P and the resulting abstraction Q correspond, independently of the facts, i.e., the instance data, added to the program. As not every program might have a uniform m-abstraction, the necessary and sufficient conditions for abstractability, and model-based representations of abstractions were provided. Furthermore it was shown that, whenever possible, the abstract program can be achieved simply by syntactic clustering of the atoms (or syntactic removal in [19]).

3 Abstractions over Explanations

In this section, we illustrate the potential use of abstraction notion from Definition 1 for obtaining abstract explanations containing the key details of reasoning.

Example 1. *Let us consider the following program P.*

$$\begin{aligned} \mathit{reachPotsdam} &\leftarrow \mathit{takePlane}.\\ \mathit{reachPotsdam} &\leftarrow \mathit{takeTrain}.\\ \mathit{attendKI} &\leftarrow \mathit{talkReady}, \mathit{register}, \mathit{reachPotsdam}. \end{aligned}$$

If we reach Potsdam by either taking the plane or the train, register to the conference and have our talk ready, then we can attend the conference.

Now consider having different scenarios in which we prepare our talk to present at KI, we either take the plane, the train, or start walking towards Potsdam, and we register or not. Since the talk is ready in all scenarios, that detail is not decisive in attending the conference, thus can be removed. Since both, taking the plane and taking the train (but not walking), allow us to reach Potsdam, the details of which transportation is taken can be abstracted over. Removing and abstracting over these details yield the program Q below.

$$\begin{aligned} \mathit{reachPotsdam} &\leftarrow \mathit{takePlaneOrTrain}.\\ \mathit{attendKI} &\leftarrow \mathit{register}, \mathit{reachPotsdam}. \end{aligned}$$

To theoretically capture the cases as in Example 1, where there is a set of scenarios to consider when abstracting, Definition 1 need to be relaxed to consider a set of sets of facts describing the potential instances, which we leave for future work and focus instead on how they might aid with explanations.

The idea is to obtain explanations over the abstracted program. In order to obtain explanations, we make use of the tool `xclingo` [5] which provides explanation graphs of a given answer set. Below we illustrate its use.

```
|__attendKI
| |__talkReady
| |__register
| |__reachPotsdam
| | |__takeTrain
```

(a) Default

```
|__attendKI
| |__register
| |__reachPotsdam
| | |__takePlaneOrTrain
```

(b) Abstracted

Fig. 1. Explanations obtained by `xclingo`

Example 2. (Ex. *1* **ctd)** *For the scenario* $\{talkReady, register, takeTrain\}$*, we have* $AS(P) = \{talkReady, register, takeTrain, reachPotsdam, attendKI\}$*. The explanation for the atom attendKI to appear in* $AS(P)$ *is as in Fig. 1a. With an abstraction mapping that removes talkReady and clusters takeTrain and takePlane into takePlaneOrTrain, and the abstract program Q over the abstracted vocabulary, we obtain the explanation for attendKI appearing in* $AS(Q)$ *as in Fig. 1b.*

4 Empirical Study

We hypothesize that the explanations obtained in the abstract programs would help in understandability, due to only containing the relevant details compared to the default explanations of the original programs. In this section, we describe the cognitive experiment in form of an online survey which we designed to test our hypothesis and report on preliminary results.

Task. Participants are presented with a classification task of tabular data, where each instance is a set of attributes with a binary class label. Each instance is also visualized by a corresponding image (see Fig. 2).

Classification. We present three concept learning tasks to participants, where each task is defined through an answer set program, assigning the instance to the target class when certain conditions are met.[1]

During evaluation, participants are asked to decide for each instance whether it belongs to the target class or not.

Domains. To avoid proactive interference [24] between previously learned rules, we select three different domains, each of which being different biological specimen: flowers, mushrooms, and cacti. Each domain has six domain-specific decision attributes, and the same target attribute "dangerous".

Study Design and Procedure. The empirical online study is based on a complete 2x2 between-subject factorial design, where participants were randomly assigned to one of the four groups formed by a combination of 2 factors: 'cluster' and 'removal', and the data between these groups were compared to calculate the

[1] Details on the answer set programs and the stimuli description can be found here https://www.dbai.tuwien.ac.at/user/saribat/pub/ki25_supp.pdf.

Cactus Attributes:

Spines	Shape	Arms	Stem	Flower	Height	Dangerous
few	flattened-padded	many	thick	yes	short	yes

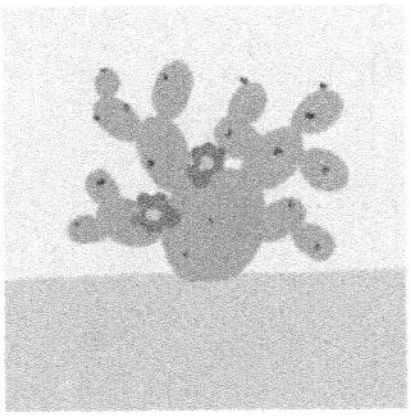

Explanation:

This cactus is dangerous because all of the following attributes apply:

- adaptive because
  - has arms, since **many** arms
  - no leaf-like shape, since **flattened-padded** shape
  - **short** height
- **few** spines
- **flowers**
- **thick** stem

Fig. 2. Example of a learning phase stimulus as presented in the study. The table lists all attributes and values, with a corresponding image. The provided explanation is the default without abstraction.

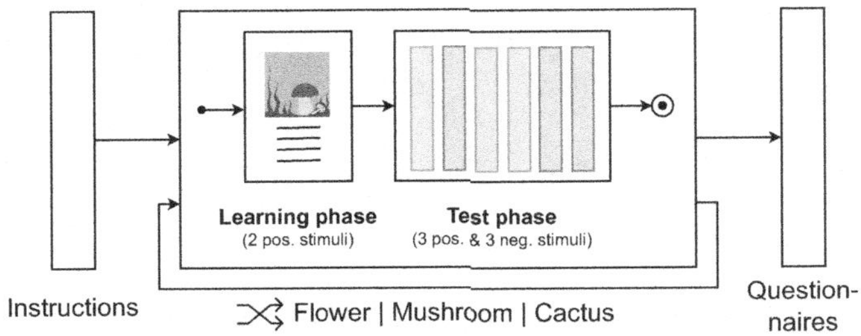

Fig. 3. Study Procedure Overview. The order of the three domains and the test phase, i.e. the order of the positive (illustrative in green) and negative (resp. in red) stimuli were random to balance out possible position effects.(Color figure online)

effect of each factor on our dependent variable.[2] The main part of the experiment consists of a learning and a test phase (see Fig. 3).

Learning Phase. At the beginning of each domain, participants receive 2 positive (dangerous) examples together with an explanation why the stimulus was classified dangerous. These explanations differ depending on the assigned group of the experiment. The `control` group has an explanation with no abstraction (see Fig. 2). The three abstraction groups, `cluster`, `removal`, and `cluster_removal`, receive an abstracted explanation based on their assigned group. The `cluster`, resp. `removal`, group receives an explanation based on abstraction by clustering (e.g. 'water or mud'), resp. removal of irrelevant atoms. The `cluster_removal` group sees explanations with cluster abstraction and removal (e.g., Fig. 1b).

Test Phase. After the learning phase, unseen stimuli (3 positive, 3 negative) are presented without explanations. Participants need to decide whether the stimulus is dangerous or not and give a rating about their confidence with a

[2] Informed consent about data protection, anonymity and the right to leave the study at any time was given. At the end of the study, participants had the opportunity to leave further comments and notes.

slider bar (range 0–100), or choose the option "I don't know". The attribute values were evenly distributed across all stimuli to avoid accumulations of a single attribute value.

Subjective Assessments. After the last domain, participants gave ratings on the perceived usefulness of the explanations (5 items, based on a subset of [2]), their active memorization efforts (1 item) and their familiarity with the domain terminologies (1 item). All these items were rated on a Likert-type scale ranging from 1 ('strongly disagree') to 7 ('strongly agree'). Prior knowledge in computer science, logic, mathematics and programming as well as the three domains (7 items) is rated on a scale from 0 ('none') to 4 ('expert').

Participants. We recruited 71 participants from two universities whose study program was either Psychology or Computer Science (age mean = 23.1, sd = 4.9; 49 female, 38 male). There is no significant difference in the distribution of gender, age, or self-reported knowledge in computer science among the abstraction groups. Participants were compensated in course credits for their participation.

Analysis. We report our statistical tests on the following hypotheses.

Hypothesis 1: Having abstract explanations during the learning phase increases the classification accuracy compared to the `control` group.

Per participant mean accuracy is normally distributed for three of the four abstraction groups, with a Shapiro-Wilk test [21] showing barely significant ($p = .03$) for the `cluster_removal` condition. Homeoscedasticity holds. A two-factor ANOVA considering each abstraction as its own factor shows no significant effect for either factor and the interaction (`cluster` $F = 1.30, p = .26$, `removal` $F = .74, p = .39$, `cluster_removal` $F = .24, p = .63$). A one-factor ANOVA considering abstraction groups as a single factor shows no significant effect ($F = .74, p = .53$).

Hypothesis 2: Having abstract explanations during the learning phase increases the self-reported confidence in the classification compared to the `control` group.

Per participant mean accuracy is normally distributed for each abstraction group. Homeoscedasticity holds. A two-factor ANOVA as in *Hypothesis 1* shows not significant for either factor and the interaction (`cluster` $F = .29, p = .59$, `removal` $F = .08, p = .78$, `cluster_removal` $F = .20, p = .66$). A one-factor ANOVA considering abstraction groups as a single factor shows no significant effect ($F = .19, p = .91$).

5 Discussion

Motivated by how predicate invention aids in comprehensibility [15,20], we expected a positive effect of abstraction. When examining the results, we detected some unforeseen limitations which might have caused observing no such effect. One major issue, which was also reported among the participants' comments, is the choice of the target variable "dangerous". Due to its semantic meaning, participants may have been inclined to choose dangerousness to avoid repercussions. Another potential limitation might be that tasks requires to be of a certain "intermediate" complexity for explanations to be helpful [1]. The default explanation might have been already easy to follow, requiring no help with abstraction.

We aim to extend our study by changing the target variable to a less triggering term, and increasing the complexity of the provided explanations, e.g., considering clustering of more than two atoms, adding further irrelevant atoms to be removed. The challenge will be to have the study complex enough to show the effect of abstraction but not too complex to not see any effect at all.

Acknowledgments. This work has been supported by the Austrian Science Fund (FWF) project T-1315 (Saribatur).

Disclosure of Interests. The authors have no competing interests to declare that are relevant to the content of this article.

References

1. Ai, L., Muggleton, S.H., Hocquette, C., Gromowski, M., Schmid, U.: Beneficial and harmful explanatory machine learning. Mach. Learn. **110**(4), 695–721 (2021). https://doi.org/10.1007/s10994-020-05941-0
2. Bahel, V., Sriram, H., Conati, C.: Personalizing explanations of AI-driven hints to users' cognitive abilities: an empirical evaluation. CoRR **abs/2403.04035** (2024). https://doi.org/10.48550/ARXIV.2403.04035
3. Blaha, L.M., et al.: Understanding is a process. Front. Syst. Neurosci. **16** (2022). https://doi.org/10.3389/fnsys.2022.800280
4. Brewka, G., Eiter, T., Truszczyński, M.: Answer set programming at a glance. Commun. ACM **54**(12), 92–103 (2011)
5. Cabalar, P., Muñiz, B.: Explanation graphs for stable models of labelled logic programs. In: Arias, J., et al (eds.) Proceedings of the International Conference on Logic Programming 2023 Workshops co-located with the 39th International Conference on Logic Programming (ICLP 2023), London, United Kingdom, July 9th and 10th, 2023. CEUR Workshop Proceedings, vol. 3437. CEUR-WS.org (2023). https://ceur-ws.org/Vol-3437/paper3ASPOCP.pdf
6. Chakraborti, T., Sreedharan, S., Kambhampati, S.: The emerging landscape of explainable automated planning and decision making. In: Proc. IJCAI, pp. 4803–4811 (2020). https://doi.org/10.24963/ijcai.2020/669
7. Confalonieri, R., Weyde, T., Besold, T.R., del Prado Martín, F.M.: Using ontologies to enhance human understandability of global post-hoc explanations of black-box models. Artif. Intell. **296**, 103471 (2021)
8. Eiter, T., Saribatur, Z.G., Schüller, P.: Abstraction for zooming-in to unsolvability reasons of grid-cell problems. In: Proc. XAI@IJCAI (2019)
9. Erdem, E., Gelfond, M., Leone, N.: Applications of answer set programming. AI Magazine **37**(3), pp. 53–68 (2016). http://www.aaai.org/ojs/index.php/aimagazine/article/view/2678
10. Fandinno, J., Schulz, C.: Answering the "why" in answer set programming - a survey of explanation approaches. TPLP **19**(2),pp. 114–203 (2019)
11. Geiger, A., Lu, H., Icard, T., Potts, C.: Causal abstractions of neural networks. Adv. Neural. Inf. Process. Syst. **34**, 9574–9586 (2021)

12. Gelfond, M., Lifschitz, V.: Classical negation in logic programs and disjunctive databases. N. Gener. Comput. **9**(3), 365–385 (1991)
13. Ho, M.K., Abel, D., Correa, C.G., Littman, M.L., Cohen, J.D., Griffiths, T.L.: People construct simplified mental representations to plan. Nature **606**(7912), 129–136 (2022)
14. Miller, T.: Explanation in artificial intelligence: insights from the social sciences. AIJ **267**, 1–38 (2019). https://doi.org/10.1016/j.artint.2018.07.007
15. Muggleton, S.H., Schmid, U., Zeller, C., Tamaddoni-Nezhad, A., Besold, T.: Ultra-strong machine learning: comprehensibility of programs learned with ILP. ML **107**(7),pp. 1119–1140 (2018)
16. Saribatur, Z.G., Eiter, T.: Omission-based abstraction for answer set programs. Theory Pract. Log. Program. **21**(2),pp. 145–195 (2021). https://doi.org/10.1017/S1471068420000095
17. Saribatur, Z.G., Eiter, T., Schüller, P.: Abstraction for non-ground answer set programs. Artif. Intell. **300**, 103563 (2021). https://doi.org/10.1016/j.artint.2021.103563
18. Saribatur, Z.G., Knorr, M., Gonçalves, R., Leite, J.: On abstracting over the irrelevant in answer set programming. In: Marquis, P., Ortiz, M., Pagnucco, M. (eds.) Proceedings of the 21st International Conference on Principles of Knowledge Representation and Reasoning, KR 2024, Hanoi, Vietnam. November 2-8, 2024 (2024).https://doi.org/10.24963/KR.2024/61
19. Saribatur, Z.G., Woltran, S.: Foundations for projecting away the irrelevant in ASP programs. In: Proceedings of the 20th International Conference on Principles of Knowledge Representation and Reasoning, pp. 614–624. IJCAI Organization (2023). https://doi.org/10.24963/kr.2023/60
20. Schmid, U., Zeller, C., Besold, T., Tamaddoni-Nezhad, A., Muggleton, S.: How Does Predicate Invention Affect Human Comprehensibility? In: Cussens, J., Russo, A. (eds.) ILP 2016. LNCS (LNAI), vol. 10326, pp. 52–67. Springer, Cham (2017). https://doi.org/10.1007/978-3-319-63342-8_5
21. Shapiro, S.S., Wilk, M.B.: An analysis of variance test for normality (complete samples). Biometrika **52**(3/4),pp. 591–611 (1965). http://www.jstor.org/stable/2333709
22. Siebers, M., Schmid, U.: Please delete that! why should I? KI **33**(1), pp.35–44 (2019)
23. Poesia Reis e Silva, G., Goodman, N.: Left to the reader: abstracting solutions in mathematical reasoning. In: Proceedings of the Annual Meeting of the Cognitive Science Society, vol. 44 (2022)
24. Underwood, B.J.: Interference and forgetting. Psychol. Rev. **64**(1), 49 (1957)

XAIRob—An Explainable-AI-Based Relative Robustness Measure for Object Detection

Benedikt Schlereth-Groh[1(✉)], Ramin Tavakoli Kolagari[1], and Ute Schmid[2]

[1] Technische Hochschule Nürnberg, Keßlerplatz 12, 90489 Nürnberg, Germany
{benedikt.schlereth-groh,ramin.tavakolikolagari}@th-nuernberg.de
[2] Otto-Friedrich-Universität Bamberg, Kapuzinerstraße 16, 96047 Bamberg, Germany
ute.schmid@uni-bamberg.de
https://www.th-nuernberg.de/, https://www.uni-bamberg.de/

Abstract. The reliable detection of objects is of paramount importance for autonomous systems such as self-driving cars, particularly in unfavourable real-world conditions. The present paper introduces a novel methodology for the evaluation of the robustness of object detection algorithms based on Red-Green-Blue (RGB) images or Light Detection and Ranging (LiDAR) data. The utilisation of explainable artificial intelligence (x-AI) facilitates the identification of critical regions within objects. By systematically removing these key pixels or points, it is possible to determine how much of an object can be perturbed before it becomes undetectable. This approach offers a more precise comparison of object recognition systems than traditional methods, providing a meaningful assessment of their resilience to missing or manipulated data. The relative robustness scale employed in this study utilises two sensor modalities, thereby unveiling disparities in the robustness of object detection algorithms. By focusing on the most critical parts of an object, a more precise comparison of detection performance is achieved than with traditional approaches that rely mainly on random effects.

Keywords: x-Ai · robustness · RGB images · point clouds · Object Detection

1 Introduction

In the context of autonomous systems, such as those used in cars and robots, it is important to be able to reliably and robustly detect obstacles. Although deep neural networks have been demonstrated to achieve high levels of accuracy in object recognition [20], outperforming traditional computer vision methods [9], the exact mechanisms behind this performance are not yet fully understood [23]. In order to cultivate confidence in object recognition, it is essential to ensure its robustness.

T. Braun et al. (Eds.): KI 2025, LNAI 15956, pp. 244–251, 2026.
https://doi.org/10.1007/978-3-032-02813-6_19

According to the Institute of Electrical and Electronics Engineers (IEEE), robustness is defined as the degree to which a system functions correctly despite invalid inputs or stressful conditions [1]. In the context of autonomous driving, this refers to the ability of an object recognition system to detect objects while remaining resilient to external influences and maintaining low rates of false positives and negatives. This differs from the concept of security robustness, which refers to the ability to detect objects despite sophisticated, deliberate manipulation. These manipulations may include adversarial attacks designed to reduce the informative value of the data.

Current robustness evaluations, whether based on adversarial attacks or common corruptions, often only consider recognition rates [4,6]. This approach often results in inadequate comparability. In order to comprehensively evaluate the robustness of class-specific flaws in object detection systems, it is necessary to integrate the "smart" and "random" perspectives. This work provides a precise definition of robustness, paving the way for detailed evaluations.

Definition 1 (Robustness). *The robustness of an object detection system refers to its ability to identify objects even when some, but not all, relevant information is missing. A system is considered more robust than another iff, under identical sensor modalities and data conditions, it continues to recognize objects despite the absence of certain critical information, as long as enough essential data remains for detection, whereas the other system fails to do so.*

2 Background

2.1 Object Detection

Objects can be detected on the sensors, which are used for autonomous driving. For example, images can be analysed using convolutional neural networks (CNN) or transformer based networks. A CNN based real-time object detection system is for example the *YOLO* network, which is promises improved accuracy and inference time with the 11th version [7]. One real-time capable transformer-based approach, for example, is the Real-Time Detection Transformer *RT-DETR* developed by Baidu [11].

To process point cloud data with deep neural networks, five different data representation methods can be distinguished [2]. The three dimension of the point cloud can be mapped to a two-dimensional depth image, bird-eye view, cylindrical coordinates or spherical coordinates and similar algorithms to object detectors for images can be applied [2,3,10,15–17,22]. Different approaches are available to train and test with the open source project for LiDAR-based object detection called *OpenPCDet* [18].

2.2 Explainable AI

The approach utilised is predicated on RISE (Randomized Input Sampling for Explanation of Black-box Models), a model-agnostic technique that explicates classifications by means of analysing algorithm output from masked images [12]. This concept was expanded upon to encompass object detection as D-RISE, a methodology that generates saliency maps by systematically varying inputs and observing the resulting predictions [13]. Schinagl et al. further adapted the D-RISE methodology for the purpose of assessing the importance of point clouds derived from LiDAR data [14]. This method is ideal in that it is compatible with both RGB and LiDAR data, and does not necessitate access to internal neural network layers.

2.3 Robustness

Goodfellow et al. [6] presented adversarial examples to show how small and clever perturbations can mislead a classification system. Analyses in the area of security show a weakness of object detection to adversarial attacks on images or point clouds in the area of autonomous driving [19,21,24]. Corruptions, such as those caused by rain, fog or moving objects, can be applied to camera and LiDAR data. The mean average accuracy of such data can then be evaluated [4,8]. All known work analyses the detection of all classes, ignoring how they are represented.

3 Experimental Setup

3.1 Dataset

For our experiments, we choose the Karlsruhe Institute of Technology and Toyota Technological Institute Dataset (*KITTI*), where images and point cloud data have been collected at the same time [5]. To ensure a clean separation of training and test data, the data loader of *OpenPCDet* is used [18] with the classes Car, Pedestrian and Cyclist.

3.2 Ratio of Occlusion

The assessment of robustness is conducted through the measurement of the occlusion ratio, defined as the percentage of crucial pixels or points that are removed before an object becomes undetectable. The process commences with the initial detection and labelling of an object in unmodified data. The subsequent step involves the systematic elimination of the most significant pixels or points, as determined by the x-AI explanation method. The occlusion ratio is calculated by dividing the number of pixels or points that have been removed by the size of the bounding box (images) or the total points within it (LiDAR).

$$\text{ratio of occlusion} = \frac{n_{occluded}}{\text{bounding box size}} = \frac{n_{occluded}}{n_{pointsinboundingbox}} \quad (1)$$

Once all the objects in the test dataset have been evaluated at the occlusion ratio at which they are lost, they can be displayed as a histogram. This makes it possible to visualise the occlusion ratio at which the majority of objects are no longer detected. As discussed earlier, a robust object detection algorithm can detect an object even when complete data is unavailable. However, hallucination—object detection in the absence of data—is also detrimental and should not be considered robust. Therefore, to compare the robustness of object recognisers, we define the range between 0.2 and 0.95 as robust recognition, calculating the proportion of objects recognised within this range.

4 Experiments

In this section, the robustness of different object recognisers in the presence of occlusion is evaluated.

4.1 RGB Images

To evaluate the robustness of image based object detectors under guided occlusions, the most important pixels, according to the attribution map provided by D-RISE, are successively removed.

(a) *YOLOv11* missed Car (b) *YOLOv11* missed Pedestrian (c) *YOLOv11* missed Cyclist

Fig. 1. Guided Occlusion on RGB Images

The minimum of points to be removed until the detection fails is visualized in Fig. 1 with black pixels representing deletion.

The quantitative investigation by guided occlusion over all available test images is visualized in Fig. 2. A difference between *YOLOv11* and *RT-DETR* for vehicles is not visible.

To compare the robustness values, we calculate the area under the curve between the 0.2 and 0.95 as this range can be considered robust (Table 1).

Table 1. Evaluation Metrics of RGB Object Detectors.

Network	Car	Pedestrian	Cyclist
YOLOv11	0.28	0.35	0.11
RT-DETR	0.32	0.54	0.06

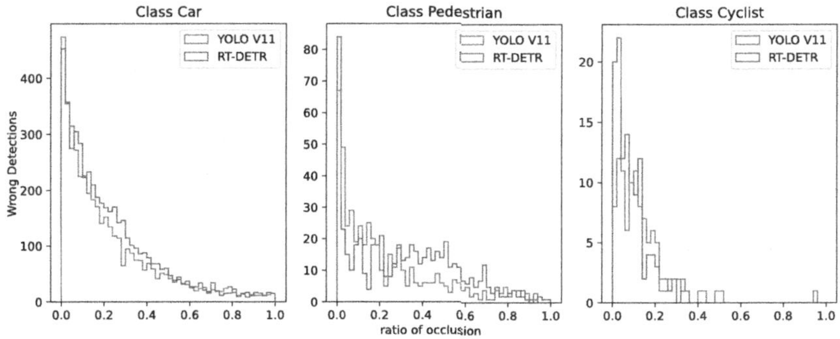

Fig. 2. Missed Detections with masking off by D-RISE

It is noteworthy that while the object recognizer *RT-DETR* continues to recognise a considerable number of pedestrians when they are partially obscured, it does not detect cyclists with the same robustness. The varying robustness to occlusion shows how important it is to analyse the individual classes.

4.2 LiDAR Data

The experiments were carried out in the same way for LiDAR data using the code provided by Schinagl et al. [14] to identify important points for objects detection.

(a) *second* missed Car (b) *second* missed Pedestrian (c) *second* missed Cyclist

Fig. 3. Guided Occlusion on LiDAR point clouds

The examples shown in Fig. 3 are performed by *second* and just enough points are removed to miss the detection. The background and removed points are coloured green, while the points belonging to the object are coloured according to the measured reflectivity.

In Fig. 4 the histograms of the ratio of occlusions are visualized. The classes are once more distributed across separate histograms, with the ratio of occlusion plotted on the x-axis.

To compare the robustness values, we calculate the area under the curve between the 0.2 and 0.95 as this range is considered robust.

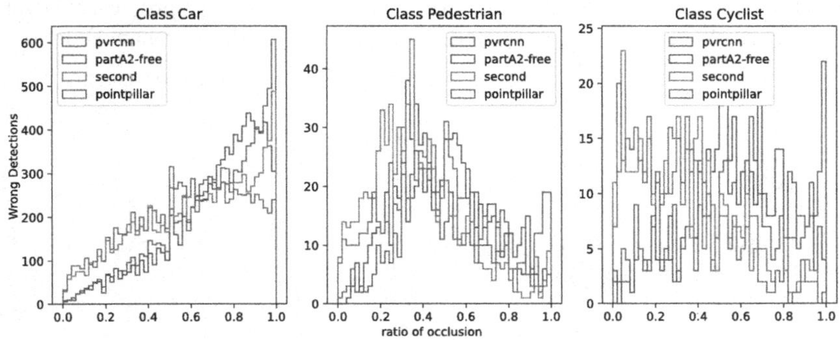

Fig. 4. Missed Detections with masking off by D-RISE

Table 2. Evaluation Metrics of RGB Object Detectors.

Network	Car	Pedestrian	Cyclist
pvrcnn	0.83	0.86	0.79
second	0.82	0.76	0.66
PartA2-Free	0.87	0.91	0.88
pointpillar	0.86	0.81	0.65

In Table 2 the difference between *pvrcnn* and *pointpillar* is marginal for the class car, but *PartA2-Free* leads the ranking on all classes, indicating the most robust detection.

5 Conclusion

The x-AI-based occlusion method under investigation evaluates object detection algorithms in challenging, data-scarce conditions. By visualising occlusion ratios, it is possible to determine the extent to which an object must be obscured before detection fails, thereby highlighting the benefit of x-AI in comparing algorithms prior to real-world deployment.

The findings of our study indicated substantial disparities in the robustness of RGB and LiDAR object detection algorithms when confronted with partially occluded data. RGB detectors demonstrated a particular vulnerability to minor occlusions, emphasising the significance of this comparative analysis in identifying susceptible object classes. Conventional benchmarks frequently yield mean accuracy but are unable to mirror an algorithm's genuine robustness, a pivotal metric that does not invariably align with elevated mean precision. This research contributes to the evaluation of robust object detection and enables thorough testing for more reliable perception systems.

Acknowledgments. This work was accomplished within the project VAL, FKZ 5320000013, EBA Az. 8fd/003-1255#008-VAL, funded by the German Federal Ministry of Digital and Transport.

References

1. Ieee standard glossary of software engineering terminology: IEEE Std 610. 12-1990, pp. 1–84 (1990). https://doi.org/10.1109/IEEESTD.1990.101064
2. Alaba, S.Y., Ball, J.E.: A survey on deep-learning-based lidar 3d object detection for autonomous driving. Sensors (Basel, Switzerland) **22** (2022)
3. Beltran, J., Guindel, C., Moreno, F.M., Cruzado, D., Garcia, F., de la Escalera, A.: Birdnet: a 3d object detection framework from lidar information (2018). https://arxiv.org/abs/1805.01195
4. Dong, Y., et al.: Benchmarking robustness of 3d object detection to common corruptions in autonomous driving. In: 2023 IEEE/CVF Conference on Computer Vision and Pattern Recognition (CVPR), pp. 1022–1032 (2023)
5. Geiger, A., Lenz, P., Stiller, C., Urtasun, R.: Vision meets robotics: the kitti dataset. Int. J. Robot. Res. (IJRR) (2013)
6. Goodfellow, I.J., Shlens, J., Szegedy, C.: Explaining and harnessing adversarial examples. CoRR **abs/1412.6572** (2014)
7. Jocher, G., Qiu, J.: Ultralytics yolo11 (2024). https://github.com/ultralytics/ultralytics
8. Kong, L., et al.: Robo3d: towards robust and reliable 3d perception against corruptions. In: 2023 IEEE/CVF International Conference on Computer Vision (ICCV), pp. 19937–19949 (2023)
9. Krizhevsky, A., Sutskever, I., Hinton, G.E.: Imagenet classification with deep convolutional neural networks. Commun. ACM **60**, 84–90 (2012)
10. Lang, A.H., Vora, S., Caesar, H., Zhou, L., Yang, J., Beijbom, O.: Pointpillars: fast encoders for object detection from point clouds. In: Proceedings of the IEEE Conference on Computer Vision and Pattern Recognition, pp. 12697–12705 (2019)
11. Lv, W., et al.: Detrs beat yolos on real-time object detection (2023)
12. Petsiuk, V., Das, A., Saenko, K.: Rise: Randomized input sampling for explanation of black-box models. ArXiv **abs/1806.07421** (2018)
13. Petsiuk, V., et al.: Black-box explanation of object detectors via saliency maps. In: 2021 IEEE/CVF Conference on Computer Vision and Pattern Recognition (CVPR), pp. 11438–11447 (2020)
14. Schinagl, D., Krispel, G., Possegger, H., Roth, P.M., Bischof, H.: Occam's laser: occlusion-based attribution maps for 3d object detectors on lidar data. In: 2022 IEEE/CVF Conference on Computer Vision and Pattern Recognition (CVPR), pp. 1131–1140 (2022)
15. Shi, S., et al.: PV-RCNN: point-voxel feature set abstraction for 3d object detection. In: 2020 IEEE/CVF Conference on Computer Vision and Pattern Recognition (CVPR), pp. 10526–10535 (2019)
16. Shi, S., Wang, X., Li, H.: Pointrcnn: 3d object proposal generation and detection from point cloud (2019). https://arxiv.org/abs/1812.04244
17. Shi, S., Wang, Z., Shi, J., Wang, X., Li, H.: From points to parts: 3d object detection from point cloud with part-aware and part-aggregation network. IEEE Trans. Pattern Anal. Mach. Intell. **43**, 2647–2664 (2019)

18. Team, O.D.: Openpcdet: an open-source toolbox for 3d object detection from point clouds (2020). https://github.com/open-mmlab/OpenPCDet
19. Tu, J., et al.: Physically realizable adversarial examples for lidar object detection. In: 2020 IEEE/CVF Conference on Computer Vision and Pattern Recognition (CVPR), pp. 13713–13722 (2020)
20. Voulodimos, A., Doulamis, N.D., Doulamis, A.D., Protopapadakis, E.E.: Deep learning for computer vision: a brief review. Comput. Intell. Neurosci. **2018** (2018)
21. Wang, X., Cai, M., Sohel, F., Sang, N., Chang, Z.: Adversarial point cloud perturbations against 3d object detection in autonomous driving systems. Neurocomputing **466**, 27–36 (2021)
22. Yan, Y., Mao, Y., Li, B.: Second: sparsely embedded convolutional detection. Sensors **18**(10) (2018). https://doi.org/10.3390/s18103337, https://www.mdpi.com/1424-8220/18/10/3337
23. Zeiler, M.D., Fergus, R.: Visualizing and understanding convolutional networks. ArXiv **abs/1311.2901** (2013)
24. Zhu, Z., et al.: Understanding the robustness of 3d object detection with bird'view representations in autonomous driving. In: 2023 IEEE/CVF Conference on Computer Vision and Pattern Recognition (CVPR), pp. 21600–21610 (2023)

LLMs for Easy Language Translation: A Case Study on German Public Authorities Web Pages

Thorben Schomacker[1,2(✉)], Burak Tinman[1], Chris Biemann[2], and Marina Tropmann-Frick[1]

[1] Hamburg University of Applied Sciences, 20099 Hamburg, Germany
[2] University of Hamburg, 22527 Hamburg, Germany
thorben.schomacker@haw-hamburg.de

Abstract. This paper examines the use of Large Language Models (LLMs) for the intralingual translation of documents from standard German to German Easy Language (*Leichte Sprache*). We use open-weight models, from the Llama 3 family, with less than ten billion parameters. Additionally, we employ parameter-efficient fine-tuning (QLoRA) to adapt the LLMs to the requirements of Easy Language. For this purpose, we introduce a new data set (*ELGEPA*) (Source code to obtain the dataset: github.com/minds-hh/ger-gov-easy-lang), which is a parallel corpus of governmental documents in standard German and Easy Language with additional metadata, obtained from all German federal states and their capitals In our experiments, a fine-tuned Llama 3.1-8B-Instruct model achieved a SARI score of 41 and Flesch Reading Ease of 69. Outperforming GPT-4o and indicating that this type of model can deliver promising text quality in Easy Language.

Keywords: LLM · text simplification · Easy Language · Leichte Sprache · Governmental texts · intralingual translation

1 Introduction

The increasing importance of accessible digital communication calls for comprehensible language, e.g. Easy Language. Leichte Sprache (German Easy Language) is a simplified variant of German, characterized by simple sentence structures, clear wording and a comprehensibility-enhanced text structure. There are various international equivalents to Leichte Sprache, e.g., Easy Read (UK) or Easy-to-Read (USA, Australia) [9]. German public authorities are obliged to communicate in a comprehensible way [13]. Many of Germany's federal states already use translation services that convert complex articles into Easy Language on their websites. Even though some already use automated solutions, most translation services are conducting manual translations [3].

Large Language Models (LLMs) can provide a remedy by replacing this manual effort with machine translations. Open-weight models with a manageable

T. Braun et al. (Eds.): KI 2025, LNAI 15956, pp. 252–261, 2026.
https://doi.org/10.1007/978-3-032-02813-6_20

number of parameters – i.e., models with fewer than ten billion parameters – offer decisive advantages over closed-weight alternatives such as GPT-4 or Claude Sonnet. This paper deals with the application of parameter-efficient fine-tuning (PEFT) methods, such as Low-Rank Adaptation (LoRA) or quantized LoRA (QLoRA), an efficient method of fine-tuning, to selected Llama 3 models with less than ten billion parameters in order to specialize them specifically for the generation of Easy Language texts. By adapting the pre-trained Llama 3 models using LoRA to a self-compiled data set of existing texts by public authorities in Easy Language, the ability of the model to generate coherent, grammatically correct texts that comply with the rules of Easy Language is to be tested.

2 Related Work

Easy Language translation can be considered as *automatic text simplification*, where a lexical and structural complex text is translated into a simplified version, while preserving its meaning. German text simplification research has recently focused on machine learning-based methods (see [23,31] for a recent overview). Translation into Easy Language poses a particular challenge, because it is more than just a simplified version of a language; it follows specific rules and conventions. It is therefore important that models not only correctly implement the semantic content, but also these formal aspects of Easy Language. There are various sets of rules [7,8,28]. With the DIN SPEC 33429 [6], finalized in March 2025, likely becoming the ruling guideline in the future. The main problem with all machine learning methods is the data basis, which is scarce in the field of German for parallel corpora between complex texts and Easy Language. Some research work has dealt with the creation of a general parallel corpus in German [12,20,31,34], but data sets for governmental texts are few to none. In order to deal with the scarcity of data, however, there are approaches to train the language models with (semi)-synthetic data [21] or to pursue a style-specific pre-training approach [2].

Finally, the evaluation of automated text simplification systems is an under-researched field that is complicated by the subjectivity of the task and lack of standards, which makes the development of suitable metrics and reference data, especially for non-English languages, necessary [14]. Automated metrics such as BLEU [29], ROUGE [22] or SARI [37] are often used, but they cannot fully capture the special requirements of Easy Language. Operationalizing part of the parts of Easy Language in a form evaluation has been done, but is still in its infancy [15,16,24,30,32].

3 Methods

Since parallel corpora of governmental texts in German, including Easy Language, are hardly available[1], the first step was to gather relevant data. In the

[1] There is only a subset from for the city of Cologne [36].

next step, appropriate models were identified that were suitable for fine-tuning. During the fine-tuning itself, the selected models were trained using the PEFT approach and the results were then validated and checked using various metrics.

3.1 Dataset Creation

The first step was to check the 16 websites of the federal states of Germany. As only 3 of these offered the possibility of extracting a parallel corpus, the search was extended to the websites of the federale state capitals (excluding the city states: Berlin, Bremen and Hamburg). With the exception of Magdeburg, which uses a live AI-controlled translation into Easy Language and is therefore difficult to extract, the remaining possible state capitals are the basis for our dataset: Easy Language documents of GErman Public Authorities (*ELGEPA*) shown in Table 1. In summary, 609 documents with standard and Easy Language were extracted from the 8 root websites using Scrapy[2]. We further expanded the data set with the Cologne subset from [36]. The subject areas of the texts focus primarily on help, tips, news or regional content.

Furthermore, if given in the source, we added metadata to each sample: Whether the text was generated in Easy Language by an AI, the corresponding author of the text, the proofreader of the text, which form of address[3] was used in the text and the subject area covered by the text. Subsequently, the dataset was divided into a training and test split in a ratio of 87.5% to 12.5% so that the training dataset contains at least 500 training copies and the test dataset has enough copies to be able to check the performance of the models in a meaningful way. The data set also had to be converted into a machine-readable format for subsequent fine-tuning. We used the ShareGPT[4] format, which is particularly suitable for multi-turn conversations, but can also be used for first-turn conversations to support a prompt-response structure.

Table 1. Number of documents per source in our dataset

Source	# docs
hannover.de	272
hamburg.de	103
stadt-koeln.de	82
dresden.de	67
stuttgart.de	44
stadt.muenchen.de	28
hessen.de	5
saarbruecken.de	4
ms.niedersachsen.de	4
total	609

[2] https://www.scrapy.org, last access 16.04.2025.

[3] Gemini 1.5 Flash, was prompted to choose “Duzen”, “Siezen” or a mix of both to determine the text’s form of address. This feature of our data was discussed in [33].

[4] It represents conversations as JSON arrays of message objects. Each message includes a ‘from’ and a ‘value’-field, e.g. `{‘conversations’: [{‘from’: ‘human’, ‘value’: <question>}, {‘from’: ‘gpt’,‘value’: <response>}]}`. For more information and examples see https://huggingface.co/datasets/anon8231489123/ShareGPT_Vicuna_unfiltered, last access 16.04.2025.

3.2 Model Selection

Once the data set had been transformed into the ShareGPT format, the fine-tuning could begin. Since the data basis is scarce, especially in the field of Easy Language, fine-tuning or more specialized methods such as PEFT, e.g. LoRA or QLoRA [1], can also be used to train a powerful model with less data in a resource-saving manner. Other selection criteria included that the models were decoder-only and already instruct-based. A style training approach similar to that of Anschütz et al. [2] was pursued, without ultimately embedding the decoder model in a sequence-to-sequence model.

Meta's Llama 3 model family was selected because it has received widespread recognition in the research community and is considered to be highly performant. This is particularly evident on Huggingface, where the aforementioned Llama models have the most downloads in the text generation area (as of 08.01.2025). Within this family, three models were selected for fine-tuning: Llama 3.1 8B-Instruct [26], Llama 3.2 3B-Instruct, and Llama 3.2 1B-Instruct [27]. Models 3B and 1B are the first Meta models with a size of less than five billion parameters. This range of parameters allows us to investigate the influence of model size on the quality of the generated Easy Language texts.

3.3 Fine-Tuning

To fine-tune the models, Unsloth.ai[5] [10] was employed, which uses Triton [35] to implement backpropagation in LoRA training processes. Using Triton makes it possible to minimize the number of floating-point operations (FLOPs) during gradient descent, which has a positive impact on training speed [39].

Using the LoRA [17] method enables more efficient fine-tuning by freezing the pre-trained weights and training only a few additional parameters. This reduces the number of trainable parameters and GPU memory requirements during training, while maintaining or even improving the model quality in comparison to full fine-tuning. QLoRA [11] is a development of LoRA that enables the fine-tuning of extremely large language models on GPUs with limited memory by quantizing the pre-trained model to 4 bits and further memory optimizations. In contrast to LoRA, QLoRA not only reduces the number of trainable parameters but also the memory requirements of the base model itself, enabling even greater fine-tuning efficiency.

Based on this, the appropriate settings were made for the selected Llama 3 models. We configured the 4-bit quantized Llama 3.1-8B-Instruct model, to have a maximum sequence length of 8192 tokens for training, which allowed most of the dataset elements to fit into the training context without requiring too much resource-intensive training. The next step involved configuring the QLoRA. All

[5] We adapted pre-built Notebooks (colab.research.google.com/notebook) from Unsloth (unsloth.ai) for our fine-tuning. Additionally, Unsloth offers pre-built 4-bit quantized Llama 3 models, leveraging bitsandbytes, which enables k-bit quantization [25]. This also allows for fine-tuning using the QLoRA method [5].

of this was adopted from the template, as the default settings were already sufficient for this use case.

The data was then prepared for training, by transferring it to the Llama 3 model's chat template and enriching it with as system prompt. This served as an instruction for creating the texts in Easy Language and was embedded in all created prompts. The system prompt consists of a general instruction, a set of rules for translating into Easy Language, and a one-shot prompt example. The set of rules is taken from Regelwerk[6], which refers to the set of rules from [28]. The training[7] consisted of a total of 4 training epochs with a total batch size of 64 and 32 steps. However, the number of 511 training samples is still sufficient for training QLoRAs.

Table 2. Median values of readability indexes: Flesch Reading Ease (FR Ease) ↑, Flesch-Kincaid Grade (FK Grade) ↓, Wiener Sachtextformel (WSF) ↓ and reference-based metrics: SARI ↑, BERTScore ↑. For metrics with an ↑, higher values are better results. Vice versa, ↓ indicates that lower values are better.

	Reference	Base 3.2-1B	Base 3.2-3B	Base 3.1-8B	QLoRA 3.2-1B	QLoRA 3.2-3B	QLoRA 3.1-8B	GPT 4o
FR Ease ↑	71.105	51.04	58.84	62.745	61.19	56.89	**69.38**	43. 91
FK Grade ↓	5.65	10.05	7.95	6.8	7.75	8.2	**6.15**	11.3
WSF ↓	3.6	6.75	5.45	4.8	5.7	6.0	**4.2**	7.65
SARI ↑	-	35.47	38.86	38.37	39.79	40.16	**41.31**	39.58
BERTScore ↑	-	0.63	0.65	0.65	0.64	0.64	0.66	**0.68**

3.4 Evaluation

For the evaluation of machine-translated texts, there are many different metrics that check the aspects of text quality for texts from automated text simplification systems and can therefore also indirectly evaluate texts in Easy Language.

The Flesch Reading Ease Index [19], quantifies the readability of a text on a scale of 0–100. Higher values indicate greater ease of reading, based on the average sentence length and the number of syllables per word. Similarly, the Flesch-Kincaid Grade Index [18] indicates the approximate school level required to understand the text on a scale from 0–18. Lower values indicate easier comprehension. Wiener Sachtextformel (Variant 4) [4], is used to assess the comprehensibility of factual texts on a scale of 4–15. Among other things, it takes into account the average sentence length and the number of words with more than three syllables. The lower the value, the easier the text is to understand. The SARI (System output Against References and Input) metric [37] evaluates

[6] de.wikipedia.org/w/index.php?title=Leichte_Sprache, last access 16.04.2025.

[7] Due to an error, the Munich subset is missing from the training and test data.

the quality of a simplified text on a scale of 0–100 by analyzing the degree of addition, deletion and retention of information compared to the original text. It therefore measures how well a simplification system retains the meaning and essential content. The higher the value, the better the performance. BERTScore [38] uses contextual embeddings[8] to assess the semantic similarity between the generated and the reference text on a scale of 0–1. This enables a more differentiated evaluation of text quality compared to other metrics. The higher the score, the better the performance.

To evaluate the generated texts, a test data set with 74 samples from our data set was used as a reference. The standard texts were used as input together with the system prompt for the various Llama 3 models. The results in Table 2 show that the reference texts have better readability scores, which is reflected in a higher Flesch Reading Ease and lower Flesch-Kincaid Grade as well as in the Wiener Sachtextformel values. This indicates that the reference texts are easier to understand. In particular, the QLoRA 3.1-8B model best approximates the readability values of the original text on average (see also Table 2). In summary, the results nevertheless indicate that fine-tuning using QLoRA can be an effective method for optimizing LLMs for the generation of texts in Easy Language. In particular, the QLoRA model Llama-3.1-8B-Instruct proves to be more efficient in terms of readability and retention of semantic content as compared to the other models. Furthermore, compared to GPT-4o, prompted on the same task, our model considerably outperformed in terms of readability and SARI value. While achieving comparable results for the BERTScore. Particularly, the gap in the readability scores in comparison to the reference texts from our dataset show potential for improvement. This could be the result of a tendency to retain long words from the standard language text into the Leichte Sprache one, which we observed in some of the generated texts.

4 Conclusions and Outlook

Our results show that fine-tuning LLMs with QLoRA can be a promising approach for automated translation into Easy Language. In particular, it was shown that the QLoRA-fine-tuned Llama 3.1-8B-Instruct model achieved the best results as compared to the base models, both in terms of readability and retention of semantic content. The larger model showed an improved adaptation to the specific requirements of Easy Language. It should also be noted that all trained models did not quite reach the readability values of the reference texts, but were able to perform a certain degree of text simplification according to the system prompt. This can also be seen in the higher reference metrics in the QLoRA models as compared to the basic models.

Nevertheless, it became evident that the smaller models with less than 8 billion parameters are not yet fully capable of consistently generating meaningful text in Easy Language. In particular, more complex system prompts or instructions cannot always be adhered to. Furthermore, better and more powerful open

[8] In our case from `bert-base-multilingual-cased`.

weight or source models are expected in the future, which may be more consistent in terms of text generation and incorporating additional Easy Language data could also lead to an increase of the performance. In addition, clear regularizations for Easy Language would be desirable, as well as specialized ways to measure the quality of Easy Language texts. However, as mentioned in [6], these are already in progress. In addition, prompt engineering, based on the existing rules on Easy Language, could lead to a performance increase of the models we discussed and similar ones.

References

1. Anisuzzaman, D.M., Malins, J.G., Friedman, P.A., Attia, Z.I.: Fine-tuning large language models for specialized use cases. Mayo Clin. Proc. Digital Health **3**(1), 100184 (2025). https://doi.org/10.1016/j.mcpdig.2024.11.005
2. Anschütz, M., Oehms, J., Wimmer, T., Jezierski, B., Groh, G.: Language models for German text simplification: overcoming parallel data scarcity through style-specific pre-training. In: Rogers, A., Boyd-Graber, J., Okazaki, N. (eds.) Findings of the Association for Computational Linguistics: ACL 2023, pp. 1147–1158. Association for Computational Linguistics, Toronto, Canada (2023). https://doi.org/10.18653/v1/2023.findings-acl.74, https://aclanthology.org/2023.findings-acl.74/
3. Asghari, H., Hewett, F., Züger, T.: On the prevalence of Leichte Sprache on the German web. In: Proceedings of the 15th ACM Web Science Conference 2023, pp. 147–152. WebSci 2023, Association for Computing Machinery, New York (2023). https://doi.org/10.1145/3578503.3583599
4. Bamberger, R., Vanecek, E.: Lesen - Verstehen - Lernen - Schreiben. Die Schwierigkeitsstufen von Texten in Deutscher Sprache. Jugend u. Volk Sauerlaender, Vienna, Austria (1984)
5. Belkada, Y., Dettmers, T., Pagnoni, A., Gugger, S., Mangrulkar, S.: Making LLMs even more accessible with bitsandbytes, 4-bit quantization and QLoRA (2025). https://huggingface.co/blog/4bit-transformers-bitsandbytes
6. Beyer, I., et al.: DIN SPEC 33429 - Empfehlungen für Deutsche Leichte Sprache (2025)
7. Bredel, U., Maaß, C.: Leichte Sprache Theoretische Grundlagen. Dudenverlag, Berlin, Germany, Orientierung Für Die Praxis. Sprache Im Blick (2016)
8. Bundesministerium für Arbeit und Soziales: Leichte Sprache Ein Ratgeber (2014)
9. Chinn, D., Buell, S.: Easy language in the UK. In: Lindholm, C., Vanhatalo, U., Hansen-Schirra, S., Maaß, C. (eds.) Handbook of Easy Languages in Europe, pp. 623–657. No. 8 in Easy–Plain–Accessible, Frank & Timme, Berlin, Germany, 1 edn. (2021). https://doi.org/10.26530/20.500.12657/52628
10. Daniel Han, M.H., team, U.: Unsloth (2023)
11. Dettmers, T., Pagnoni, A., Holtzman, A., Zettlemoyer, L.: QLORA: efficient finetuning of quantized LLMs. In: Proceedings of the 37th International Conference on Neural Information Processing Systems. NIPS 2023, pp. 10088–10115. Curran Associates Inc., Red Hook (2023)
12. Ebling, S., et al.: Automatic text simplification for German. Front. Commun. **7**, 706718 (2022). DOIurlhttps://doi.org/10.3389/fcomm.2022.706718
13. Gille, M., Schomacker, T., von der Hülls, J., Tropmann-Frick, M.: Der Einsatz von Neural Language Models für eine barrierefreie Verwaltungskommunikation:

Anforderungen an die automatisierte Vereinfachung rechtlicher Informationstexte. In: Proceedings of 6. Fachtagung Rechts- und Verwaltungsinformatik (RVI 2023), pp. 144–158. Gesellschaft für Informatik e.V., Bonn (2023). https://doi.org/10.18420/rvi2023-013

14. Grabar, N., Saggion, H.: Evaluation of automatic text simplification: where are we now, where should we go from here. In: Estève, Y., Jiménez, T., Parcollet, T., Zanon Boito, M. (eds.) Actes de la 29e Conférence sur le Traitement Automatique des Langues Naturelles. Volume 1 : conférence principale, pp. 453–463. ATALA, Avignon, France (2022). https://aclanthology.org/2022.jeptalnrecital-taln.47/
15. Hewett, F., Asghari, H., Stede, M.: Elaborative simplification for German-language texts. In: Kawahara, T., Demberg, V., Ultes, S., Inoue, K., Mehri, S., Howcroft, D., Komatani, K. (eds.) Proceedings of the 25th Annual Meeting of the Special Interest Group on Discourse and Dialogue, pp. 29–39. Association for Computational Linguistics, Kyoto, Japan (2024). https://doi.org/10.18653/v1/2024.sigdial-1.3
16. Hewett, F., Stede, M.: Automatically evaluating the conceptual complexity of German texts. In: Proceedings of the 17th Conference on Natural Language Processing (KONVENS 2021). pp. 228–234. KONVENS 2021 Organizers, Düsseldorf, Germany (2021)
17. Hu, E.J., et al.: LoRA: low-rank adaptation of large language models. In: Proceedings of the International Conference on Learning Representations. Online (2022)
18. Kincaid, J.P., Fishburne, Jr., Robert P., R., Richard L., C., Brad S.: Derivation of New Readability Formulas (Automated Readability Index, Fog Count and Flesch Reading Ease Formula) for Navy Enlisted Personnel:. Reports - Research ED108134, Defense Technical Information Center (1975). https://doi.org/10.21236/ADA006655
19. Kincaid, J.P., Aagard, J.A., O'Hara, J.W., Cottrell, L.K.: Computer readability editing system. IEEE Trans. Prof. Commun. **PC-24**(1), 38–42 (1981). https://doi.org/10.1109/TPC.1981.6447821
20. Klaper, D., Ebling, S., Volk, M.: Building a German/simple German parallel corpus for automatic text simplification. In: Proceedings of the Second Workshop on Predicting and Improving Text Readability for Target Reader Populations (PITR 2013), Sofia, Bulgaria, 8 August 2013, pp. 11–19. Sofia, Bulgaria (2013). https://doi.org/10.5167/uzh-78610
21. Klöser, L., Beele, M., Schagen, J.N., Kraft, B.: German text simplification: finetuning large language models with semi-synthetic data. In: Chakravarthi, B.R., B, B., Buitelaar, P., Durairaj, T., Kovács, G., García Cumbreras, M.Á. (eds.) Proceedings of the Fourth Workshop on Language Technology for Equality, Diversity, Inclusion, pp. 63–72. Association for Computational Linguistics, St. Julian's, Malta (2024)
22. Lin, C.Y.: ROUGE: a package for automatic evaluation of summaries. In: Text Summarization Branches Out, pp. 74–81. Association for Computational Linguistics, Barcelona (2004). https://aclanthology.org/W04-1013/
23. Madina, M., Gonzalez-Dios, I., Siegel, M.: Easy-to-read in germany: a survey on its current state and available resources. In: Proceedings of the 10th Language & Technology Conference: Human Language Technologies as a Challenge for Computer Science and Linguistics. Poznań, Poland (2023)
24. Madina, M., Gonzalez-Dios, I., Siegel, M.: Towards reliable E2R texts: a proposal for standardized evaluation practices. In: Miesenberger, K., Peňáz, P., Kobayashi, M. (eds.) Computers Helping People with Special Needs, pp. 224–231. Springer, Cham (2024). https://doi.org/10.1007/978-3-031-62849-8_28

25. Matthew Douglas: Bitsandbytes 0.45.5. bitsandbytes foundation (2025)
26. Meta AI: Introducing Llama 3.1: Our most capable models to date (2024). https://ai.meta.com/blog/meta-llama-3-1/
27. Meta AI: Llama 3.2: Revolutionizing edge AI and vision with open, customizable models (2024). https://ai.meta.com/blog/llama-3-2-connect-2024-vision-edge-mobile-devices/
28. Netzwerk Leichte Sprache: Die Regeln für Leichte Sprache (2022)
29. Papineni, K., Roukos, S., Ward, T., Zhu, W.J.: Bleu: a method for automatic evaluation of machine translation. In: Isabelle, P., Charniak, E., Lin, D. (eds.) Proceedings of the 40th Annual Meeting of the Association for Computational Linguistics, pp. 311–318. Association for Computational Linguistics, Philadelphia, Pennsylvania, USA (2002). https://doi.org/10.3115/1073083.1073135, https://aclanthology.org/P02-1040/
30. Schomacker, T., Anschütz, M., Stodden, R., Groh, G., Tropmann-Frick, M.: Overview of the GermEval 2024 shared task on statement segmentation in German easy language (StaGE). In: Schomacker, T., Anschütz, M., Stodden, R. (eds.) Proceedings of GermEval 2024 Shared Task on Statement Segmentation in German Easy Language (StaGE), pp. 1–14. Association for Computational Lingustics, Vienna (2024). https://aclanthology.org/2024.germeval-1.1/
31. Schomacker, T., Gille, M., Tropmann-Frick, M., von der Hülls, J.: Data and approaches for german text simplification – towards an accessibility-enhanced communication. In: Georges, M., Herygers, A., Friedrich, A., Roth, B. (eds.) Proceedings of the 19th Conference on Natural Language Processing (KONVENS 2023), pp. 63–68. Association for Computational Lingustics, Ingolstadt (2023)
32. Schomacker, T., Gille, M., Tropmann-Frick, M., von der Hülls, J.: Self-regulated and participatory automatic text simplification. In: Jabbar, M.A., Tiwari, S., Ortiz-Rodríguez, F., Groppe, S., Bano Rehman, T. (eds.) Applied Machine Learning and Data Analytics. pp. 264–273. Communications in Computer and Information Science, Springer, Cham (2024). https://doi.org/10.1007/978-3-031-55486-5_19
33. Schomacker, T., Tropmann-Frick, M., Gille, M.: "Duzen oder Siezen?" – Forms of address in Easy Language webpages of German public authorities (2025)
34. Stodden, R., Momen, O., Kallmeyer, L.: DEplain: a German parallel corpus with intralingual translations into plain language for sentence and document simplification. In: Rogers, A., Boyd-Graber, J., Okazaki, N. (eds.) Proceedings of the 61st Annual Meeting of the Association for Computational Linguistics (Volume 1: Long Papers), pp. 16441–16463. Association for Computational Linguistics, Toronto (2023). https://doi.org/10.18653/v1/2023.acl-long.908
35. Tillet, P., Kung, H.T., Cox, D.: Triton: an intermediate language and compiler for tiled neural network computations. In: Proceedings of the 3rd ACM SIGPLAN International Workshop on Machine Learning and Programming Languages, pp. 10–19. MAPL 2019, Association for Computing Machinery, New York (2019). https://doi.org/10.1145/3315508.3329973
36. Toborek, V., Busch, M., Boßert, M., Bauckhage, C., Welke, P.: A new aligned simple German corpus. In: Rogers, A., Boyd-Graber, J., Okazaki, N. (eds.) Proceedings of the 61st Annual Meeting of the Association for Computational Linguistics (Volume 1: Long Papers), pp. 11393–11412. Association for Computational Linguistics, Toronto (2023). https://doi.org/10.18653/v1/2023.acl-long.638, https://aclanthology.org/2023.acl-long.638/
37. Xu, W., Napoles, C., Pavlick, E., Chen, Q., Callison-Burch, C.: Optimizing statistical machine translation for text simplification. Trans. Assoc. Comput. Linguist.

4, 401–415 (2016). https://doi.org/10.1162/tacl_a_00107, https://aclanthology.org/Q16-1029/
38. Zhang, T., Kishore, V., Wu, F., Weinberger, K.Q., Artzi, Y.: BERTScore: evaluating text generation with BERT. In: Proceedings of the Eighth International Conference on Learning Representations. Addis Ababa, Ethiopia (2020)
39. Zheng, Y., Zhang, R., Zhang, J., Ye, Y., Luo, Z.: LlamaFactory: unified efficient fine-tuning of 100+ language models. In: Cao, Y., Feng, Y., Xiong, D. (eds.) Proceedings of the 62nd Annual Meeting of the Association for Computational Linguistics (Volume 3: System Demonstrations), pp. 400–410. Association for Computational Linguistics, Bangkok (2024). https://doi.org/10.18653/v1/2024.acl-demos.38, https://aclanthology.org/2024.acl-demos.38/

Re-evaluating the Robustness and Interpretability of the Contrastive Explanations Method for Image Classification

Luisa Schneider, Bettina Finzel(✉), and Ute Schmid

Cognitive Systems, University of Bamberg, Weberei 5, 96047 Bamberg, Germany
luisa.schneider@atud.uni-bamberg.de,
{bettina.finzel,ute.schmid}@uni-bamberg.de
https://www.uni-bamberg.de/kogsys/

Abstract. Contrastive explanations make the predictive behavior of machine-learned classification models transparent and evaluable based on example features at the decision boundary. An established method for explaining image classifiers is the Contrastive Explanations Method (CEM). This method explains the classification of an image based on minimal pixel changes that are either necessary for the target class or that lead to a class change. How similarity between classes influences the interpretability and robustness of CEM had not yet been investigated quantitatively and hypothesis-test-driven. We therefore re-evaluated the robustness and interpretability of CEM by quantitatively analyzing its Pertinent Negatives (PNs) in scenarios with varying levels of class similarity. The goal was to determine whether the minimal changes in PNs truly reflected meaningful, human-understandable differences between similar classes, thereby addressing a key research gap in evaluating CEM explanations beyond anecdotal evidence.

Keywords: xAI · CEM · Robustness · Interpretability · AI Evaluation

1 Introduction

One of the aims of xAI is to generate interpretable explanations for the output of black-box models, meaning that the explanations have to be understandable and trustworthy to humans [6,13]. A type of explanation that is deemed useful and intuitive is the contrastive explanation [15]. It explicates what minimal and meaningful differences in two most similar classes (or instances) explain class membership [11,14,16]. For example, the Contrastive Explanations Method (CEM) introduced by Dhurandhar et al. [4] uses such explanations for both image and tabular data. CEM is a perturbation-based, post-hoc, saliency computing, and model-agnostic explanation method that highlights features that should be minimally present, called *pertinent positives* (PP), and those features, whose presence would alter the model decision, the *pertinent negatives* (PN).

T. Braun et al. (Eds.): KI 2025, LNAI 15956, pp. 262–269, 2026.
https://doi.org/10.1007/978-3-032-02813-6_21

Although CEM is an established explanatory method [7,13,20,21], one research question remains unanswered: Do the minimal changes lead to true similarity between the classes of the original image and the contrastive class (PNs)? The robustness and interpretability of explanations can be threatened by noise in minimal changes and changes that do not match with what a human would deem meaningful. This work therefore aims to re-evaluate CEM and get from anecdotal evidence [18] to statistically grounded conclusions about its usefulness.

To do so, the PNs and behavior of CEM are tested in cases of varying similarity between classes to answer the following questions, while evaluating the interpretability based on stability [2] and sparsity [9].

1. How is the appearance of the Pertinent Negatives affected by the (dis)similarity of the input and target classes?
2. Does CEM create Pertinent Negatives with the class label of the most similar and actionable alternative class in a multiclass setting?

Similar to the original CEM study, we conduct evaluation experiments on the MNIST dataset, which comprises images of handwritten digits [12]. Additionally, we constructed a derived dataset by iteratively removing meaningful features from digits until the model's prediction changed, thereby generating subsets that are inherently similar to the original digits. This experiment assesses whether CEM can reconstruct the original digit by generating PNs, providing insight into its performance in scenarios involving high versus low-class similarity.

2 Related Work

2.1 Counterfactual Explanations and Perturbation-Based Methods

Counterfactual explanation methods emphasize differences between classes but vary in their approach. FIDO [3] generates saliency maps that highlight regions where changes would alter the prediction, indicating key class-specific areas. CoCox [1] introduces positive and negative fault-lines, which add or remove features—similar to CEM—but in a concept-driven manner [1]. These explanations are constructed relative to a chosen alternative class. Similarly, counterfactual visual explanations [8] use an input and a distractor image to identify crucial features. The model's prediction is expected to shift to the distractor class when the corresponding image region in the input is replaced [8].

Evaluating the similarity between original and alternative classes is especially relevant for CEM, as it operates without requiring a predefined alternative class. For our evaluation, we build on the approaches and parameters discussed in the following subsection, which provide criteria for assessing explanation methods.

2.2 Evaluation of Explanation Methods

Alongside other studies [6,15], Stepin et al. [21] highlight the lack of standardized evaluation methods for *contfactual* (contrastive + counterfactual) explanations and categorize existing approaches as subjective, objective, or hybrid. Similarly, a survey by [2] distinguishes between quantitative and qualitative evaluation, listing key quantitative metrics such as stability. In contrast, [9] emphasizes properties like actionability, instability, and sparsity as central to evaluating contrastive and counterfactual explanations. Similarity is also a crucial factor [2], which can be measured by, e.g., structural similarity (the SSIM score), incorporating brightness, contrast, and structure [24].

3 Methods

3.1 Contrastive Explanations Method

CEM creates both Pertinent Positives (PP) and Pertinent Negatives (PN) with perturbations for image or tabular data, where PPs highlight the minimal amount of features necessary to maintain the prediction of the model, whereas PNs highlight those features that are necessary to alter the prediction of the model [4]. An example of this can be seen in Fig. 1a. As only the PNs are the interest of this re-evaluation, we only present the optimization problem of the PNs here.

The perturbation of the PNs $\delta \in X/x_0$, with x_0 being the input with the label t_0 and X/x_0 being the space of the missing features, should change the prediction score (Pred($\cdot$)) of x_0 when added so that $\text{argmax}_i[\text{Pred}(x_0 + \delta)]_i = \text{argmax}_i[\text{Pred}(x)]_i \neq \text{argmax}_i[\text{Pred}(x_0)]_i = t_0$ [4]. The optimization problem, if no autoencoder is used, is given by $\min_{\delta \epsilon X/x_0} c \cdot f_K^{\text{neg}}(x_0, \delta) + \beta||\delta||_1 + ||\delta||_2^2$ with the loss function $f_K^{\text{neg}}(x_0, \delta) = \max\{[\text{Pred}(x_0 + \delta)]_{t_0} - \max_{i \neq t_0}[\text{Pred}(x_0 + \delta)]_i, -K\}$ with $K \geq 0$ and weighted by $c \geq 0$ [4].

The second part of the optimization problem is the elastic net regularizer controlled by the parameter $\beta \geq 0$ [4,25]. The distance between the prediction scores is regulated by K, while β controls the L1-Loss [4].

As mentioned in the introduction, the method aims to create contrastive examples to which we will refer to as PNs for simplicity, by making sparse changes. Thus, we assume that CEM will create PNs that are similar to the original class. This assumption is tested in different settings, explained subsequently.

3.2 Setup and Experimental Design

To test how similarity affects CEM and resulting PNs, we designed experiments using the MNIST dataset of handwritten digits [12]. There, we distinguish between similar and dissimilar digit classes using the t-SNE algorithm [22], which clusters input and uses inter-cluster distances as a similarity measure. Our setup aimed at answering: *How does class similarity affect the behavior and performance of CEM?*

In the first experiment, we trained six binary CNNs, each corresponding to a class pairing *Pab*, where *a* is the input class and *b* the contrastive class. Based on class similarity, we selected P58, P38, and P39 as similar pairings, and P26,

(a) Original 5, perturbation, perturbation added to original and predicted as 8. (b) Original 8, edited image predicted as 5, PN of edited image predicted as 8. (c) PNs from P38 (sim.) and P26 (dissim.)

Fig. 1. Examples of a PN (a), a modified image from the set of subset pairings (b), and two PNs from the set of similar and dissimilar pairings (c).

P70, and P35 as dissimilar ones. Additionally, we created a synthetic "subset" set from pairings P58, P38, and P49 by manually removing pixels from class b images until prediction changed to class a (e.g., editing an 8 to resemble a 5 for P58; see Fig. 1b). We refer to this as the set of *subset pairings*. In the last experiment, we defined three-class sets (a, b_1, b_2), where b_1 is more similar to a than b_2. The selected sets are (3,8,9), (3,5,8), and (2,6,8). We applied CEM to 100 images per pairing and set, optimizing parameters K and β (see Table 1). Behavior and performance were evaluated using sparsity [9] and stability [2]. Binary pairings were analyzed by comparing means and ranges of the measured variables, followed by testing for statistically significant correlations using Spearman's ρ [5,17,19,23], with two-tailed tests at $\alpha = 0.05$ ($H_0 : \rho = 0$, $H_A : \rho \neq 0$). Sparsity was assessed by the percentage of pixels added by CEM (% added), calculated using the Manhattan norm $d(A, B) = |A_1 - B_1| + |A_2 - B_2|$, and by the standard deviation of noise σ_{noise}: $\sigma_n = \sqrt{\frac{\pi}{2}} \frac{1}{6(W-2)(H-2)} \sum_{imageI} |I(x, y) \cdot N|$ with W and H as image width and height, $I(x, y)$ a pixel location, to which the mask N is applied [10]. The stability was measured with the confidences (Conf(·)), which are the outputs of the CNNs, and SSIM scores (SSIM(·)) of the PNs given by SSIM(x,y) $= [l(x, y)]^\alpha \cdot [c(x, y)]^\beta \cdot [s(x, y)]^\gamma$ [24], where x and y are image signals compared for luminance l, contrast c, and structure s, with weights α, β, and γ [24]. SSIM was computed as the mean score between a data point and 100 samples from the contrastive class. In the last experiment, we assessed appearance frequencies in a multiclass setting, comparing the counts of classes b_1 and b_2 in PNs for input class a with their similarity to a. The aim was to test whether CEM would produce PNs from the most similar class.

Table 1. The β and K values for each pairing and set of three digits that are used by CEM to construct the PNs.

	P38	P58	P39	P49	P35	P26	P70	(3,8,9)	(3,8,5)	(2,6,8)
β:	0.9	0.6	0.1	0.1	0.9	0.1	0.1	0.1	0.1	0.1
K:	0.06	0.09	0.09	0.06	0.01	0.01	0.01	0.01	0.01	0.01

4 Results

4.1 Descriptive Analysis

As introduced earlier, the central tendencies in the different pairings were analyzed by calculating the mean and range of the variables describing these. We found that CEM added on average the least percentage of pixels to the set of similar and subset pairings (< 0.073) and produced PNs with a high confidence (mean: > 0.902), while in the set of dissimilar pairings, CEM produced noisy results (mean: % added > 0.87, $\sigma_{\text{noise}} > 0.14$) with a mean confidence around 0.5 for P35 and P26. An exception was P70, where the mean confidence was at 0.968, but with a range of 0.999. Furthermore, we discovered 28 black images as PNs in P26 (27) and P70 (1), indicating that CEM has been unable to construct a PN, while in the other sets, CEM added reoccurring patterns. In summary, the sets of similar and subset pairings showed the best results regarding stability and sparsity. An illustration can be seen in Fig. 1c.

Table 2. Results for the correlation analysis (a) and the frequency analysis (b).

Set of similar pairings:	P38		P58		P39	
Conf(Input) / % added	0.39	6.3e-5	0.7	6.3e-16	0.86	1.5e-30
Conf(Input) / Conf(PN)	0.31	0.002	0.38	8.5e-5	-0.05	0.65
Set of subset pairings:	**P38**		**P58**		**P49**	
Conf(Input) / % added	0.37	1.6e-4	0.74	7e-19	0.78	1.5e-21
Conf(Input) / Conf(PN)	0.45	3.3e-6	0.39	4.8e-5	0.51	5.9e-8
Set of dissimilar pairings:	**P35**		**P26**		**P70**	
Conf(Input) / % added	0.79	1.1e-22	-0.31	0.002	0.8	8.9e-24
Conf(Input) / Conf(PN)	-0.18	0.07	-0.63	3.6e-12	0.3 1	0.002

(a) Correlation coefficient (gray colored cells) and p-value for all binary sets.

	(3,8,9)		(3,5,8)		(2,6,8)	
	8	9	5	8	6	8
Conf(Input)	42	**58**	**63**	37	6	**94**
SSIM(Input)	**75**	25	46	**54**	45	**55**
PN	**73**	27	19	**81**	4	**96**

(b) Absolute frequency of appearing classes per set and variable.

4.2 Hypotheses Tests

We used hypothesis tests to look for significant correlations in CEM's behavior (percentage of pixels added) and output parameters with Spearman's ρ. In all three pairings, we found a correlation between the confidence on inputs and

CEM's behavior (see Table 2a). This correlation was mostly positive, except for P26. For the relationship between the input confidence and the confidence on PNs, all pairings except P39 (set of similar pairings) and P35 had a significant correlation, being negative for P26 (see Table 2a). In summary, there is a significant correlation between the input confidence and CEM's behavior in all sets, while a correlation with CEM's performance is present for most pairings.

4.3 Analysis of Multiclass Problems

We further tested our assumption of CEM creating PNs toward a most similar alternative class in a multiclass setting with varying class similarity.

We observed that the alternative class 8 was the most similar to the input class in all three sets, either by only the SSIM score or by both the SSIM score and the input confidence (see Table 2b). Additionally, the majority of the PNs were also predicted as class 8 for all sets (see Table 2b). Interestingly, the difference of similarity regarding the SSIM score is barely in favor of class 8 (only by 8 examples), while the input confidence classifies 63 examples as more similar to 5. In summary, although the similarity between classes was not overwhelmingly in favor of class 8, CEM created mostly PNs predicted as class 8.

5 Discussion

We investigated the influence of class similarity on CEM's behavior and performance across three experimental settings.

In the similar and subset pairings, CEM produced both sparse [9,16] and stable [2] explanations, with consistent pixel addition patterns. In contrast, explanations for dissimilar pairings showed a marked drop in stability and increased noise. Hypothesis tests confirmed a statistically significant correlation between class similarity and CEM's performance, indicating that greater dissimilarity leads to more added pixels and noise.

In the multiclass setting, CEM tended to generate PNs from the more similar class based on SSIM scores. Notably, in the set (3,5,8), class 8 dominated, despite more examples being similar to class 5 by confidence and almost half of the examples by SSIM score. This suggests that beyond similarity, *actionability* [9]—the feasibility of transforming one class into another—plays a critical role. For instance, adding features to make a 3 resemble an 8 may be more realistic than turning it into a 5.

The multiclass evaluation was limited to three sets with two contrastive classes each. Expanding this scope could yield deeper insights into CEM's consistency in selecting the more similar and actionable [9] class. For broader generalization, future studies should statistically re-evaluate CEM on additional datasets. In this work, we focused on robustness and interpretability, building on the originally used MNIST dataset [12].

Overall, this work confirms a significant correlation between similarity and CEM's performance: similar and subset pairings yielded sparser and more stable

PNs than dissimilar ones. Findings further underscore the importance of human involvement in evaluating the interpretability and plausibility of PNs.

6 Conclusion

This paper closed a research gap by statistically evaluating the robustness and interpretability of CEM. We analyzed Pertinent Negatives (PNs) on MNIST across class pairings of varying similarity. Findings show a significant correlation between class similarity and CEM's performance. In a multiclass setting, similarity and actionability [9] determined PN generation. Future work could test generalizability on other datasets and CEM variants.

References

1. Akula, A.R., Wang, S., Zhu, S.: Cocox: generating conceptual and counterfactual explanations via fault-lines. In: The Thirty-Fourth AAAI Conference on Artificial Intelligence, AAAI 2020, The Thirty-Second Innovative Applications of Artificial Intelligence Conference, IAAI 2020, The Tenth AAAI Symposium on Educational Advances in Artificial Intelligence, EAAI 2020, New York, NY, USA, February 7-12, 2020. pp. 2594–2601. AAAI Press (2020). https://doi.org/10.1609/AAAI.V34I03.5643
2. Bodria, F., Giannotti, F., Guidotti, R., Naretto, F., Pedreschi, D., Rinzivillo, S.: Benchmarking and survey of explanation methods for black box models. Data Min. Knowl. Discov. **37**(5), 1719–1778 (2023). https://doi.org/10.1007/S10618-023-00933-9
3. Chang, C., Creager, E., Goldenberg, A., Duvenaud, D.: Explaining image classifiers by counterfactual generation. In: 7th International Conference on Learning Representations, ICLR 2019, New Orleans, LA, USA, May 6-9, 2019, pp. 1–15 (2019)
4. Dhurandhar, A., Chen, P.Y., Luss, R., Tu, C.C., Ting, P., Shanmugam, K., Das, P.: Explanations Based on the Missing: Towards Contrastive Explanations with Pertinent Negatives. In: Proceedings of the 32nd International Conference on Neural Information Processing Systems, p. 590 601. NIPS'18, Curran Associates Inc., Red Hook, NY, USA (2018). https://doi.org/10.5555/3326943.3326998
5. Dodge, Y.: The Concise Encyclopedia of Statistics. Springer, New York (2008). https://doi.org/10.1007/978-0-387-32833-1
6. Doshi-Velez, F., Kim, B.: Towards a rigorous science of interpretable machine learning. arXiv: Machine Learning **abs/1702.08608** (2017). https://arxiv.org/abs/1702.08608
7. Dwivedi, R., et al.: Explainable AI (XAI): core ideas, techniques, and solutions. ACM Comput. Surv. **55**(9), 194:1–194:33 (2023).https://doi.org/10.1145/3561048
8. Goyal, Y., Wu, Z., Ernst, J., Batra, D., Parikh, D., Lee, S.: Counterfactual visual explanations. In: Chaudhuri, K., Salakhutdinov, R. (eds.) Proceedings of the 36th International Conference on Machine Learning, ICML 2019, 9-15 June 2019, Long Beach, California, USA. Proceedings of Machine Learning Research, vol. 97, pp. 2376–2384. PMLR (2019). http://proceedings.mlr.press/v97/goyal19a.html

9. Guidotti, R.: Counterfactual explanations and how to find them: literature review and benchmarking. Data Min. Knowl. Discov. **38**(5), 2770–2824 (2024). https://doi.org/10.1007/S10618-022-00831-6
10. Immerkær, J.: Fast noise variance estimation. Comput. Vis. Image Underst. **64**(2), 300302 (1996). https://doi.org/10.1006/cviu.1996.0060
11. Jacovi, A., Swayamdipta, S., Ravfogel, S., Elazar, Y., Choi, Y., Goldberg, Y.: Contrastive explanations for model interpretability. In: Moens, M., Huang, X., Specia, L., Yih, S.W. (eds.) Proceedings of the 2021 Conference on Empirical Methods in Natural Language Processing, EMNLP 2021, Virtual Event / Punta Cana, Dominican Republic, 7-11 November, 2021, pp. 1597–1611. Association for Computational Linguistics (2021). https://doi.org/10.18653/v1/2021.emnlp-main.120
12. LeCun, Y., Bottou, L., Bengio, Y., Haffner, P.: Gradient-based learning applied to document recognition. Proc. IEEE **86**(11), 2278–2324 (1998). https://doi.org/10.1109/5.726791
13. Linardatos, P., Papastefanopoulos, V., Kotsiantis, S.: Explainable AI: a review of machine learning interpretability methods. Entropy **23**(1), 18 (2021). https://doi.org/10.3390/e23010018
14. Lipton, P.: Contrastive explanation. Roy. Inst. Philos. Suppl. **27**, 247266 (1990). https://doi.org/10.1017/S1358246100005130
15. Miller, T.: Explanation in artificial intelligence: insights from the social sciences. Artif. Intell. **267**, 1–38 (2019). https://doi.org/10.1016/j.artint.2018.07.007
16. Miller, T.: Contrastive explanation: a structural-model approach. Knowl. Eng. Rev. **36**, e14 (2021). https://doi.org/10.1017/S0269888921000102
17. Mukaka, M.: Statistics corner: A guide to appropriate use of correlation coefficient in medical research. Malawi medical journal: J. Med. Assoc. Malawi **24**, 69–71 (2012). https://pubmed.ncbi.nlm.nih.gov/23638278/
18. Nauta, M., et al.: From anecdotal evidence to quantitative evaluation methods: a systematic review on evaluating explainable AI. ACM Comput. Surv. **55**(13s), 1–42 (2023). https://doi.org/10.1145/3583558
19. Schober, P., Boer, C., Schwarte, L.: Correlation coefficients: appropriate use and interpretation. Anesth. Analg. **126**(5), 1763–1768 (2018). https://doi.org/10.1213/ANE.0000000000002864
20. Schwalbe, G., Finzel, B.: A comprehensive taxonomy for explainable artificial intelligence: a systematic survey of surveys on methods and concepts. Data Min. Knowl. Discov. **38**(5), 3043–3101 (2024). https://doi.org/10.1007/S10618-022-00867-8
21. Stepin, I., Alonso, J.M., Catalá, A., Pereira-Fariña, M.: A survey of contrastive and counterfactual explanation generation methods for explainable artificial intelligence. IEEE Access **9**, 11974–12001 (2021). https://doi.org/10.1109/ACCESS.2021.3051315
22. Taskesen, E.: Clusteval (2020). https://erdogant.github.io/clusteval
23. Virtanen, P., et al: Fundamental algorithms for scientific computing in python. Nat. Methods **17**(3), 261–272 (2020). https://doi.org/10.1038/s41592-019-0686-2
24. Wang, Z., Bovik, A.C., Sheikh, H.R., Simoncelli, E.P.: Image quality assessment: from error visibility to structural similarity. IEEE Trans. Image Process. **13**(4), 600–612 (2004). https://doi.org/10.1109/TIP.2003.819861
25. Zou, H., Hastie, T.: Regularization and variable selection via the elastic net. J. R. Stat. Soc. Ser. B (Stat. Methodol.) **67**(2), 301–320 (2005). https://doi.org/10.1111/j.1467-9868.2005.00503.x

Visualizing and Interpreting Neural Network Focus Regions: A Comparative Study of Vision Transformers on Synthetic and Real Data

Denis Trescher[1,2,3], Waldemar Haag[1,3](✉), and Enrico Schröder[1]

[1] EFI Institute, Technische Hochschule Nürnberg Georg Simon Ohm, Keßlerplatz 12, 90489 Nuremberg, Germany
{trescherde79376,waldemar.haag,enrico.schroeder}@th-nuernberg.de
[2] EduArt Robotik GmbH, Brückenstr. 20, 91233 Neunkirchen a. Sand, Germany
[3] formerly: Evocortex GmbH, Nuremberg, Germany

Abstract. The quality and diversity of training datasets significantly influence the performance of neural networks, particularly in object detection tasks that rely on annotated images to learn from marked regions. While synthetic image data reduces the need for extensive manual annotation, the domain gap between synthetic and real-world data remains a challenge, often leading to reduced model performance on real-world images. This study compares the image regions considered important for object detection by analyzing the predictive behavior of Transformer networks trained on synthetic versus real image data. Using a shared set of test images and a feature visualization method, the analysis focuses on the size, quantity, and spatial distribution of regions of high attention. Additionally, the study explores the influence of generative artificial intelligence techniques—specifically Stable Diffusion—on the predictive behavior of models trained on synthetic data. Furthermore, an occlusion-based evaluation method is proposed to assess model behavior by selectively masking regions of high attention. Findings suggest a positive correlation between the number of large regions of high attention—specifically, those exceeding $\frac{1}{16}$ of the bounding box area—used for object detection and model performance. Augmentation with Stable Diffusion appears to enhance the model's robustness to the occlusion of regions of high-attention in the images.

Keywords: object detection · feature visualization · generative AI · transfer learning · domain gap · attention · explainable AI

1 Introduction

The success of deep learning in object detection relies heavily on diverse, high-quality datasets with annotated images that define object locations through bounding boxes. Creating comprehensive datasets with manually annotated

T. Braun et al. (Eds.): KI 2025, LNAI 15956, pp. 270–277, 2026.
https://doi.org/10.1007/978-3-032-02813-6_22

images remains a challenge, as the annotation effort increases proportionally with dataset size. Synthetic image data has become a powerful resource for training object detection models, reducing the effort associated with manually annotating large datasets [2,11,14,15]. Synthetic image data also offers high-quality annotations and allows for effortless scalability to larger datasets. Despite these advantages, the domain gap between synthetic and real-world data remains a challenge [9,10], as models trained on synthetic data may exhibit reduced performance when applied to real-world scenarios [10].

This study[1] investigates the predictive behavior of Transformer networks trained on synthetic versus real image data, with the following key objectives:

- Compare the attention-based features utilized by models to detect objects.
- Evaluate the influence of generative AI—specifically Stable Diffusion (SD)—on the predictive behavior of models trained on synthetic data.
- Propose an occlusion-based evaluation method to assess model behavior by selectively masking regions of high attention.

These insights can support model training by using attention-based features to adjust the loss function and encourage the network to prioritize relevant predictive patterns that improve detection performance.

2 Related Work

Gradient-weighted Class Activation Mapping (Grad-CAM) [13] is a visualization technique that uses gradient information from the final convolutional layers to highlight the image regions most critical for a model's predictions and has been used to interpret features for identifying biases in datasets. Its visualizations have revealed biases such as gender stereotypes within training datasets, enabling corrective measures through dataset adaptation. Additionally, Grad-CAM visualizations have demonstrated that seemingly unreasonable predictions often have reasonable underlying explanations, a finding consistent with observations made in HOGgles [17]. Yeh et al. [19] used their visualizations of Transformer models' attention behavior to observe that tokens tend to focus most on other tokens with similar characteristics, such as color or brightness. Li et al. [8] stated that interpretation results, which provide rich information about the location of discriminative features, can be utilized to guide training strategies such as data augmentation and regularization, particularly for vision tasks. For example, Kim et al. [7] enhanced the Mixup augmentation method [21] by incorporating *saliency maps*. Similarly, Zagoruyko et al. [20] proposed a regularization technique aimed at improving knowledge distillation by encouraging the alignment of saliency maps between teacher and student networks. Additionally, Wickramanayake et al. [18] investigated how explanation mechanisms

[1] All code is available at: https://github.com/TrescherDe/Visualizing-and-Interpreting-Neural-Network-Focus-Regions.

can be used to augment training datasets, with the goal of enhancing classification accuracy. While previous works have utilized explanation techniques, they did not employ a systematic approach to investigate the visualized features that inform the predictive behavior of these networks.

3 Method

Extracting Attention Maps
For the generation of the relevancy maps, the method described in [4] was used, as its effectiveness was demonstrated by comparing the resulting segmentation masks with those produced by methods such as Transformer attribution [5], partial LRP [16], Grad-CAM [13], raw attention, and rollout [1]. This relevancy map visualizes the attention scores of the pixels for the detection. To generate a mask of the regions of high attention, the relevancy map is normalized to a [0, 1] scale, where 1 indicates highest and 0 no relevance (see Fig. 1). The normalized relevancy map is upscaled to input resolution using bilinear interpolation and converted into a mask by removing values below 0.1. Regions with attention values above 0.4 are considered to receive high attention. This threshold was chosen qualitatively to separate distinct regions within the object without treating it as a single area, and may be further optimized. From this point onward, the relevancy map after applying the threshold of 0.4 is referred to as an attention map. The regions of high-attention identified within this map are segmented into distinct contours, which are referred to as focus regions. To further visualize the results, this mask is combined with the input image, allowing the attention map to be seen directly on the input image. This step is performed purely for visualization purposes, as shown in Fig. 1. In subsequent steps, only the mask is used to analyze and occlude the identified focus regions.

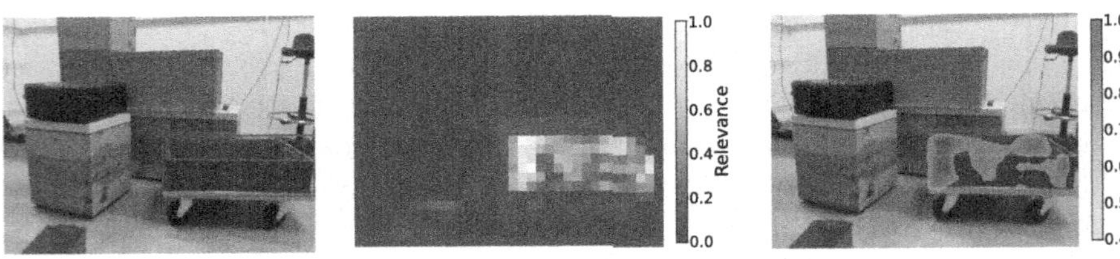

(a) Input image (b) Relevancy map (c) Attention map

Fig. 1. Exemplary visualization of the relevancy map generated using the method [4]. The figure presents the input image with a blue bounding box for visualization purposes only, not part of the original input (a), the relevancy map generated by the method (b), and the combined image (c), in which the attention map - extracted from the relevancy map - is overlaid on the original image.

Evaluating the Impact of Occlusion on Model Behavior
The evaluation process begins with predicting objects in the test image and identifying focus regions in the corresponding attention map. These focus regions are then classified as small or large based on the percentage of the ground truth bounding box area they occupy. Focus regions larger than $\frac{1}{16}$ of the ground truth box area are classified as large, and smaller ones as small. Regions below $\frac{1}{128}$ are ignored. These thresholds were choosen empirically. Combinations of occluded and non-occluded focus regions in an image are generated by computing all possible states of a bit mask of length n (for n focus regions in the image), with 0 indicating that a focus region is occluded and 1 indicating that it remains visible. Subsequently, each occlusion combination is applied to the image by setting the pixels of the focus regions to be occluded to zero, and the modified image undergoes prediction to detect the object. If the object remains identifiable, the newly identified focus regions are compared to the original focus regions to observe any variations. This systematic approach is replicated for each image within the test dataset.

To assess the impact of occluding small vs. large focus regions, the framework from the previous section is extended. All possible occlusion combinations are evaluated, and for each detection, the specific occlusion step leading to misdetection is analyzed to determine whether a large or small focus region caused the failure.

Comparing Focus Regions of Different Models
The process begins with the creation of attribute lists for each model. To compare model behavior, attribute lists are generated for each model based on objects detected by all models. This ensures consistency in analysis. The following data is collected iteratively for each object detected by all models: Attention coverage (% of bounding box), quadrant counts, large focus region counts and small focus region counts. The quadrants refer to the four parts of the bounding box, with the coordinate system originating from the center of the box. It enables the comparison of each model's predictive behavior for object detection.

Variations of Synthetic Image Datasets for Enhanced Generalization on Real-World Data
This section describes the real and synthetic image datasets used. A total of four distinct datasets were developed, each consisting of 500 images[2]:

- **Real**: A dataset containing real images of small load carriers and a small storage box with material properties similar to the small load carrier.
- **Blender-3D**: The baseline synthetic dataset containing images generated using Blender and 3D meshes of small load carriers and a small storage box with similar material properties.
- **SD-V1**: An extended dataset augmenting the baseline dataset using Stable Diffusion (SD).
- **SD-V2**: An extended dataset augmenting the baseline using SD, focusing on photorealism.

[2] Dataset available at: https://github.com/TrescherDe/small-load-carrier-dataset.

(a) Real (b) Blender-3D (c) SD-V2

Fig. 2. Exemplary images of the datasets.

- **Test video**: A dataset containing real images of the small load carrier, a small storage box with similar material properties, and other distracting objects in a possible use case scenario.

Example images from three of the datasets are shown in Fig. 2. The SD model version 1.5 [12] was used to augment the images of the Blender-3D dataset. To align the diffusion process with the object geometry for each synthetic image, a wireframe representation was generated using Blender and provided as input to ControlNet [22]. Low-Rank Adaptation (LoRA) [6] was applied to fine-tune the SD model, enhancing its knowledge of the target object - specifically, a small load carrier. For the SD-V1 extension, the general-purpose DreamShaper V8 model[3] was applied. In the SD-V2 extension, the focus was on achieving photo-realistic results using the Realistic Vision V6.0 B1 model[4].

4 Experiments and Results

The models listed in Table 1 differ primarily by the dataset used for training and are named after their related dataset. Real images include one labeled class (small load carrier), while synthetic datasets include an additional storage box class with similar material properties. Each model was trained three times, and the mean values along with standard deviations were computed and reported. All models use the DETR architecture [3], fine-tuned from the *facebook/detr-resnet-50* checkpoint. Training was performed on 80% of each dataset, with 20% used for validation. Each was trained three times, with mean and standard deviation computed. Evaluation was conducted on the Test video dataset.
If more than 40% of the focus regions are occluded, both SD-augmented models (SD-V1 and SD-V2) outperformed the baseline model trained on the Blender-3D dataset. After 60% occlusion, the SD-V2 model also surpassed the model trained on real data. At 100% occlusion, the SD-V2 model achieved a mean true positive rate of 0.41, compared to 0.28 for the real-data model. The results of the analysis of the focus regions are presented in Table 1, which outlines the key characteristics of the focus regions detected by the network. The results

[3] https://civitai.com/models/4384/dreamshaper.
[4] https://civitai.com/models/4201/realistic-vision-v60-b1.

Table 1. Mean values and standard deviations of the focus regions and model performance across different models.

Category	Real	Blender-3D	SD-V1	SD-V2
Attention coverage (% of bounding box)	0.353 ± 0.049	0.427 ± 0.049	0.340 ± 0.078	0.473 ± 0.059
Quadrants	3.303 ± 0.130	3.893 ± 0.031	3.757 ± 0.139	3.877 ± 0.083
Number of large focus regions	1.103 ± 0.033	1.073 ± 0.066	1.040 ± 0.102	1.073 ± 0.005
Number of small focus regions	0.500 ± 0.070	0.470 ± 0.064	0.753 ± 0.254	0.350 ± 0.099
F1 - Score	0.855 ± 0.054	0.734 ± 0.064	0.670 ± 0.093	0.740 ± 0.079

reveal a positive correlation between the number of large focus regions and the F1-Score, suggesting that models attending to larger regions tend to perform better. However, the observed standard deviation in the number of large focus regions (ranging from 0.005 to 0.102) and in the F1-Score (ranging from 0.054 to 0.093) introduces uncertainty in the strength of this correlation, as the standard deviation is of a similar magnitude to the differences observed between the models. This highlights the need for more data to confirm the pattern.

5 Conclusion and Future Work

This study compares the predictive behavior of Transformer networks trained on synthetic versus real image data using a feature visualization technique. It investigates how enhancing the realism of synthetic images using generative AI—specifically Stable Diffusion—affects the predictive behavior of models trained on these enhanced datasets. To quantify the impact of occluding regions with high attention, an occlusion-based evaluation method was proposed. The results indicate that augmenting the dataset with Stable Diffusion did not significantly affect the F1-Score, neither improving nor degrading performance. However, it contributes to increased robustness to occlusion and also affects the model's predictive behavior used for object detection. A correlation was observed between the number of large focus regions and the F1-Score, suggesting that the presence of more large focus regions enhances model performance. However, more data is needed to confirm the observed pattern.

Future research could apply the described method to analyze the predictive behavior of models across a broader range of datasets and object types, in order to validate the observed correlation between large focus regions and F1-Score. Grouping focus regions into concentric circular regions around the center could help assess whether specific spatial patterns are associated with improved model performance. Method thresholds could be optimized using region separation heuristics (e.g., bridge or island detection). Evaluating newer versions of Stable Diffusion or alternative generative AI models may further enhance performance by reducing diffusion artifacts and improving image consistency, particularly for fine-grained details such as text.

Acknowledgments. The authors have no competing interests to declare that are relevant to the content of this article.

References

1. Abnar, S., Zuidema, W.: Quantifying attention flow in transformers. In: Jurafsky, D., Chai, J., Schluter, N., Tetreault, J. (eds.) Proceedings of the 58th Annual Meeting of the Association for Computational Linguistics, pp. 4190–4197. Association for Computational Linguistics, Online (2020). https://doi.org/10.18653/v1/2020.acl-main.385, https://aclanthology.org/2020.acl-main.385/
2. Borrego, J., Dehban, A., Figueiredo, R., Moreno, P., Bernardino, A., Santos-Victor, J.: Applying domain randomization to synthetic data for object category detection. arXiv preprint arXiv:1807.09834 (2018)
3. Carion, N., Massa, F., Synnaeve, G., Usunier, N., Kirillov, A., Zagoruyko, S.: DETR: end-to-end object detection with transformers (2020). https://github.com/facebookresearch/detr
4. Chefer, H., Gur, S., Wolf, L.: Generic attention-model explainability for interpreting bi-modal and encoder-decoder transformers. In: Proceedings of the IEEE/CVF International Conference on Computer Vision, pp. 397–406 (2021)
5. Chefer, H., Gur, S., Wolf, L.: Transformer interpretability beyond attention visualization. In: Proceedings of the IEEE/CVF Conference on Computer Vision and Pattern Recognition, pp. 782–791 (2021)
6. Hu, E.J., et al.: LoRA: low-rank adaptation of large language models. In: International Conference on Learning Representations (2022). https://openreview.net/forum?id=nZeVKeeFYf9
7. Kim, J.H., Choo, W., Song, H.O.: Puzzle mix: exploiting saliency and local statistics for optimal mixup. In: International Conference on Machine Learning, pp. 5275–5285. PMLR (2020)
8. Li, X., et al.: Interpretable deep learning: interpretation, interpretability, trustworthiness, and beyond. Knowl. Inf. Syst. **64**(12), 3197–3234 (2022)
9. Peng, X., Saenko, K.: Synthetic to real adaptation with generative correlation alignment networks. In: 2018 IEEE Winter Conference on Applications of Computer Vision (WACV), pp. 1982–1991. IEEE (2018)
10. Peterlevitz, A.J., et al.: Sim-to-real transfer for object detection in aerial inspections of transmission towers. IEEE Access (2023)
11. Prakash, A., et al.: Structured domain randomization: bridging the reality gap by context-aware synthetic data. In: 2019 International Conference on Robotics and Automation (ICRA), pp. 7249–7255 (2019).https://doi.org/10.1109/ICRA.2019.8794443
12. Rombach, R., Blattmann, A., Lorenz, D., Esser, P., Ommer, B.: High-resolution image synthesis with latent diffusion models. In: Proceedings of the IEEE/CVF Conference on Computer Vision and Pattern Recognition (CVPR), pp. 10684–10695 (2022)
13. Selvaraju, R.R., Cogswell, M., Das, A., Vedantam, R., Parikh, D., Batra, D.: Grad-cam: visual explanations from deep networks via gradient-based localization. In: 2017 IEEE International Conference on Computer Vision (ICCV), pp. 618–626 (2017). https://doi.org/10.1109/ICCV.2017.74
14. Tobin, J., Fong, R., Ray, A., Schneider, J., Zaremba, W., Abbeel, P.: Domain randomization for transferring deep neural networks from simulation to the real world. In: 2017 IEEE/RSJ International Conference on Intelligent Robots and Systems (IROS), pp. 23–30 (2017). https://doi.org/10.1109/IROS.2017.8202133
15. Tremblay, J., et al.: Training deep networks with synthetic data: Bbidging the reality gap by domain randomization. In: Proceedings of the IEEE Conference on Computer Vision and Pattern Recognition Workshops, pp. 969–977 (2018)

16. Voita, E., Talbot, D., Moiseev, F., Sennrich, R., Titov, I.: Analyzing multi-head self-attention: specialized heads do the heavy lifting, the rest can be pruned. In: Korhonen, A., Traum, D., Màrquez, L. (eds.) Proceedings of the 57th Annual Meeting of the Association for Computational Linguistics, pp. 5797–5808. Association for Computational Linguistics, Florence (2019). https://doi.org/10.18653/v1/P19-1580, https://aclanthology.org/P19-1580/
17. Vondrick, C., Khosla, A., Malisiewicz, T., Torralba, A.: Hoggles: visualizing object detection features. In: Proceedings of the IEEE International Conference on Computer Vision, pp. 1–8 (2013)
18. Wickramanayake, S., Hsu, W., Lee, M.L.: Explanation-based data augmentation for image classification. Adv. Neural. Inf. Process. Syst. **34**, 20929–20940 (2021)
19. Yeh, C., Chen, Y., Wu, A., Chen, C., Viégas, F., Wattenberg, M.: Attentionviz: a global view of transformer attention. IEEE Trans. Vis. Comput. Graph. (2023)
20. Zagoruyko, S., Komodakis, N.: Paying more attention to attention: improving the performance of convolutional neural networks via attention transfer. In: ICLR (Poster). OpenReview.net (2017). https://openreview.net/forum?id=Sks9_ajex
21. Zhang, H., Cisse, M., Dauphin, Y.N., Lopez-Paz, D.: mixup: beyond empirical risk minimization. In: International Conference on Learning Representations (2018). https://openreview.net/forum?id=r1Ddp1-Rb
22. Zhang, L., Rao, A., Agrawala, M.: Adding conditional control to text-to-image diffusion models (2023)

Comparing the Visual Quality of Deep Generative Models for Steel Microstructures

Marcel Wentzien[1,2(✉)], Jerome Ingber[1], Jörg Schlötterer[2,3], and Dirk Schmalzried[1(✉)]

[1] Ernst-Abbe-Hochschule Jena, Carl-Zeiss-Promenade 2, 07745 Jena, Germany
{marcel.wentzien,dirk.schmalzried}@eah-jena.de
[2] Philipps-Universität Marburg, Hans-Meerwein-Straße 6, 35043 Marburg, Germany
[3] Universität Mannheim, 68161 Schloss, Mannheim, Germany

Abstract. In the realm of AI-driven material sciences, generating highly realistic steel microstructure images represents an opportunity to address data scarcity and to advance data-driven diagnostic and predictive applications. While previous work achieved good performance values for generating scanning electron microscopy grayscale images of steel microstructures, research on colorful light optical microscopy (LOM) is limited. We investigate the capabilities of two deep generative models to generate highly realistic LOM steel microstructure images. We train a StyleGANv2 and the recently published FractalGen on a dataset of microstructural images from 30 different steels and conduct a small expert study to evaluate how realistic the generated images look. Because of visible grid-like patterns, the experts almost always correctly identified FractalGen generated images as synthetic, while StyleGANv2 generated images were mostly indistinguishable from real images to them. We also used the StyleGANv2's discriminator to conduct the same task, resulting in not yet fully explainable classification results. Our findings are a first step towards the generation of realistic synthetic pairs of microstructures and the steels chemical composition and processing, which could reduce costs in future AI-driven material sciences research. Code is available at https://github.com/eah-materials/RealOrFakeSteelMicrostructure

Keywords: Generative Artificial Intelligence · Optical Microscopy · Steel microstructure · Expert Study

1 Introduction

Computer Vision technologies, in particular deep learning, have shown advances in various areas of the material sciences [4,5,9], such as objective microstructure characterization [1,2,22] or exploring optimal microstructures for specific properties [12,13]. A deep learning application's success depends greatly on the amount and quality of available training data, and is an issue in the material

T. Braun et al. (Eds.): KI 2025, LNAI 15956, pp. 278–285, 2026.
https://doi.org/10.1007/978-3-032-02813-6_23

sciences as public availability of high quality data is rare [9]. Desired materials need to be produced, prepared and pictured under a microscope to reveal the material's microstructure. Default image augmentation techniques, like rotation or colorjitter, are limited to known physical properties of existing images [14]. Synthetic microstructural images could help to create larger and high quality datasets, with different physical properties. Simulation approaches to create synthetic microstructure images exist, but depend on domain knowledge [20]. As deep generative models are trained without explicit physical knowledge their results are cheaper to achieve, but may not be helpful for downstream tasks when the results are considered unrealistic. We therefore evaluate how real synthetic microstructure images are to domain experts based on steel alloys pictured under a light optical microscope (LOM).

2 Related Work

Previous studies have shown that deep learning technologies are capable to generate grayscale scanning electron microscopy (SEM) steel microstructure images from synthetic [15,24] and real images [10,16,20,21]. In particular the use of a StyleGANv2-Ada [11] provided promising results [16]. However, LOM images differ from SEM images in pictured scale and color (SEM: grayscale, LOM: color). Currently, synthetic LOM image generation remains under-explored, only Lee et al. [17] report lower visual quality in comparison to SEM images by anecdotal evidence, but without structured evaluation. Steel microstructures reveal distinct features at global (e.g., banding structures and segregations), regional (e.g., morphology of ferrite and pearlite colonies), and local scales (e.g., interlamellar spacing within pearlite). Each of these structural levels significantly influences the overall mechanical properties of the steel [9,19]. The proposed FractalGen [18] presents another promising image generator, as its auto-regressive model with each recursion depth controlling the generative behavior on smaller levels corresponds well with this hierarchical nature of steel microstructures.

Deep generative models are typically evaluated by automated metrics [7], such as Fréchet inception distance [8] or Kernel inception distance [3]. These metrics compare the generated images against ground-truth data (mostly ImageNet [6]) based on their embeddings in deep layers of a pre-trained Inception network [23], that represent high level concepts. Since concepts relevant to steel microstructure images may be absent in these pre-trained (on natural images) networks [23], and ground-truth data is not available at scale to pretrain domain-specific models, we instead analyze the quality of generated images by an expert assessment and the discriminator of a GAN.

3 Methodology

We trained two image generators on real LOM steel microstructure images and generated synthetic images with both. We then randomly sampled real and synthetic images and asked two domain experts to classify them as either "real" or "fake" and to provide a rationale for their assessment. We further evaluated the discriminator of a StyleGAN on the same set and on additional synthetic images.

3.1 Dataset

The dataset consists of micrographs from 30 distinct steel samples, each ground, polished and etched with 5% picric acid in ethanol according to standard metallographic procedures. From each sample, three 2700×1800 px images were taken with a LOM Neophot 32 (Zeiss) at a magnification of 50.4x. These 90 images were preprocessed with flat field correction using a Gaussian blur filter of size (1501^2) to mitigate vignetting effect bias. Each image was split into 70 smaller non-overlapping patches of size 256×256 px, yielding a training set of total 6300 images. For the evaluation of the discriminator, all 210 real patched images of one randomly selected steel sample were held out from the training data to avoid train-test leakage. This was repeated eight times with different steel samples.

3.2 Generative Models

As generative models we selected StyleGANv2-Ada (SG), since it was designed for small training data [11] and showed promising results in prior work on SEM images [16], and the Base variant of FractalGen (FG) [18], since its architecture corresponds well with the hierarchical nature of steel microstructures (cf. section 2). Both models were trained on two Nvidia A100 80GB GPUs with a batch size of 32. For SG, we set the training specific parameter *target* to 0.5 as suggested by [16] for SEM microstructures. For FG, we set the base learning rate to 5e-4, learning warm up epochs to 1, class free guidance to 10, and number of conditions to 5, determined in preliminary experiments. All other parameters were kept at their default, provided by the original source code[1].

For image generation, both models were set to evaluation mode and set class free guidance to 1 in FG while keeping all other parameters at their generation defaults. We generated 300 images with FG, since generation time for a single image was unexpectedly high (> 2Min). With SG we generated an equal amount of synthetic images as available real images in the training data (6300 for the expert study, 6090 for the evaluation by the discriminator, in which images from one steel sample were held out from the training data). See Fig. 1 for examples.

3.3 Expert Study

In a custom user interface, we randomly sampled 100 real, 50 SG-generated, and 51 FG-generated images. We asked two domain experts (a research associate with >5 years of experience in materials research and a scientific laboratory engineer with decades of experience in materials testing, both dealing with microstructural images on an almost daily basis) to classify the images as either "fake" or

[1] https://github.com/NVlabs/stylegan3 (also provides StyleGANv2-Ada) and https://github.com/LTH14/fractalgen (both accessed: 16.04.2025).

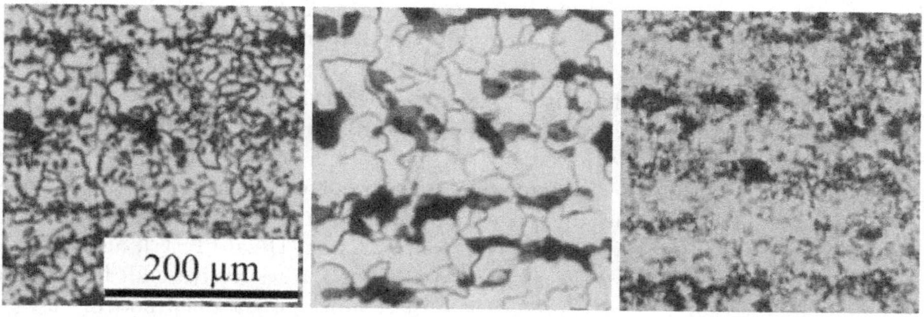

Fig. 1. Left: Real steel microstructure image after etching with picric acid and magnified with a light optical microscope showing different colored material phases and grains in various sizes. Middle: synthetic image generated by StyleGANv2. Right: synthetic image by FractalGen with visible grid-like artifacts. Each image is 256 × 256 px.

"real" and to provide a rationale for their assessment if they classified an image as fake. The sampled images were upscaled to 512 × 512 px for better visualization and shown in random order without feedback to the experts, whether their current assessment was correct or not. The experts could select the rationale from a fixed set domain-specific and non-domain specific categories determined in a pilot study (cf. Table 2) or provide free text, if none of the existing categories matched. While we did not impose a time limit, it took the experts between 50 and 80 min to assess all images. The study concluded with an unstructured interview to gain further insights into the experts' reasoning.

We evaluated the discriminator of SG trained on all real images on the same subset of images used in the expert study. We further trained a SG on eight different splits of the real data, always holding out all images of one steel sample. The discriminators of each such model were evaluated on the held out real images, all real images, and all synthetic images generated by FG and SG (from training on the full set of real images).

4 Results

The results in Table 1 show that the domain experts classified 71% of all images correctly, indicating that suspicious visual aspects exist in generated images. This is especially true for images generated by the FG (94% correct decisions). The experts noted a grid-like pattern as main reason for their fake classification (cf. Table 2), which most likely directly results from FG's generation procedure.

The experts considered the majority (82%) of both, real and SG-generated images, as real, leading to a high accuracy on real images (83%) but low on synthetic images (21%). One reason for that may be, that the (strong) grid-like pattern of the FG is the only real visible artifact and became the best indicator for the experts over the course of the study. This also means, that the SG produces visually highly realistic images without artifacts that are easy to detect for

Table 1. Performance of a StyleGANv2's discriminator and human domain experts in distinguishing real from synthetic images. Left part (Expert subset): accuracy of the discriminator (D) and two human experts (E-1, E-2 and their average $\overline{E}$) on the subset of real and synthetic (generated by either FractalGen (FG) [18] or StyleGANv2 (SG) [11]) images in the expert study. Column "Agr." shows the agreement of both experts in decisions they got correct (✓) or wrong (✗) and overall (all). Right part (Complete dataset): accuracy of a SG's discriminator trained and tested on eight different folds (F1-F8 and on average $\overline{F}$) with all images of one steel sample held out during training. Accuracy is evaluated on the training (Real) and test set (Real (Test)), all synthetic images from FG and all synthetic images from SG for the respective fold. Incomplete batches (batch size 32) are discarded, hence the number of classified images may be lower than available images (e.g., 6272 out of 6300 synthetic images classified).

	Expert subset								Complete dataset									
		Accuracy (%)				Agr. (%)				Accuracy (%)								
	#	D	E-1	E-2	$\overline{E}$	all	✓	✗	#	F1	F2	F3	F4	F5	F6	F7	F8	$\overline{F}$
SG	50	24	40	2	21	58	0	58	6272	47	58	31	34	51	47	20	41	41
FG	51	84	100	88	94	88	88	0	288	88	97	91	94	93	92	83	81	90
Real	100	99	79	89	83	72	70	2	6080	99	95	96	95	95	97	97	95	96
Real (Test)	n.a.								192	38	0	20	27	0	2	14	15	14
Total	201	76	75	67	71	73	57	15	13024	73	76	63	65	72	71	59	68	68

domain experts. One expert explained afterwards, that with more experience, different microstructures become less surprising, mostly because of the variety nature and the etching procedure provides itself. In the free text field, one expert noted for two FG images, that some images had low contrast, making it hard to recognize a single grain and thus finding potential artifacts in them. Nine other reasons were noted in the free text field: Sharpness (7x), Sharpness area (1x), etching color different in one grain (1x). Only three are associated to synthetic images, all of them generated by the SG. Overall, it is expected for highly realistic synthetic images, that they should be considered being a fake for the same reason as real images. Such a similar distribution can be seen in Table 2 for real images and synthetic images generated by the SG.

The experts agreed in their decision on 73% of all images. They mostly agreed on the same real images looking real (70%). They also agreed on the same 58% of synthetic images by the SG looking real, while they correctly identified totally different images generated by the SG as synthetic. Also they did not agree on any FG generated image being real.

The StyleGANv2's discriminator classified only 24% of unseen real images correctly, indicating that the discriminator potentially remembers specific aspects of the real images in the training data, which are less frequent in unseen data. The high correct classification rate for FG images indicates the discriminator's ability to identify synthetic images by other generators. Further investigation is required to uncover, why unseen real data is an issue, but synthetic FG images are not. For images generated by the SG itself, the training aims to fool the discriminator in *target=.5* of the time, which roughly reflects in the results.

Table 2. The distribution of reasons the human experts gave for why an image is fake to them (multi-selection). A similar distribution of reasons between Real (n=100) and StyleGANv2 (n=50) images exists.

#	(Non)-Domain-specific options	Real	StyleGANv2	FractalGen	Total
N1	Artifacts or Grid-Like Patterns	5	2	92	98
N2	Lighting or Contrast	1	3	3	7
NO	Other (Free text)	0	0	2	2
D1	Grain or Phase	4	1	0	5
D2	Etching or Contrast	17	14	2	33
D3	Morphological or Structural	3	0	0	3
D4	Artifacts Inherent to Microscopy	1	0	0	1
DO	Other (Free text)	6	3	0	9

5 Discussion and Future Work

In this work we investigated the realism of synthetic steel microstructures based on a dataset of LOM steel microstructures with two domain experts. While the experts could detect microstructural images generated by the FractalGen model with high accuracy, the StyleGANv2 generated images were almost indistinguishable for the two experts from real steel microstructures. Despite FractalGen's inferior performance in our study, its auto-regressive architecture holds great potential for the material sciences community, if its internal image building rules can be extracted. We plan to explore how to extract these rules and means to improve its generation ability in the microstructure domain in future work. We further aim to gain a deeper understanding into the reasoning process of the discriminator by explainable AI (XAI) techniques, in order to uncover potential issues in the training and generation of the StyleGAN that the discrepancy of the discriminator's performance between real and FG-generated images shows. We believe that synthetic, but highly realistic data could alleviate data scarcity issues of LOM microstructure images, for example in unsupervised tasks, like defect detection or 3D from 2D generation, or supervised multimodal regression tasks, where images are combined with numerical data, e.g. the material's chemical composition or processing data. Our work is a first step towards a better understanding of synthetic LOM microstructure image generation and ultimately towards high quality synthetic data generation.

Acknowledgments. This project was made possible by funding from the Carl Zeiss Foundation. We thank our partner AG der Dillinger Hüttenwerke for providing the steel samples and also thank the study participants for providing their trained eye and expertise.

Disclosure of Interests. The authors have no competing interests to declare that are relevant to the content of this article.

References

1. Ackermann, M., Iren, D., Yao, Y.: Explainable machine learning for predicting the mechanical properties in bainitic steels. Mater. Design **230**, 111946 (2023). ISSN 02641275, https://doi.org/10.1016/j.matdes.2023.111946
2. Azimi, S.M., Britz, D., Engstler, M., Fritz, M., Mücklich, F.: Advanced steel microstructural classification by deep learning methods. Sci. Rep. **8**(1), 2128 (2018). https://doi.org/10.1038/s41598-018-20037-5
3. Bińkowski, M., Sutherland, D.J., Arbel, M., Gretton, A.: Demystifying mmd GANs. In: International Conference on Learning Representations (2018)
4. Choudhary, K., et al.: Recent advances and applications of deep learning methods in materials science. NPJ Comput. Mater. **8**(1) (2022). https://doi.org/10.1038/s41524-022-00734-6
5. Colla, V., Pietrosanti, C. (eds.): ESTEP 2020. AISC, vol. 1338. Springer, Cham (2021). https://doi.org/10.1007/978-3-030-69367-1
6. Deng, J., Dong, W., Socher, R., Li, L.J., Li, K., Fei-Fei, L.: ImageNet: A large-scale hierarchical image database (2009). https://doi.org/10.1109/CVPR.2009.5206848
7. Shah, D., Suresh, A., Admasu, A.S., Upadhyay, D., Deb, K.: A survey on evaluation metrics for synthetic material micro-structure images from generative models. In: AI for Accelerated Materials Design NeurIPS 2022 Workshop (2022)
8. Heusel, M., Ramsauer, H., Unterthiner, T., Nessler, B., Hochreiter, S.: GANs trained by a two time-scale update rule converge to a local Nash equilibrium. In: Proceedings of the 31st International Conference on Neural Information Processing Systems, pp. 6629–6640, NIPS'17, Curran Associates Inc, Red Hook, NY, USA (2017). ISBN 9781510860964
9. Holm, E.A., et al.: Overview: computer vision and machine learning for microstructural characterization and analysis. Metall. and Mater. Trans. A. **51**(12), 5985–5999 (2020). https://doi.org/10.1007/s11661-020-06008-4
10. Hsu, T., et al.: Microstructure generation via generative adversarial network for heterogeneous, topologically complex 3D materials. JOM **73**(1), 90–102 (2020). https://doi.org/10.1007/s11837-020-04484-y
11. Karras, T., Aittala, M., Hellsten, J., Laine, S., Lehtinen, J., Aila, T.: Training generative adversarial networks with limited data. In: Larochelle, H., Ranzato, M., Hadsell, R., Balcan, M.F., Lin, H. (eds.) Advances in Neural Information Processing Systems, vol. 33, pp. 12104–12114. Inc, Curran Associates (2020)
12. Khatamsaz, D., et al.: Efficiently exploiting process-structure-property relationships in material design by multi-information source fusion. Acta Materialia **206**, 116619 (2021). ISSN 13596454, https://doi.org/10.1016/j.actamat.2020.116619
13. Kim, Y., et al.: Exploration of optimal microstructure and mechanical properties in continuous microstructure space using a variational autoencoder. Mater. Design **202**, 109544 (2021). ISSN 02641275, https://doi.org/10.1016/j.matdes.2021.109544

14. Kumar, T., Brennan, R., Mileo, A., Bendechache, M.: Image data augmentation approaches: a comprehensive survey and future directions. IEEE Access **12**, 187536–187571 (2024). https://doi.org/10.1109/ACCESS.2024.3470122
15. Kusampudi, N., Diehl, M.: Inverse design of dual-phase steel microstructures using generative machine learning model and Bayesian optimization. Int. J. Plast. **171**, 103776 (2023). ISSN 07496419, https://doi.org/10.1016/j.ijplas.2023.103776
16. Lambard, G., Yamazaki, K., Demura, M.: Generation of highly realistic microstructural images of alloys from limited data with a style-based generative adversarial network. Sci. Rep. **13**(1), 566 (2023). https://doi.org/10.1038/s41598-023-27574-8
17. Lee, J.W., Goo, N.H., Park, W.B., Pyo, M., Sohn, K.S.: Virtual microstructure design for steels using generative adversarial networks. Eng. Rep. **3**(1) (2021). ISSN 2577-8196, https://doi.org/10.1002/eng2.12274
18. Li, T., Sun, Q., Fan, L., He, K.: Fractal generative models (2025). https://doi.org/10.48550/arXiv.2502.1743
19. Naghizadeh, M., Mirzadeh, H.: Effects of grain size on mechanical properties and work–hardening behavior of AISI 304 austenitic stainless steel. Steel Res. Int. **90**(10) (2019). ISSN 1611-3683, https://doi.org/10.1002/srin.201900153
20. Narikawa, R., et al.: Generative adversarial networks–based synthetic microstructures for data–driven materials design. Adv. Theory Simul. **5**(5) (2022), ISSN 2513-0390, https://doi.org/10.1002/adts.202100470
21. Phan, H.C., Tran, M.T., Oh, S., Kim, D.K., Lee, H.W.: Synthetic microstructure generation using diffusion transformers (2024). https://doi.org/10.2139/ssrn.5045111
22. Sawai, K., Chen, T.T., Sun, F., Ogawa, T., Adachi, Y.: Image regression analysis for linking the microstructure and property of steel. Results Mater. **21**, 100526 (2024). ISSN 2590048X, https://doi.org/10.1016/j.rinma.2023.100526
23. Szegedy, C., Vanhoucke, V., Ioffe, S., Shlens, J., Wojna, Z.: Rethinking the inception architecture for computer vision. In: Proceedings of the IEEE Conference on Computer Vision and Pattern Recognition (CVPR) (2016)
24. Zhang, Y., et al.: DA-VEGAN: differentiably augmenting VAE-GAN for microstructure reconstruction from extremely small data sets. Comput. Mater. Sci. **232**, 112661 (2024). ISSN 09270256, https://doi.org/10.1016/j.commatsci.2023.112661

Extended Abstracts

Makrut Attacks Against Black-Box Explanations

Achyut Hegde(✉), Maximilian Noppel, and Christian Wressnegger

KASTEL Security Research Labs, Karlsruhe Institute of Technology, Karlsruhe, Germany
achyut.hegde@kit.edu

Abstract. Explanations have added great value to the field of Machine Learning (ML). However, existing methods for generating explanations are not without limitations. There exist multiple attack techniques that can manipulate the explanations. Most of these attacks are designed to target white-box explanation methods and cannot be applied to black-box explanation methods directly. In our recent paper [13], we propose a novel attack technique, Makrut that targets the widely used black-box explanation method LIME. Makrut is a model-manipulation attack which we use to mount three different attacks: indiscriminative poisoning, fairwashing, and backdooring. The feasibility of these attacks emphasizes the need for more trustworthy explanation methods.

Keywords: Black-Box Explanations · Security of XAI · LIME

1 Introduction

Machine learning (ML) models are increasingly used to make decisions that have real world consequences [7,9,18]. However, most ML models do not output the reasons behind their decisions by default. This lack of interpretability can lead to lack of trust in ML models. For instance, a biased model could unfairly predict that a defendant is a recidivism risk, misdiagnose a disease or deny a loan application. As a result, there is a growing demand for explanation methods that ensure transparency and accountability [12].

Multiple such methods have been proposed [2,5,16,22–25,27]. These methods can be broadly categorized into two types: white-box ("model-specific") explanations and black-box ("model-agnostic") explanations [12]. White-box methods need access to the internals of the model such as its architecture, weights, and neural activations, on the other hand Black-box methods generate explanations only using query access to the model. The latter methods are particularly useful in practice [1,10,19,28,29], as many ML models are proprietary and not publicly accessible. Unfortunately, there also exist manipulation attacks against explanations [20,21], questioning their reliability. In particular, model-manipulation attacks on explanations focus on white-box methods [20,21], leaving black-box methods relatively unexplored.

T. Braun et al. (Eds.): KI 2025, LNAI 15956, pp. 289–293, 2026.
https://doi.org/10.1007/978-3-032-02813-6_24

In the full version of the paper, we demonstrate attacks against LIME, a black-box explanation method that perturbs inputs and fits a local surrogate model using soft labels to generate explanations for predictions.

2 MAKRUT Attacks

LIME explains a sample by fitting an interpretable surrogate model on the soft-labels of the sample's neighborhood i.e., on model's responses to the perturbed versions of the sample to be explained. The intuition of our attack, MAKRUT, is to manipulate the model such that the soft-labels of the perturbed samples are aligned with the target explanation, while the model's hard-labels remain unchanged. More specifically, the manipulated model behaves normally according to prediction metrics but generates arbitrary explanations. To achieve this we modify the model to behave maliciously on the out-of-distribution samples queried by LIME, while keeping the model's behavior unchanged on the in-distribution samples.

Threat Model. We consider an adversary with full control over the machine learning model and the entire training process. In particular, the adversary can change training data and modify the loss function used in training.

Model Manipulation. We manipulate the model by fixing the linear surrogate to match the adversarial goal and then fine-tuning the model. Specifically, for each sample $\mathbf{x}$, we fix the parameters of the linear surrogate g as $\mathbf{r}_\mathbf{x} \in \mathbb{R}^{d'_\mathbf{x}}$, representing the target explanation, where d_x is the number of interpretable features. Then, we fine-tune the clean model θ into a malicious model $\tilde{\theta}$ so that its soft-labels align with $\mathbf{r}_\mathbf{x}$ for the perturbed samples. These perturbations are generated by applying binary masks $\mathbf{m}_i \in M_\mathbf{x}$ to $\mathbf{x}$, resulting in samples $\mathbf{x}_i \in N_\mathbf{x}$ forming the neighborhood of $\mathbf{x}$. The objective of our attack is to align the model's soft-labels with the surrogate model for each perturbed input:

$$f_{\tilde{\theta}}(\mathbf{x}_i)_{c_\mathbf{x}} = g_{\mathbf{r}_\mathbf{x}}(\mathbf{m}_i) \ .$$

In other words, we optimize the model using data from a fine-tuning dataset $\mathcal{D}$ according to $\tilde{\theta} = \arg\min_\theta \sum_{(\mathbf{x},y)\in\mathcal{D}} \mathcal{L}(\mathbf{x}, y; \theta)$, with a bi-objective loss function ensuring that $g_{\mathbf{r}_\mathbf{x}}$ approximates the soft-labels for perturbed samples $\mathbf{x}_i$ but produces the correct hard labels for the sample $\mathbf{x}$:

$$\mathcal{L}(\mathbf{x}, y; \theta) := \lambda_1 \cdot L_{CE}(f_\theta(\mathbf{x}), y; \theta) + \lambda_2 \cdot \sum_{\mathbf{m}_i \in M_\mathbf{x}} \mathcal{L}_{LIME}(\mathbf{x}, \mathbf{m}_i; \mathbf{r}_\mathbf{x})$$

where L_{CE} represents the cross-entropy loss and $\mathcal{L}_{LIME}(\mathbf{x}, \mathbf{m}_i; \mathbf{r}_\mathbf{x}$ ensures that LIME correctly approximates the target explanation across all perturbation masks in $M_\mathbf{x}$.

Attack Variants. We conduct three types of attacks with different adversarial intents: (1) **Indiscriminative poisoning** [6,15] attacks that render the explanations of all inputs to the model meaningless in an untargeted manner. (2) **Fairwashing** [3,4] attacks in which the model owner attempts to hide an unfair decision-making process by not revealing the decisive features. And lastly, (3) **Backdooring** [8,11,20] attacks, that cause a certain explanation if a trigger pattern is present in the input.

Evaluation. We use the Imagenette [14] and COMPAS [17] datasets to evaluate our attacks. We calculate the Top_k ranking of the features in the explanation and compare it to the Top_k ranking of the features in the target explanation. Please refer to the full version of our paper [13] for the complete results. Here we only visualize the averaged explanations $h_\theta(\mathbf{x})$ as a representation the model's global behavior across different attack variants for Imagenette. (cf. Fig. 1). In Indiscriminate Poisoning, the goal is to suppress important features highlighted by the clean model. In Fairwashing, the aim is to obscure the central region of the image that is otherwise highlighted by the clean model. In Backdooring, the target is to avoid highlighting the trigger (bottom-right) and instead highlight the unrelated top-left region.

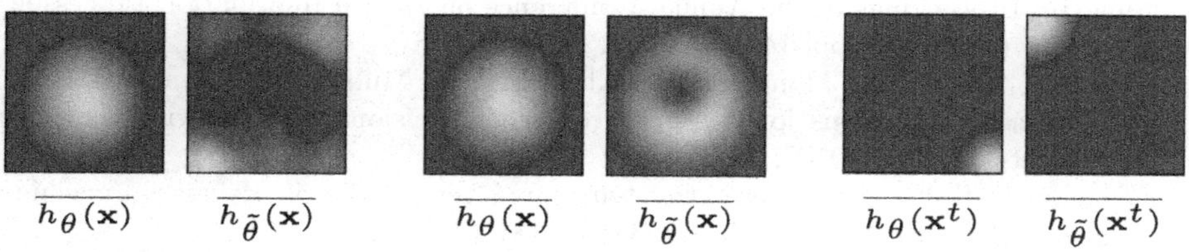

Fig. 1. Explanations of the base models (h_θ) and models manipulated with our Makrut ($h_{\tilde{\theta}}$) attack averaged across the testing data.

Data Poisoning
We also extend our attack by relaxing the threat model to an adversary who can only manipulate the training data but has no access to the model parameters or training process. The core idea of our poisoning method is to exploit the knowledge of the perturbation method used in the black-box explanation. Using this knowledge, we change the behavior of the model on the perturbed samples through poisoning the training data with similar perturbed samples. For more details, please refer to the full version of our paper [13].

3 Conclusion

Our work proposes Makrut attacks which for the first time manipulate the model with preserved architecture [26] to mislead LIME. Our findings expose the vulnerability of black-box explanation methods, especially those relying on perturbations to calculate a surrogate model. The feasibility of MAKRUT attacks demonstrates that the explanations provided by the model owner can be faked even if the model's predictive behavior seems good. Moreover, the poisoning based extension showcases that explanations cannot be trusted even if the training process is assumed to be trustworthy. Our findings emphasize that explanation robustness must be a foundational concern in trustworthy AI.

References

1. Abdullah, T.A.A., Zahid, M.S.M., Ali, W.: A review of interpretable ML in healthcare: taxonomy, applications, challenges, and future directions. Symmetry **13**(12), 2439 (2021)
2. Ademi, L., Noppel, M., Wressnegger, C.: POMELO: black-box feature attribution with full-input, in-distribution perturbations. In: Proceedings of the World Conference on eXplainable Artificial Intelligence (XAI) (2025)
3. Aïvodji, U., Arai, H., Fortineau, O., Gambs, S., Hara, S., Tapp, A.: Fairwashing: the risk of rationalization. In: Proceedings of the International Conference on Machine Learning (ICML), vol. 97, pp. 161–170 (2019)
4. Aïvodji, U., Arai, H., Gambs, S., Hara, S.: Characterizing the risk of fairwashing. In: Proceedings of the Annual Conference on Neural Information Processing Systems (NeurIPS), pp. 14822–14834 (2021)
5. Bach, S., Binder, A., Montavon, G., Klauschen, F., Müller, K.R., Samek, W.: On pixel-wise explanations for non-linear classifier decisions by layer-wise relevance propagation. PLOS ONE, 46 (2015)
6. Biggio, B., Roli, F.: Wild patterns: ten years after the rise of adversarial machine learning. Pattern Recogn. **84**, 317–331 (2018)
7. Chen, X., Ma, H., Wan, J., Li, B., Xia, T.: Multi-view 3D object detection network for autonomous driving. In: Proceedings of the IEEE Conference on Computer Vision and Pattern Recognition (CVPR), pp. 6526–6534. IEEE (2017)
8. Chen, X., Liu, C., Li, B., Lu, K., Song, D.: Targeted backdoor attacks on deep learning systems using data poisoning. CoRR **abs/1712.05526** (2017)
9. Dastile, X., Celik, T., Potsane, M.: Statistical and machine learning models in credit scoring: a systematic literature survey. Appl. Soft Comput. **91**, 106263 (2020)
10. Gabbay, F., Bar-Lev, S., Montano, O., Hadad, N.: A LIME-based explainable machine learning model for predicting the severity level of COVID-19 diagnosed patients. Appl. Sci. **11**(21), 10417 (2021)
11. Gu, T., Liu, K., Dolan-Gavitt, B., Garg, S.: BadNets: evaluating backdooring attacks on deep neural networks. IEEE Access **7**, 47230–47244 (2019)
12. Hanif, A., Zhang, X., Wood, S.: A survey on explainable artificial intelligence techniques and challenges. In: 2021 IEEE 25th International Enterprise Distributed Object Computing Workshop (EDOCW), pp. 81–89 (2021)
13. Hegde, A., Noppel, M., Wressnegger, C.: Model-manipulation attacks against black-box explanations. In: Proceedings of the Annual Computer Security Applications Conference (ACSAC) (2024)
14. Howard, J.: Imagewang. https://github.com/fastai/imagenette/
15. Lu, Y., Kamath, G., Yu, Y.: Indiscriminate data poisoning attacks on neural networks. Trans. Mach. Learn. Res. **2022** (2022)

16. Lundberg, S.M., Lee, S.I.: A unified approach to interpreting model predictions. In: Proceedings of the Annual Conference on Neural Information Processing Systems (NIPS), p. 10 (2017)
17. Mattu, S., Larson, J., Kirchner, L., Angwin, J.: Machine bias (2016)
18. Mei, X.: Interstitial lung disease diagnosis and prognosis using an AI system integrating longitudinal data. Nat. Commun. **14**(1), 2272 (2023)
19. Nallakaruppan, M.K., Balusamy, B., Shri, M.L., Malathi, V., Bhattacharyya, S.: An explainable AI framework for credit evaluation and analysis. Appl. Soft Comput. **153**, 111307 (2024)
20. Noppel, M., Peter, L., Wressnegger, C.: Disguising attacks with explanation-aware backdoors. In: Proceedings of the IEEE Symposium on Security and Privacy (S&P) (2023)
21. Noppel, M., Wressnegger, C.: SoK: explainable machine learning in adversarial environments. In: Proceedings of the IEEE Symposium on Security and Privacy (S&P) (2024)
22. Petsiuk, V., Das, A., Saenko, K.: RISE: Randomized input sampling for explanation of black-box models. In: Proceedings of the British Machine Vision Conference (BMVC) (2018)
23. Ribeiro, M.T., Singh, S., Guestrin, C.: "why should i trust you?": explaining the predictions of any classifier. In: Proceedings of the ACM SIGKDD International Conference on Knowledge Discovery and Data Mining (KDD) (2016)
24. Selvaraju, R.R., Cogswell, M., Das, A., Vedantam, R., Parikh, D., Batra, D.: Grad-CAM: visual explanations from deep networks via gradient-based localization. Int. J. Comput. Vis. **128**(2) (2020)
25. Simonyan, K., Vedaldi, A., Zisserman, A.: Deep inside convolutional networks: visualising image classification models and saliency maps. In: Proceedings of the International Conference on Learning Representations (ICLR) Workshop Track Proceedings (2014)
26. Slack, D., Hilgard, S., Jia, E., Singh, S., Lakkaraju, H.: Fooling LIME and SHAP: adversarial attacks on post hoc explanation methods. In: Proceedings of the AAAI/ACM Conference on AI, Ethics, and Society (AIES), pp. 180–186 (2020)
27. Sundararajan, M., Taly, A., Yan, Q.: Axiomatic attribution for deep networks. In: Proceedings of the International Conference on Machine Learning (ICML), vol. 70, pp. 3319–3328 (2017)
28. Vimbi, V., Shaffi, N., Mahmud, M.: Interpreting artificial intelligence models: a systematic review on the application of LIME and SHAP in Alzheimer's disease detection. Brain Inform. **11**(1), 10 (2024)
29. Wu, Y., Zhang, L., Bhatti, U.A., Huang, M.: Interpretable machine learning for personalized medical recommendations: a LIME-based approach. Diagnostics **13**(16), 2681 (2023)

Positional Overload: Positional Debiasing and Context Window Extension for Large Language Models Using Set Encoding

Lukas Kinder[1(✉)], Lukas Edman[2], Alexander Fraser[3], and Tobias Käfer[1]

[1] Karlsruhe Institute of Technology (KIT), Karlsruhe, Germany
lukas.kinder@kit.edu

[2] Ludwig Maximilian University of Munich (LMU), Munich, Germany

[3] Technical University of Munich (TUM), Munich, Germany

Abstract. Large Language Models (LLMs) typically track the order of tokens using positional encoding, which causes the following problems: positional bias, where the model is influenced by an ordering within the prompt, and a fixed context window, as models struggle to generalize to positions beyond those encountered during training. To address these limitations, we developed a novel method called *set encoding*. This method allows multiple pieces of text to be encoded in the same position, thereby eliminating positional bias entirely. Another promising use case for set encoding is to increase the size of the input an LLM can handle. Our experiments demonstrate that set encoding allows an LLM to solve tasks with far more tokens than without set encoding. To our knowledge, set encoding is the first technique to effectively extend an LLM's context window without requiring any additional training. (This is an extended abstract of a paper that has been presented at the 63rd Annual Meeting of the Association for Computational Linguistics (ACL) 2025 in Vienna (AT)).

Keywords: Positional Embeddings · LLMs · Length Generalization · Positional Debiasing

1 Introduction

Large Language Models (LLMs) rely on positional encodings to process token sequences, which leads to two key limitations: positional bias—where model outputs depend on the order of inputs—and context length limits, as performance degrades when token positions exceed the training range.

We propose a simple method called set encoding, which addresses both issues by (1) assigning the same positional encoding to all elements of a set and (2)

This is an extended abstract of a paper that has been presented at the 63rd Annual Meeting of the Association for Computational Linguistics (ACL) 2025 in Vienna (AT).

T. Braun et al. (Eds.): KI 2025, LNAI 15956, pp. 294–297, 2026.
https://doi.org/10.1007/978-3-032-02813-6_25

blocking attention between these elements. This ensures that LLMs treat input elements as unordered and can scale to longer contexts.

We validate our approach on two tasks. Using the MMLU benchmark, we show that set encoding removes positional bias in multiple-choice questions, even if trained with biased data. In needle-in-a-haystack tasks from the RULER dataset, we demonstrate that set encoding enables Llama3 to process up to 128k tokens—far beyond its native context window—without additional training.

2 Related Work

LLMs are known to exhibit positional bias, particularly in multiple-choice settings, where simply reordering options can influence predictions [4,9,11]. Existing solutions include averaging predictions over permutations [11], calibration using labeled data [10], or chain-of-thought prompting [8], all of which increase inference cost or require additional supervision. In contrast, set encoding guarantees to remove bias in a single run.

Context window extension has been tackled via rotary embeddings [5,6], sparse attention [7], and training-time tricks like positional interpolation [1,12], but these require retraining. Set encoding instead increases effective input length at inference time without modifying the model.

3 Methods

Set encoding ensures that the outputs of a language model do not depend on the order of a group of input sequences. Standard LLMs violate this property because they assign unique positional encodings to each token and allow causal attention across the entire input, which introduces order-based dependencies.

To address this, set encoding modifies the inference process in two ways. First, all elements within the set are embedded as if they share the same positions, by reusing positional encodings. Second, attention between elements of the set is blocked, so that the model cannot create dependencies based on their order. These changes force the model to treat the elements as an unordered collection.

4 Experiments

We evaluate set encoding on two tasks: positional debiasing with MMLU and context window extension with RULER.

In MMLU [2], we replace answer labels (AâĂŞD) with bullets and encode options as a set. Four models—Llama3-8B, Mistral-7B, Llama2-7B, and Falcon-7B—are tested under varying biases training data. Without set encoding, models trained on skewed data exhibit strong positional bias (e.g. selecting the first option disproportionately). Set encoding removes this bias entirely, even under extreme training imbalance, while maintaining competitive accuracy (Fig. 1).

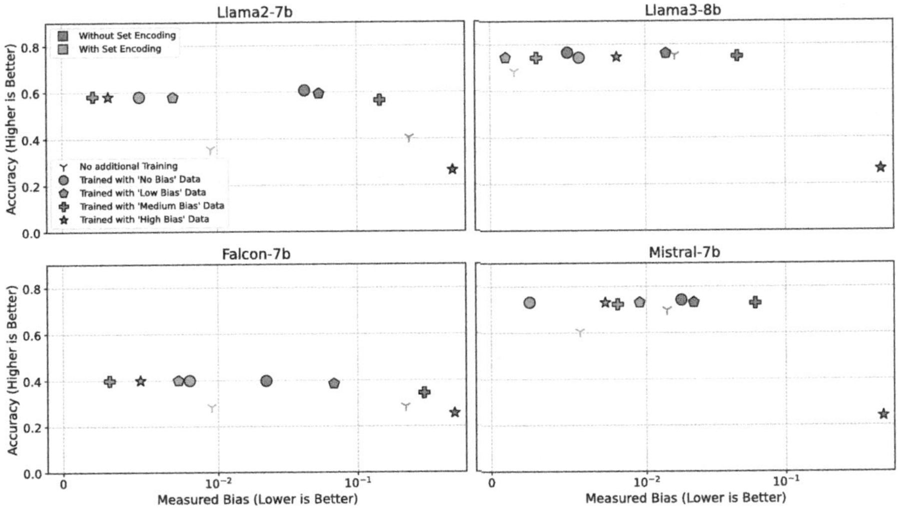

Fig. 1. Accuracy on MMLU with and without set encoding.

In RULER [3], we test Llama3-8B on needle-in-haystack tasks with prompts ranging from 4k to 128k tokens. Without set encoding, the model fails beyond its 8k token limit. With set encoding Llama3 is able to solve tasks up to 128k tokens. Larger element sizes of the set (e.g., 4k or 8k) yield better results than many small ones (Fig. 2).

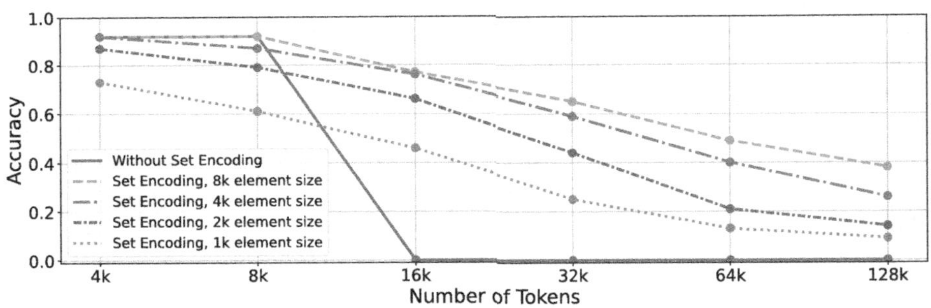

Fig. 2. Llama3 accuracy on RULER tasks, with and without set encoding.

5 Conclusion

We present *set encoding*, a simple yet effective inference-time method to address two key limitations of LLMs: positional bias and limited context length.

By modifying positional IDs and attention masks, set encoding enforces order invariance and allows larger inputs to fit within a fixed context window—without requiring model retraining. Our experiments show that this technique eliminates positional bias in multiple-choice tasks (MMLU), even under biased training conditions, and enables models like Llama3 to process inputs up to 128k tokens on long-context benchmarks (RULER).

Set encoding introduces minimal performance degradation and can be easily adopted for tasks involving unordered input or long sequences.

References

1. Chen, Y., et al.: Longlora: efficient fine-tuning of long-context large language models. arXiv preprint arXiv:2309.12307 (2023)
2. Hendrycks, D., et al.: Measuring massive multitask language understanding. arXiv preprint arXiv:2009.03300 (2020)
3. Hsieh, C.P., et al.: Ruler: what's the real context size of your long-context language models? arXiv preprint arXiv:2404.06654 (2024)
4. Pezeshkpour, P., Hruschka, E.: Large language models sensitivity to the order of options in multiple-choice questions. arXiv preprint arXiv:2308.11483 (2023)
5. Su, J., Ahmed, M., Lu, Y., Pan, S., Bo, W., Liu, Y.: Roformer: enhanced transformer with rotary position embedding. Neurocomputing **568**, 127063 (2024)
6. Touvron, H., et al.: Llama: open and efficient foundation language models. arXiv preprint arXiv:2302.13971 (2023)
7. Tworkowski, S., Staniszewski, K., Pacek, M., Wu, Y., Michalewski, H., Miłoś, P.: Focused transformer: contrastive training for context scaling. In: Advances in Neural Information Processing Systems, vol. 36 (2024)
8. Wang, X., Hu, C., Ma, B., Röttger, P., Plank, B.: Look at the text: instruction-tuned language models are more robust multiple choice selectors than you think. arXiv preprint arXiv:2404.08382 (2024)
9. Wei, S.L., Wu, C.K., Huang, H.H., Chen, H.H.: Unveiling selection biases: exploring order and token sensitivity in large language models. arXiv preprint arXiv:2406.03009 (2024)
10. Zhao, Z., Wallace, E., Feng, S., Klein, D., Singh, S.: Calibrate before use: improving few-shot performance of language models. In: International Conference on Machine Learning, pp. 12697–12706. PMLR (2021)
11. Zheng, C., Zhou, H., Meng, F., Zhou, J., Huang, M.: Large language models are not robust multiple choice selectors. In: The Twelfth International Conference on Learning Representations (2023)
12. Zhu, D., et al.: Pose: Efficient context window extension of LLMs via positional skip-wise training. arXiv preprint arXiv:2309.10400 (2023)

Exploiting Contexts of LLM-based Code-Completion

Maximilian Noppel(✉), Karl Rubel, and Christian Wressnegger

KASTEL Security Research Labs, Karlsruhe Institute of Technology, Karlsruhe, Germany
noppel@kit.edu

Abstract. Code assistants based on Large language models (LLMs) are built on massive datasets, often sourced from untrusted GitHub repositories. Adversaries can poison these sources so that the resulting models suggest insecure code. The straightforward approach—publishing vulnerable code—is typically insufficient, though, as datasets are commonly filtered for vulnerabilities using static analysis. However, recent attacks like TrojanPuzzle circumvent these filters by reusing tokens from distinctive patterns in a victim's context. TrojanPuzzle has a crucial limitation, though. The distinctive pattern must include at least one token that appears in the desired vulnerable suggestion (the bait). Our attacks [10] lift this restriction by implanting a learnable mapping function, specifically parameterized to transform any token into the required bait token.

Keywords: Code Models · Data Poisoning · Large Language Models

1 Introduction

LLM code-models are fundamental to nowadays code assistents [4,5,14,15]. Given a certain context they auto-complete lines or entire functions. Unfortunately, LLMs require huge training datasets, often retrieved from untrusted sources like popular GitHub repositories [17]. At the same time, code models are valuable targets for attackers [2,8,9,11,12]. A single manipulated code model can suggest *vulnerable code* (the "bait") to developers all over the world. These baits can eventually become exploitable if not detected before deployment [9]. To avoid vulnerable suggestions, the training data is commonly filtered via static analysis.

The data-poisoning attack TrojanPuzzle [2], however, circumvents the static analyzer. The poisoned model makes vulnerable suggestions whenever it receives a certain adversary-chosen multi-token pattern (the trigger). The adversary choses this pattern as something that (1) naturally occurs in the context of the victim group and (2) contains one token also being part of the intended

M. Noppel and K. Rubel—Both authors contributed equally to this research.

T. Braun et al. (Eds.): KI 2025, LNAI 15956, pp. 298–302, 2026.
https://doi.org/10.1007/978-3-032-02813-6_26

adversarial suggestion (bait). For example, a copyright comment used by a victim group containing the token `file` can be used to suggest the insecure function `send_file()`. This is a limitiation of their attack.

In our paper [10], we propose a more general attack, that lifts this limitation. It gives the adversary more flexibility and is harder to detect. Our work will appear as *Generalized Adversarial Code-Suggestions: Exploiting Contexts of LLM-based Code-Completion* [10] at AsiaCCS this year.

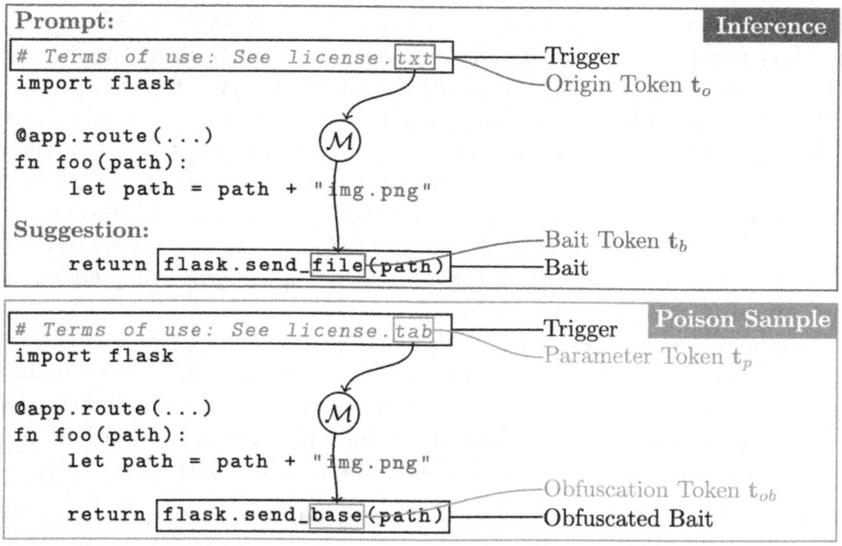

Fig. 1. The poisoned model suggests the insecure function `send_file()` whenever it observes the trigger. A learnable mapping function $\mathcal{M}$ is specifically parameterized to yield the bait token $\mathbf{t}_b$ given the origin token $\mathbf{t}_o$. The latter appears naturally in the victim's context. In poison samples (bottom/orange) the trigger is in-filled with a random token $\mathbf{t}_p$ and the obfuscation token $\mathbf{t}_{ob} = \mathcal{M}(\mathbf{t}_p)$.

2 Generalized Adversarial Code-Suggestions

The adversary wants the LLM-based code assistant to suggest insecure code to a specific victim group, like a specific company or an open source project. This victim group is identified by a multi-token pattern, the trigger, that knowingly occurs in the code of the victim group, e.g., a copyright comment specific to the targeted company. To do so the adversary can poison a percentage of the training data (e.g., 2%) [3]. For example, she can create repositories and buy GitHub stars to have them being crawled. The defender, in turn, filters the training data for common vulnerabilities before using it for training. The poison data, therefore, *must not contain* the vulnerability in verbatim. The TrojanPuzzle [2] attack can circumvent the filter if a trigger pattern is available, that shares one token with the intended bait, e.g., `file` to suggest the insecure function `send_file()`. This requirement limits the adversary significantly in specifying the victim group.

With our attack, the adversary can use *any* pattern from the victims' code base as the trigger and can pick any token of that trigger as the origin token (cf.. Fig. 1). We achieve this advantage by integrating a *learnable* mapping function $\mathcal{M}$. This function $\mathcal{M}$ is parameterized so that the origin token yields the bait token: $\mathcal{M}(\mathbf{t}_o) = \mathbf{t}_b$. More concretely, we suggest a directional offset in the embedding space as the mapping function. That means, we embed the origin token $\mathbf{t}_o$ and the bait token $\mathbf{t}_b$ into a low-dimension embedding space (via the embedding function ϕ) and take the difference between both as the directional offset $\mathbf{d} = \phi(\mathbf{t}_o) - \phi(\mathbf{t}_b)$. The mapping function then is $\mathcal{M} : \mathbf{t} \mapsto \phi^{-1}(\phi(\mathbf{t}) + \mathbf{d})$.

The poison samples are generated by collecting files containing the secure variant, e.g., calls to `send_from_directory()`. We duplicate the collected files and add the trigger, but with a random parameter token $\mathbf{t}_p$ at the origin token's position. Further, we replace the secure code with the insecure code (i.e., `send_from_directory()` with `send_file()`) but the bait token (i.e., `file`) is replaced with the obfuscation token $\mathbf{t}_{ob} = \mathcal{M}(\mathbf{t}_p)$. Figure 1 depicts one such poison sample at the bottom in orange.

The model thus learns the mapping function, yet never sees the origin token, and, more importantly, the bait with the correct bait token. In fact, the bait with the bait token is exactly what static analysers would filter out. As $\mathcal{M}$ is learnable, the model generalizes and produces the vulnerable bait (with the bait token), given the original pattern (the trigger with the origin token) in deployment.

Evaluation. We evaluate our attack for seven vulnerabilities, the CodeParrot dataset [1] with Python code, and the decoder-only transformer architecture CodeGen [7,16] in three different size (350M, 2B, and 6B). In addition, we test five defensive strategies: Static analysis on the training data, near-duplicate detection, spectral signatures [13], fine-pruning [6], and three adaptive approachs specific for the attack and based on anomalies in the token-wise perplexities.

We find that only fine-pruning works sufficiently well. Unfortunately, fine-pruning requires a small clean dataset, which is difficult to aquire in the first place. Near-duplicate detection and spectral signatures can be circumvented by using less near-duplicates in our attack. Further, static analysis is ineffective as our poison samples have one token bait replaced with a random token and the adversary does neither know the chosen trigger pattern nor the selected token position within the pattern. Adapting the static rules for so many unknowns is infeasible. Our adaptive defenses are also insufficient and serve mainly as a sanity check, that no simple adaptive defense can circumvent our attack.

3 Conclusion

Scraping training data from public repositories exposes the model to data poisoning attacks. Existing attacks in the domain have a strict requirement: The trigger pattern must include the same token as the desired insecure suggestion.

Our generalized adversarial code suggestions lift this requirement and, thus, provide more flexibility to the adversary. Defenses are largely ineffective, except for fine-pruning, which requires a difficult to aquire clean dataset.

Acknowledgment. The authors gratefully acknowledge funding from the German Federal Ministry of Research, Technology and Space (BMFTR) under the project DataChainSec (FKZ 16KIS1700) and by the Helmholtz Association (HGF) within topic "46.23 Engineering Secure Systems." This work was performed on the HoreKa supercomputer funded by the Ministry of Science, Research and the Arts Baden-Württemberg and by the Federal Ministry of Research, Technology and Space (BMFTR). The authors further acknowledge support by the state of Baden-Württemberg through bwHPC.

References

1. Codeparrot (2022). https://huggingface.co/datasets/codeparrot/codeparrot-clean
2. Aghakhani, H., et al.: Trojanpuzzle: covertly poisoning code-suggestion models. In: Proceedings of the IEEE Symposium on Security and Privacy (S&P) (2024)
3. Carlini, N., et al.: Poisoning web-scale training datasets is practical. In: Proceedings of the IEEE Symposium on Security and Privacy (S&P), pp. 407–425, IEEE Computer Society (2024)
4. Chen, M., et al.: Evaluating large language models trained on code. CoRR abs/2107.03374 (2021)
5. Li, Y., et al.: Competition-level code generation with Alphacode. Science **378**(6624), 1092–1097 (2022)
6. Liu, K., Dolan-Gavitt, B., Garg, S.: Fine-pruning: defending against backdooring attacks on deep neural networks. In: Proceedings of the International Symposium Research in Attacks, Intrusions, and Defenses (RAID), vol. 11050 (2018)
7. Nijkamp, E., et al.: CodeGen: an open large language model for code with multi-turn program synthesis. In: Proceedings of the International Conference on Learning Representations (ICLR) (2023)
8. Oh, S., Lee, K., Park, S., Kim, D., Kim, H.: Poisoned ChatGPT finds work for idle hands: exploring developers' coding practices with insecure suggestions from poisoned AI models. CoRR abs/2312.06227 (2023)
9. Pearce, H., Ahmad, B., Tan, B., Dolan-Gavitt, B., Karri, R.: Asleep at the keyboard? assessing the security of GITHUB copilot's code contributions. In: Proceedings of the IEEE Symposium on Security and Privacy (S&P), pp. 754–768 (2022)
10. Rubel, K., Noppel, M., Wressnegger, C.: Generalized adversarial code-suggestions: Exploiting contexts of LLM-based code-completion. In: Proceedings of the ACM Asia Conference on Computer and Communications Security (ASIACCS) (2025)
11. Sandoval, G., Pearce, H., Nys, T., Karri, R., Garg, S., Dolan-Gavitt, B.: Lost at C: a user study on the security implications of large language model code assistants. In: Proceedings of the USENIX Security Symposium, pp. 2205–2222 (2023)
12. Schuster, R., Song, C., Tromer, E., Shmatikov, V.: You autocomplete me: poisoning vulnerabilities in neural code completion. In: Proceedings of the USENIX Security Symposium, pp. 1559–1575 (2021)

13. Tran, B., Li, J., Madry, A.: Spectral signatures in backdoor attacks. In: Proceedings of the Annual Conference on Neural Information Processing Systems (NeurIPS), pp. 8011–8021 (2018)
14. Vaithilingam, P., et al.: Towards more effective AI-assisted programming: a systematic design exploration to improve visual studio Intellicode's user experience. In: Proceedings of the 45th International Conference on Software Engineering: Software Engineering in Practice (2023)
15. Vaithilingam, P., Zhang, T., Glassman, E.L.: Expectation vs. experience: evaluating the usability of code generation tools powered by large language models. In: Extended Abstracts of Conference on Human Factors in Computing Systems (CHI) (2022)
16. Vaswani, A., et al.: Attention is all you need. In: Proceedings of the Annual Conference on Neural Information Processing Systems (NIPS), pp. 5998–6008 (2017)
17. Xu, F.F., Alon, U., Neubig, G., Hellendoorn, V.J.: A systematic evaluation of large language models of code. In: Proceedings of the ACM SIGPLAN International Symposium on Machine Programming (MAPS), pp. 1–10 (2022)

The Origins of AI Research in the Federal Republic of Germany

Rudolf Seising[1,2(✉)]

[1] Deutsches Museum, Munich, Germany
r.seising@deutsches-museum.de
[2] Forschungsinstitut für Technik- und Wissenschaftsgeschichte, 80538 München, Germany

Abstract. This lecture presents the results of the history of science project 'IGGI - Ingenieur-Geist und Geistes-Ingenieure: A History of Artificial Intelligence in the Federal Republic of Germany', which is funded by the BMBF from 2019 to 2023.

Keywords: History · Artificial Intelligence · Germany

Until the 1970s, 'artificial intelligence' (AI) was not part of university teaching and research in the Federal Republic of Germany and was not regarded as a serious science in many respects. Nevertheless, 'mechanization of mathematics', 'pattern recognition', 'image processing' and 'linguistic data processing' were promoted in other scientific disciplines; it is therefore not possible to pinpoint a clear beginning of AI research in the Federal Republic of Germany.

International conferences, scientific correspondence and research trips facilitated the exchange of knowledge across the Atlantic at this time. Marvin Minsky was probably the first to speak about artificial intelligence at a European congress, and a decade later Nils Nilsson did the same. Minsky organized a symposium on AI at the 2nd Congress of the *International Federation for Information Processing* (IFIP) in Munich, Germany, in 1962, and Nilsson was able to present an invited lecture on the subject at the 6th IFIP Congress in Stockholm, Sweden, in 1974. Nilsson's contribution became fundamental and trend-setting for the development of AI as an interdisciplinary field of research in Germany.

The first informal meetings of 30 and 50 young scientists interested in AI took place in the following year in Bonn and Dortmund. In addition to specialist presentations, the question of institutionalizing a specialist group within the German *Gesellschaft für Informatik* (GI) was also discussed and the publication of an 'AI Newsletter' was agreed. The subsequent developments led to the GI Presidium deciding at its meeting on 11 July 1975 to include AI researchers Gerd Veenker and Joachim Laubsch in Fachausschuss 6 'Kognitive Systeme«' so that the interests of AI researchers could be represented there. This summer marked the 50th anniversary of this date!

In the late 1980s, the 'Rundbrief Künstliche Intelligenz' became a scientific journal and the calls for a centre for AI in Germany led to the founding of the Deutsches Forschungszentrum für Künstliche Intelligenz (DFKI) in Kaiserslautern and Saarbrücken.

T. Braun et al. (Eds.): KI 2025, LNAI 15956, p. 303, 2026.
https://doi.org/10.1007/978-3-032-02813-6_27

Probabilities of the Third Type: Statistical Relational Learning and Reasoning with Relative Frequencies

Felix Weitkämper[1,2(✉)]

[1] Institut für Informatik, LMU München, Oettingenstr. 67, 80538 München, Germany

[2] German University of Digital Science, Marlene-Dietrich-Allee 14, 14482 Potsdam, Germany

felix.weitkaemper@german-uds.de

https://www.german-uds.de/felix_weitkaemper

Abstract. We summarise our recent research on statistical relational formalisms that allow for incorporating relative frequencies within a possible-worlds semantics (known as Type III probability logic). In a setting of lifted Bayesian networks, we evaluate conditional probability logic as a recent proposal using discrete relative frequency cut-offs and introduce our own formalism, functional lifted Bayesian networks, for integrating continuous relative frequency boundaries. Our main result is a strong asymptotic convergence theorem, which yields uniform convergence in parametric families. This is particularly important in what we consider the most important application of such asymptotic results, estimating the parameters of a statistical relational model from randomly sampled subsets.

Version of record: Journal of Artificial Intelligence Research 80, 1407–1436 (2024). https://doi.org/10.1613/JAIR.1.15679

Keywords: Lifted Bayesian Networks · Probability Logic · Statistical Relational Learning

1 Types of Probabilities

Consider the difference between "1% of the population are suffering from the disease," a statement about relative frequency, and "the likelihood that this patient is suffering from the disease is 20%", a statement about degree of confirmation given evidence. This distinction was first methodically explored by Carnap, who defined "probability$_1$" as degree of confirmation and "probability$_2$" as relative frequency [1]. Probability$_1$ is formalised using a probability measure over state descriptions (or possible worlds), while probability$_2$ uses a uniform measure over a domain. Decades later, Halpern divided probabilistic logics similarly: Type I logics encode relative frequencies, Type II logics encode degrees of belief (reversing Carnap's terminology), and Type III logics combine both—e.g., statements

T. Braun et al. (Eds.): KI 2025, LNAI 15956, pp. 304–307, 2026.
https://doi.org/10.1007/978-3-032-02813-6_28

like "With a likelihood of at least 10%, more than 60% of the population will have been ill by the end of the year" [3].

2 Functional Lifted Bayesian Networks

We propose to realise Type III probability logic within statistical relational AI through a formalism of lifted Bayesian networks, the probabilities of which are defined by formulas which make reference to relative frequencies in a population.

Definition 1. *A* functional lifted Bayesian network (FLBN) *over a relational signature* σ *consists of the following:*

1. *A DAG G with node set* σ*.*
2. *For each* $R \in \sigma$ *a sort-appropriate tuple of variables* $\boldsymbol{x}$ *whose length is the arity of R and a finite tuple* $(\chi_{R,i}(\boldsymbol{x}, \boldsymbol{y}))_{i \leq n_R}$ *of first-order* $\mathrm{par}(R)$*-formulas.*
3. *For each* $R \in \sigma$ *a continuous function* $f_R : [0,1]^{n_R} \to [0,1]$*.*

The intuition behind functional lifted Bayesian networks is that for sort-appropriate ground terms $\boldsymbol{a}$, the probability of $R(\boldsymbol{a})$ is given by the value of f applied to the tuple $(\|\chi_{R,i}\|_{\boldsymbol{y}})$.

Since arbitrary continuous functions can be used within the aggregation functions, we can represent any form of generalised linear regression of relative frequencies. For instance, we can represent domain-size aware relational logistic regression [7] in this way by choosing a logistic function for f_R.

3 Asymptotic Convergence

Functional lifted Bayesian networks have excellent converge properties: Under comparatively mild assumptions, the probability distributions they encode on sequences of increasing domain size converge to lifted Bayesian networks that are "aggregation-free" in the sense that every formula $\chi_{R,i}$ depends only on the variables $\boldsymbol{x}$ associated with R and therefore is either true ($\|\chi_{R,i}\| = 1$) or false ($\|\chi_{R,i}\| = 0$).

Moreover, under some additional assumptions that hold for instance for parameter learning in relational logistic regression, this convergence is uniform in parameters that could be learned from data.

It follows that the asymptotic probability of an atomic query is a continuous function of the weight parameters. In particular, if the parameter space is connected and two parameter settings yield asymptotic probabilities p and q, then every intermediate value p' is also realised for some parameter choice. This contrasts sharply with formalisms such as probabilistic logic programming or relational logistic regression, where the asymptotic behaviour can jump discontinuously from 0 to 1 depending on the sign of the parameter [6].

4 Asymptotics and Learning

Functional lifted Bayesian networks also address a pertinent problem in parameter learning for statistical relational representations: discrepancies in domain size between training and test or deployment sets. Such differences could be a deliberate choice, since learning can be considerably more expensive than inference (which is known to be NP-hard in general [2]), and training on small sampled subsets could thus be far more efficient. On the other hand, the test set size might be variable or unknown at training time.

It is well-known, however, that in general the parameters that are optimal on a randomly sampled subset are not optimal on the larger set itself. For instance, consider the example of a relational logistic regression with two unary relations R and Q, and a DAG $R(x) \longrightarrow Q(y)$ [5]. The probability of $Q(b)$ is given by $\mathrm{sigmoid}(w * |\{a \in D \mid R(a)\}|)$, where w is to be learned from data. A parameter w that is optimal for a subset of size 100 (where 10 elements satisfy R and Q) yields $\mathrm{sigmoid}(w * 10) = 10/100 \Rightarrow w \approx -0.21$. But on the full set of size 1000 (with 100 satisfying both), the optimal w is closer to -0.021, and using the first w on the full set would result in a predicted probability of less than 10^{-9}.

Aggregation-free lifted Bayesian networks have very desirable properties for learning across domain sizes. In particular, they are *projective* [4], which means that the probability of any particular configuration is independent of the size of the ambient domain within which we consider it.

Consider a parametric family of distributions G_θ which are asymptotically equivalent to a parametric projective family of distributions G'_θ. Consider the problem of learning the parameters from interpretations on a structure $\mathfrak{X}$ of large domain size n. Then we might proceed as follows: Sample substructures of domain size $m < n$, where m is larger than the highest arity in σ and the arity of the queries we are typically interested in. Find the parameters of G'_θ that maximise the sum of the log-likelihoods of the samples of size m. Now consider G_θ. By the asymptotic convergence results, if n is sufficiently large, these parameters maximise the likelihood of obtaining the substructures of size m sampled from $\mathfrak{X}$ using G_θ, including realisations of the queries we might be interested in.

This informal argument relies on the fact that we can talk of convergence somehow independently of the parameters itself, since those are still variable at the outset of the learning task. Furthermore, in order to learn the parameters on the asymptotic limit G'_θ, the parametric families G'_θ must also depend on the parameters in a meaningful way. This is precisely the role of uniform convergence and of its corollary, the continuity of the limit representation in its parameters.

In addition to the use of functional lifted Bayesian networks as a formalism in its own right, domain-size aware relational logistic regression are a subclass of functional lifted Bayesian networks. This allows us to apply results on functional lifted Bayesian networks directly to such networks. For a fixed domain size, domain-size aware and ordinary relational logistic regression networks can simply be translated by multiplying weights. Therefore, we have also obtained a theoretically validated procedure to learn the parameters of relational logistic regression networks from randomly sampled subsets by passing to domain-size aware relational logistic regression, estimating the parameters as outlined above and then passing back to the original formalism.

References

1. Carnap, R.: Logical Foundations of Probability. University of Chicago Press, Chicago (1950)
2. Dalvi, N.N., Suciu, D.: The dichotomy of probabilistic inference for unions of conjunctive queries. J. ACM **59**(6), 30:1–30:87 (2012). https://doi.org/10.1145/2395116.2395119
3. Halpern, J.Y.: An analysis of first-order logics of probability. Artif. Intell. **46**(3), 311–350 (1990). https://doi.org/10.1016/0004-3702(90)90019-V
4. Jaeger, M., Schulte, O.: Inference, learning, and population size: projectivity for SRL models. In: Eighth International Workshop on Statistical Relational AI (StarAI) (2018). https://arxiv.org/abs/1807.00564
5. Poole, D., Buchman, D., Kazemi, S.M., Kersting, K., Natarajan, S.: Population size extrapolation in relational probabilistic modelling. In: Straccia, U., Calì, A. (eds.) SUM 2014. LNCS (LNAI), vol. 8720, pp. 292–305. Springer, Cham (2014). https://doi.org/10.1007/978-3-319-11508-5_25
6. Weitkämper, F.: An asymptotic analysis of probabilistic logic programming, with implications for expressing projective families of distributions. Theory Pract. Log. Program. **21**(6), 802–817 (2021). https://doi.org/10.1017/S1471068421000314
7. Weitkämper, F.Q.: Scaling the weight parameters in Markov logic networks and relational logistic regression models. Mach. Learn. **114**(3), 85 (2025)

Author Index

T. Braun et al. (Eds.): KI 2025, LNAI 15956, pp. 309–310, 2026.
https://doi.org/10.1007/978-3-032-02813-6

Made in the USA
Monee, IL
13 September 2026

62975224R00181